Corrosion Resistance Enhancement of the Materials Surface

Corrosion Resistance Enhancement of the Materials Surface

Editors

Costica Bejinariu
Nicanor Cimpoesu

MDPI • Basel • Beijing • Wuhan • Barcelona • Belgrade • Manchester • Tokyo • Cluj • Tianjin

Editors
Costica Bejinariu
Department of Materials
Engineering and Industrial
Safety
"Gheorghe Asachi" Technical
University of Iasi
Iasi
Romania

Nicanor Cimpoesu
Materials Science
"Gheorghe Asachi" Technical
University of Iasi
Iasi
Romania

Editorial Office
MDPI
St. Alban-Anlage 66
4052 Basel, Switzerland

This is a reprint of articles from the Special Issue published online in the open access journal *Materials* (ISSN 1996-1944) (available at: www.mdpi.com/journal/materials/special_issues/ corrosion_resistance_enhancement).

For citation purposes, cite each article independently as indicated on the article page online and as indicated below:

LastName, A.A.; LastName, B.B.; LastName, C.C. Article Title. *Journal Name* **Year**, *Volume Number*, Page Range.

ISBN 978-3-0365-8029-6 (Hbk)
ISBN 978-3-0365-8028-9 (PDF)

© 2023 by the authors. Articles in this book are Open Access and distributed under the Creative Commons Attribution (CC BY) license, which allows users to download, copy and build upon published articles, as long as the author and publisher are properly credited, which ensures maximum dissemination and a wider impact of our publications.

The book as a whole is distributed by MDPI under the terms and conditions of the Creative Commons license CC BY-NC-ND.

Contents

About the Editors vii

Preface to "Corrosion Resistance Enhancement of the Materials Surface" ix

Diana Nicoleta Avram, Corneliu Mircea Davidescu, Iosif Hulka, Mircea Laurentiu Dan,
Elena Manuela Stanciu and Alexandru Pascu et al.
Corrosion Behavior of Coated Low Carbon Steel in a Simulated PEMFC Environment
Reprinted from: *Materials* 2023, 16, 3056, doi:10.3390/ma16083056 1

Ioana Maria Carmen Ienașcu, Adina Căta, Adriana Aurelia Chis, Mariana Nela Ştefănuţ,
Paula Sfîrloagă and Gerlinde Rusu et al.
Some *Brassicaceae* Extracts as Potential Antioxidants and Green Corrosion Inhibitors
Reprinted from: *Materials* 2023, 16, 2967, doi:10.3390/ma16082967 17

V. Encinas-Sánchez, A. Macías-García, M. T. de Miguel, F. J. Pérez and J. M. Rodríguez-Rego
Electrochemical Impedance Analysis for Corrosion Rate Monitoring of Sol–Gel Protective
Coatings in Contact with Nitrate Molten Salts for CSP Applications
Reprinted from: *Materials* 2023, 16, 546, doi:10.3390/ma16020546 33

Liang Yu, Chen Zhang, Yuan Liu, Yulong Yan, Pianpian Xu and Yanli Jiang et al.
Comparing the Corrosion Resistance of 5083 Al and $Al_2O_3 3D/5083$ Al Composite in a Chloride
Environment
Reprinted from: *Materials* 2022, 16, 86, doi:10.3390/ma16010086 47

Costică Bejinariu, Viorel Paleu, Ciprian Vasile Stamate, Ramona Cimpoesu, Margareta
Coteată and Gheorghe Badarau et al.
Microstructural, Corrosion Resistance, and Tribological Properties of Al_2O_3 Coatings Prepared
by Atmospheric Plasma Spraying
Reprinted from: *Materials* 2022, 15, 9013, doi:10.3390/ma15249013 67

Ibrahim H. Zainelabdeen, Fadi A. Al-Badour, Rami K. Suleiman, Akeem Yusuf Adesina,
Necar Merah and Fadi A. Ghaith
Influence of Friction Stir Surface Processing on the Corrosion Resistance of Al 6061
Reprinted from: *Materials* 2022, 15, 8124, doi:10.3390/ma15228124 85

Guangyuan Yang, Fuwei Liu, Ning Hou, Sanwen Peng, Chunqing He and Pengfei Fang
Preparation of One-Dimensional Polyaniline Nanotubes as Anticorrosion Coatings
Reprinted from: *Materials* 2022, 15, 3192, doi:10.3390/ma15093192 103

Ana Maria Roman, Victor Geantă, Ramona Cimpoeșu, Corneliu Munteanu, Nicoleta Monica
Lohan and Georgeta Zegan et al.
In-Vitro Analysis of FeMn-Si Smart Biodegradable Alloy
Reprinted from: *Materials* 2022, 15, 568, doi:10.3390/ma15020568 113

Michael Kahl and Teresa D. Golden
Corrosion Resistance of Electrochemically Synthesized Modified Zaccagnaite LDH-Type Films
on Steel Substrates
Reprinted from: *Materials* 2021, 14, 7389, doi:10.3390/ma14237389 131

Catalin Panaghie, Ramona Cimpoeșu, Bogdan Istrate, Nicanor Cimpoeșu, Mihai-Adrian
Bernevig and Georgeta Zegan et al.
New Zn_3Mg-xY Alloys: Characteristics, Microstructural Evolution and Corrosion Behavior
Reprinted from: *Materials* 2021, 14, 2505, doi:10.3390/ma14102505 145

Costica Bejinariu, Diana-Petronela Burduhos-Nergis and Nicanor Cimpoesu
Immersion Behavior of Carbon Steel, Phosphate Carbon Steel and Phosphate and Painted Carbon Steel in Saltwater
Reprinted from: *Materials* **2021**, *14*, 188, doi:10.3390/ma14010188 . . . . **161**

About the Editors

Costica Bejinariu

Professor Ph.D. Eng. Costica Bejinariu is a full professor and researcher at the "Gheorghe Asachi" Technical University of Iasi, Romania, with over 35 years of academic experience. He has been Vice Dean of the Faculty of Materials Science and Engineering and doctoral supervisor since 2009, with five completed doctoral theses and 10 ongoing doctoral students. His main field of expertise is materials engineering. He has published a total of 25 books, chapters, and monographs, over 250 research papers with more than 1000 citations, and an H index of 19-WoS/21-Scopus. He worked on over 50 contracts for scientific research and development, for five of which he was project director and for two of which he was responsible. Furthermore, he owns 12 invention patents in the field of materials engineering, which were presented at numerous invention salons and won medals. He is an active member of professional associations in the field, a reviewer for scientific journals, and a member of scientific committees at different congresses. He is a member of the Committee of Materials Engineering, Domain Engineering, and Materials Science of the National Council for Attestation of University Titles, Diplomas, and Certificates (CNATDCU), which is an independent consulting entity at a national level in Romania.

Nicanor Cimpoesu

Prof. habil. phd. eng. Nicanor Cimpoesu has 18 years of experience in the research and development of materials, including non-ferrous and ferrous materials and shape memory alloys (SMAs), including five national research projects as coordinator. He has published 155 ISI-indexed articles and proceedings papers and has an H-index of 17. He is part of a team with 18 years of experience in the study of SMAs, 15 years of electron microscopy and EDS/EBSD analysis, and 8 years in the biodegradable metallic materials field.

Preface to "Corrosion Resistance Enhancement of the Materials Surface"

From the perspective of engineers, physicists, medical doctors, researchers, and scientists, we analyze and discuss different topics on corrosion resistance. The high potential of the enhancement of a material's surface (using metallic, polymer, or ceramic layers or by active functionalization through laser or ion beam, mechanical, or chemical transformation) makes them suitable for many medical or industrial applications. Actual activity in the domain presents a few problems connected to the obtaining and processing of metallic alloys, the modification of the surface state, and the characterization, modeling, and simulation or prototyping technologies. This Special Issue of *Materials* intends to focus on the most recent advances in obtaining materials with active surfaces used in the industrial, automotive, chemical, or medical fields with enhanced performances.

Costica Bejinariu and Nicanor Cimpoesu
Editors

Article

Corrosion Behavior of Coated Low Carbon Steel in a Simulated PEMFC Environment

Diana Nicoleta Avram [1], Corneliu Mircea Davidescu [1,2], Iosif Hulka [2], Mircea Laurentiu Dan [1], Elena Manuela Stanciu [3,*], Alexandru Pascu [3] and Julia Claudia Mirza-Rosca [4]

[1] Faculty of Industrial Chemistry and Environmental Engineering, Politehnica University Timisoara, 2 Piata Victoriei, 300006 Timisoara, Romania
[2] Renewable Energy Research Institute—ICER, Politehnica University Timisoara, 138 Gavril Musicescu Street, 300774 Timisoara, Romania
[3] Materials Engineering and Welding Department, Transilvania University of Brasov, 29 Eroilor Blvd., 500036 Brasov, Romania; alexandru.pascu@unitbv.ro
[4] Department of Mechanical Engineering, Las Palmas de Gran Canaria University, 35017 Las Palmas de Gran Canaria, Spain
* Correspondence: elena-manuela.stanciu@unitbv.ro

Abstract: Here, potential metallic bipolar plate (BP) materials were manufactured by laser coating NiCr-based alloys with different Ti additions on low carbon steel substrates. The titanium content within the coating varied between 1.5 and 12.5 wt%. Our present study focussed on electrochemically testing the laser cladded samples in a milder solution. The electrolyte used for all of the electrochemical tests consisted of a 0.1 M Na_2SO_4 solution (acidulated with H_2SO_4 at pH = 5) with the addition of 0.1 ppm F^-. The corrosion resistance properties of the laser-cladded samples was evaluated using an electrochemical protocol, which consisted of the open circuit potential (OCP), electrochemical impedance spectroscopy (EIS) measurements, and potentiodynamic polarization, followed by potentiostatic polarization under simulated proton exchange membrane fuel cell (PEMFC) anodic and cathodic environments for 6 h each. After the samples were subjected to potentiostatic polarization, the EIS measurements and potentiodynamic polarization were repeated. The microstructure and chemical composition of the laser cladded samples were investigated by scanning electron microscopy (SEM) combined with energy-dispersive X-ray spectroscopy (EDX) analysis.

Keywords: electrochemical evaluation; protective coatings; PEMFC; corrosion; laser cladding

1. Introduction

Without the natural greenhouse effect (GHG) regulating the planet's atmospheric temperature, life as it exists today would not be possible. The atmosphere retains enough solar energy to maintain an average global temperature of 14 °C due to the natural existence of GHGs on Earth. Carbon dioxide (CO_2), methane (CH_4), ozone (O_3), nitrous oxide (N_2O), and steam (H_2O) are the main greenhouse gases that are naturally recirculated in the atmosphere. Since the Industrial Revolution, increases in the concentrations of the main GHGs have been recorded [1]. Human interference has brought considerable changes to the natural landscape, because of activities such as deforestation and burning fossil fuels for transportation and heating, agricultural, and industrial activities. This has augmented the natural greenhouse effect, with CO_2 being added in voluminous amounts in the Earth's atmosphere [2–4]. The U.S. Energy Information Administration (EIA) reported that, from 1980 to 2012, CO_2 emissions and primary energy consumption grew at an average annual increase of 1.7% and 2%, respectively. In this period, CO_2 emissions grew by 75% and the global primary consumption of energy increased by 85% [5]. Nowadays, replacing conventional energy sources with cleaner energy alternatives has become a priority. Anthropogenic GHG emissions can be reduced with the use of renewable energy sources, such as wind, solar, hydropower, geothermal, biomass, and fuel cells [6,7].

Among these green technologies, the fuel cell is regarded as a promising solution to the energy crisis and a key to combat the global warming phenomena. The advantages of using fuel cells include a high power density, silent operation, good performance, and low or even zero emissions if the fuel is from a renewable source. Among the different types of fuel cells, PEMFCs have attracted a lot of interest in the scientific community because the technology can be implemented in different sectors, from portable electronics to power generation systems [7–9]. Over the past years, intense research has been done on implementing the PEMFCs systems in the automotive sector. Several prototypes of fuel cell vehicles (FCV) have been successfully manufactured, such as Toyota Mirai, Hyundai ix35, Daimler F-Cell, and Honda FCX-Clarity. The great potential of fuel cell applications has also extended to various vehicles such as buses, trucks, forklifts, and airport transportation vehicles. However, the commercialization of such systems has two main obstacles, which are their low durability and high price. Although substantial advancements have been made in recent years, these previously mentioned hurdles still impose a challenge in the current technologies [7,10].

One of the key components that could determine a better performance for PEMFCs is represented by the bipolar plate. BP interconnects the cathode of a cell with the anode of the adjacent cell, forming a stack. To achieve adequate power output, several individual cells are connected in series to form a fuel cell stack. This means that BPs represent a considerable fraction of the weight and cost of the stack. Other functions of the BPs include (i) homogenous distribution of reactant gases within the flow channels to the active sites, (ii) management of water, and (iii) heat transfer, ensuring the electrical flow within the cell and offering mechanical stability to the stack. To fulfill all of the requirements, BP materials must combine a suitable mechanical strength with a high corrosion resistance, together with a good electrical and thermal conductivity [11,12]. High density graphite, polymer-based composites and metals are the main materials used for the development of BPs. Among the three, metallic BPs have attracted a lot of interest in the research community due to their superior mechanical and physical properties. Various metals and metallic alloys, such as stainless steel [13,14], aluminum [15], mild steel [16–18], copper [12], titanium [13,19], aluminum alloys [20], copper alloys [21], titanium alloys [22], and nickel alloys [23,24], have been studied as possible candidates for BPs materials. Mild steels have the advantage of being inexpensive materials with excellent mechanical properties, but with a weak corrosion resistance in the acidic PEMFCs environment. Different surface modification techniques or applying protective coatings can increase the corrosion resistance of low-carbon steels [12]. In their research, Bai et al. [16,17] reported that chromized-AISI 1045 could be a good alternative BP material for PEMFC applications. Various test analyses have been used to evaluate the performance of chromized-AISI 1045, and the results show that it is comparable to that of stainless steel or graphite. Previous work [25] has included studies on surface modification AISI-1020 with a chromized coating containing carbides and nitrides, with promising results for PEMFC applications. Yuan et al. [18] reported that the corrosion resistance of nickel coated AISI 1045 was improved after low temperature pack chromizing and that it is a good alternative material for PEMFC applications.

In our previous studies [26,27], we successfully laser cladded NiCr(Ti)-based coatings onto low carbon steel plates and electrochemically tested them in a very acidic PEMFC environment (aggressive conditions). The NiCr-based superalloy was selected as the base material due to its good corrosion resistance [28], while titanium was added as a reinforcing material into the composition (up to 10 wt% Ti addition). From the literature and applications, strong acidic solutions used as an electrolyte have a great influence on the corrosion of metallic BPs. Lædre et al. [29] showed that very low pH values can change the composition and thickness of the oxide layer in such a way that it may not exist in an operating PEMFC. This is also supported by the research of Feng et al. [30], who concluded that using mild conditions, instead of aggressive conditions (solutions such as 1 or 0.5 M H_2SO_4), for electrochemical testing could simulate the PEMFC working environment better, while the research of Hinds et al. [31] showed that in situ measurements of pH on the exhaust water from fuel cell stacks was in the range of 3–4 for the cathode and 5–7 for the anode.

In this regard, our present study focused on electrochemically testing the laser cladded materials in a milder solution. The electrolyte used for all of the experiments was Na_2SO_4 0.1 M (acidulated with H_2SO_4 at pH = 5) + 0.1 ppm F^-. Furthermore, the titanium content within the coating from 1.5 to 12.5 wt% Ti was added, and the variation in titanium was studied in terms of the microstructure and corrosion resistance. The morphology and chemical composition of the laser cladded samples were investigated by SEM combined with EDX analysis. The electrochemical protocol consisted of OCP and EIS measurements, potentiodynamic polarization, followed by potentiostatic polarization under simulated PEMFC anodic and cathodic environments for 6 h each. After the samples were subjected to potentiostatic polarization, the EIS measurements and potentiodynamic polarization were repeated.

2. Materials and Methods

2.1. Materials and Deposition Process

Feedstock powders of MetcoClad 625F (NiCr-based powder) and Metco 4010A (Ti powder, 99% purity) from Oerlikon Metco Switzerland were used as the laser-cladding materials. AISI 1010 plates with dimensions of 60 × 25 × 5 mm were used as the substrates for laser cladding NiCr-based alloys (designated B in our manuscript) with different Ti additions. The titanium content within the coating varied as follows: 1.5, 3, 5, 7, 10, and 12.5 wt% Ti. The chemical compositions of the feedstock powders and for the low carbon steel were the same as those presented by the manufacturers [32–34].

The equipment used for powder deposition and the deposition parameters were as presented in our previous study [26]. In summary, a diode laser with λ = 975 nm (Coherent F1000) mounted on a robotic arm (Cloos 6-axes) was used for laser cladding of feedstock powders fed from a processing head (Precitec WC 50) onto the AISI 1010 substrate. The carrier gas (argon) was used in a feeding system (Termach AT-1200 HPHV) to transport the powders to the processing head. The process parameters used for laser cladding deposition included preheating the feedstock powder, and using an 859 W laser power with a 40 cm/min deposition speed and 15.1 L/min argon flow. Ten partially overlapped tracks were deposited onto the AISI 1010 substrate with an 45% overlap degree between the subsequent tracks. The schematic representation is shown in Figure 1.

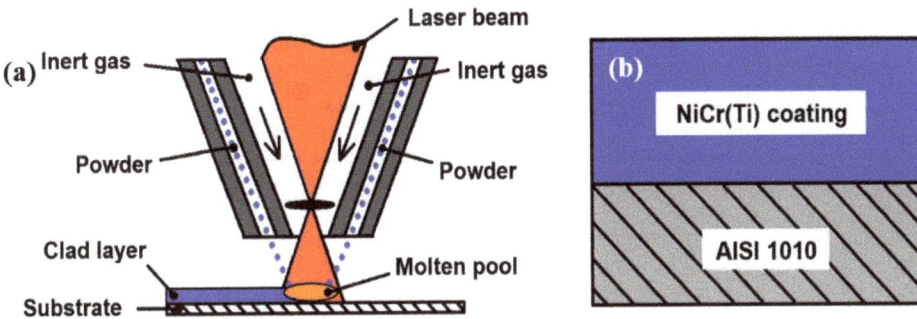

Figure 1. Schematic representation of (**a**) the laser-cladded process and (**b**) protective coatings.

2.2. Characterization of Protective Coatings

The microstructure of the laser-cladded coatings was investigated in cross section at a low magnification (1000×) using a scanning electron microscope (FESEM Zeiss Sigma 300 VP, Carl Zeiss, Jena, Germany). To reveal the microstructure of the coatings produced by laser cladding at a high magnification (30,000×), the samples were examined by FESEM (Quanta FEG 250, FEI, Hillsboro, OR, USA) using backscattered electron detector (BSD).

Elemental analysis was performed as well using energy dispersive X-ray spectroscopy (EDX) with an Apollo SSD detector (EDAX Inc. Mahwah, NJ, USA).

A Future-Tech FM-700 microhardness tester with a Vickers indenter (Future-tech Corp, Kanagawa, Japan) was used for the microhardness measurements. Ten HV_{01} indentations were performed on each coating in the cross-section starting from the top towards the substrate; two measurements were performed in the heat affected zone (HAZ) and another two were performed in the substrate. For the measurements, a load of 100 gf and a dwell time of 15 s were employed for each indentation.

A three-electrode system was used for the electrochemical evaluation of the laser cladded samples in a solution consisting of 0.1 M Na_2SO_4 (acidulated with H_2SO_4 at pH = 5) with the addition of 0.1 ppm F^-. The corrosion behavior of the samples was tested using a potentiostat/galvanostat SP150 (BioLogic Science Instruments, Seyssinet-Pariset, France). The samples were placed in a corrosion cell with an exposed area of 1 cm^2, together with a platinum mesh and an Ag/AgCl electrode acting as the counter and reference electrode, respectively. Different grades of SiC abrasive papers (grit size from 120 to 2400) and a 3 µm diamond suspension were used to grind and polish the samples until they had a mirror-like surface. A solution of 10% oxalic acid was used to electrolytically etch the samples for 12 s at 3 V to visualize the microstructure of the laser-cladded coatings. Afterward, distilled water and ethanol were used to clean the samples.

The electrochemical evaluation protocol consisted of OCP and EIS measurements, followed by potentiodynamic and potentiostatic polarization. After the samples were subjected to potentiostatic polarization, the EIS measurements and potentiodynamic polarization were repeated. EIS measurements were employed with the impedance module of SP150, as described in our previous paper [26], in the frequency range of 10^{-1}–10^5 Hz with a 10 mV AC voltage amplitude. A ZView 3.4 software (Scribner Associates, Inc., Southern Pines, NC, USA) was used to fit all of the experimental EIS data to their corresponding equivalent electrical circuit (EEC). Potentiodynamic polarization curves were recorded in the potential range of ±0.250 V at a scan rate of 1 $mV \cdot s^{-1}$. Potentiostatic polarization curves were employed to simulate the cathodic and the anodic PEMFC environments. In this measurement, potentials of +0.736 V and −0.493 V were applied to the electrochemical configuration, respectively, and they were maintained for 6 h each. Afterwards, the EIS measurements and potentiodynamic polarization were repeated in the same procedure, as previously described.

3. Results and Discussions

3.1. Microstructure and EDX Analysis of Cladded Coatings

Representative SEM images of the laser-cladded samples were taken from the cross-section for coatings without Ti and with the addition of 5 and 10 wt% Ti, and are presented in Figure 2a–c. It can be noticed that there were no significant defects, such as cracks or pores, across the coatings. The etchant revealed a dendritic structure for all of the laser cladded coatings. Furthermore, it can be seen that as the Ti content increasesd within the coating, the size of the dendrites increased.

In our previous study, it was observed that the coatings consisted mainly of γ-Ni-Cr phase with Ti present at 51.59° 2θ, which dissolved into the FCC Ni-Cr phase. Moreover, no other peaks associated with other precipitates of Ni-Cr alloys such M6C, MC and M23C6 carbides were identified due to their low intensity [26]. Generally, these precipitates developed at grain boundaries and inter-dendritic areas during solidification and within the coating, increasing the cracking susceptibility at the fusion zone, reducing its corrosion resistance and ductility [35]. To investigate the precipitates, EDX spot analysis was performed on an un-etched laser-cladded coating at high magnification representative micrographs, as presented in Figure 3. It is well known that SEM analysis using BSD is used to detect elastically scattered electrons, delivering micrographs with information about the composition. Thus, the number of back scattered electrons reaching the detector is strongly related to the atomic number of the elements. The higher the atomic number, the brighter the element appears in the image.

Generally, three main areas could be observed in the cross-sectional micrographs of the investigated coatings (i) a gray area, which represents the Ni-Cr matrix; (ii) a lighter phase rich in Nb and Mo, which represent laves; and (iii) dark areas with a high Si content (for the coating without Ti) and high Ti content (for the coatings manufactured with Ti addition), which represent MC carbides. During deposition, Ti powder particles dissolve in the molten pool and react with C atoms, leading to the formation of TiC when solidification takes place [36]. Ti particles precipitated during solidification generally in spherical and rhombus shapes distributed randomly within the coatings. Lei et al. also reported that a rhombus morphology for TiC precipitates was formed during the deposition of TiC/NiCrBSiC composite carbides using laser cladding [37]. From the images, it can be observed that the size of carbides increased with the addition of Ti within the coatings.

3.2. Microhardness of the Coatings

Figure 4 represents the microhardness comparison of the investigated coatings. The microhardness values were measured across the cross-section of the coatings, starting from the top of the coatings down to the substrate. It can be seen from the graph that the coatings manufactured under the same process parameters have great differences in their function after the addition of Ti. The increase in hardness values were attributed to the formation of hard phases within the coatings. With the increase in Ti content, the number of hard phases increased, which directly led to an increase in hardness. Consequently, the coating without the addition of Ti presented the lowest hardness, while the coating with the addition of 12.5% Ti presented the highest hardness values. The fluctuation in the measurements was caused by the randomly distributed hard phases within the coatings. HAZ had a somewhat higher hardness than the substrate, caused by the metallurgical changes that occurred as a result of the fast heating−cooling process that took place during cladding.

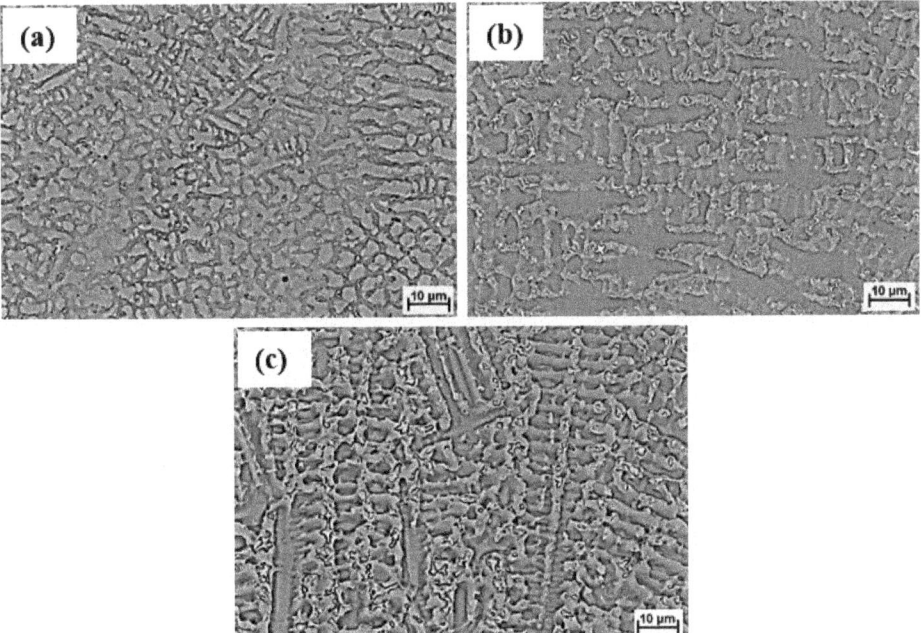

Figure 2. SEM images of laser-cladded coatings with different Ti contents after etching with oxalic acid: (**a**) B, (**b**) B + 5% Ti, and (**c**) B + 10% Ti.

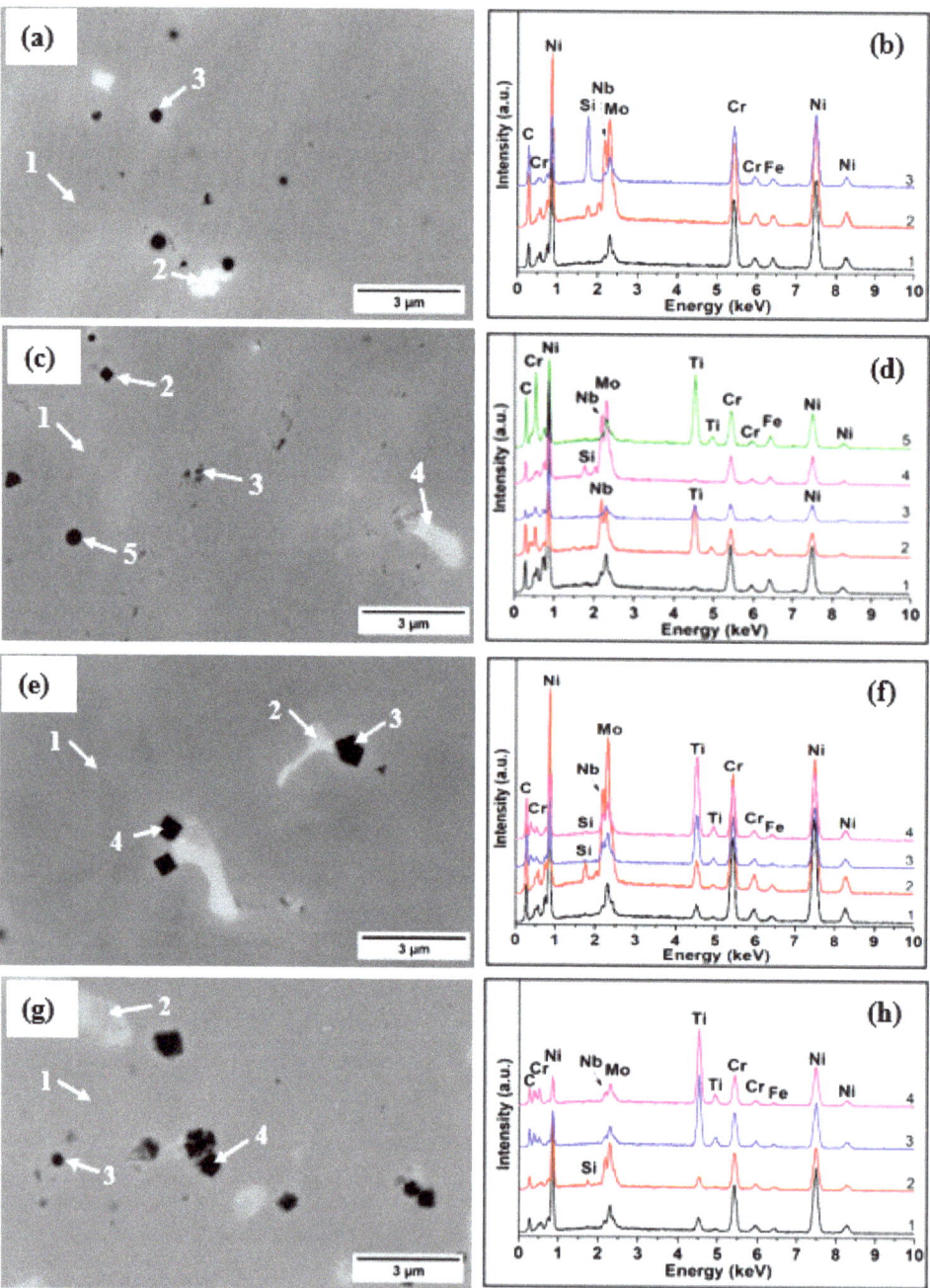

Figure 3. SEM images with the corresponding EDX spectra of coatings with different Ti contents: (**a**,**b**) B, (**c**,**d**) B + 1.5% Ti, (**e**,**f**) B + 5% Ti, and (**g**,**h**) B + 12% Ti.

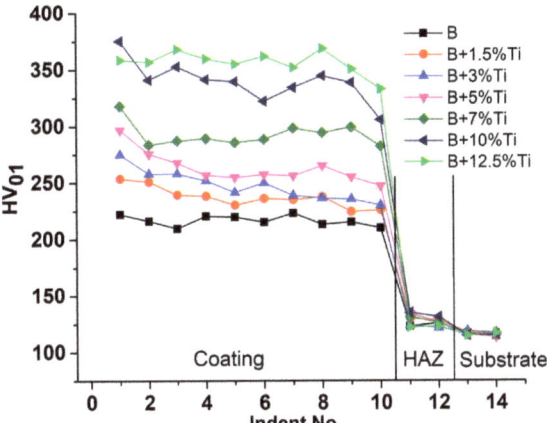

Figure 4. Microhardness values measured in the cross-section from the top of the coatings to the substrate.

3.3. Electrochemical Evaluation before the Durability Test

Corrosion resistance in a simulated PEMFC environment is an important criterion when evaluating the performance of a potential BPs material. Potentiodynamic polarization curves of the substrate and laser cladded coatings with various Ti contents were recorded at a scan rate of 1 mV·s^{-1}, before submitting the samples to a potentiostatic polarization test. Figure 5 presents the Tafel plots of the substrate and laser-cladded samples tested in Na$_2$SO$_4$ 0.1 M + 0.1 ppm F$^-$ at room temperature, before measuring the current transients in anodic and cathodic environments. From the potentiodynamic curves, it can be observed that all of the coatings presented an enhanced corrosion resistance compared with the substrate. Comparing the coatings, a slight negative shift in the corrosion potential with the increase in Ti content was observed.

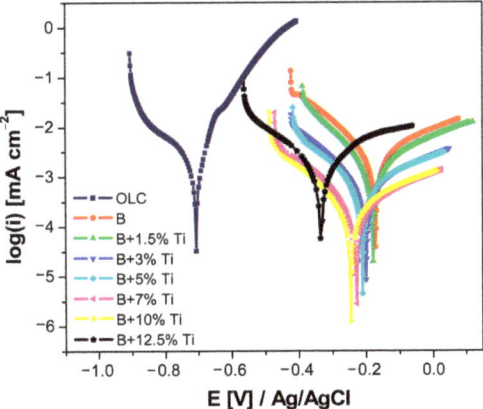

Figure 5. Potentiodynamic polarization curves recorded on all laser-cladded samples, before potentiostatic polarization, at a scan rate of 1 mV·s^{-1}.

Using the Tafel extrapolation method, the corrosion current density (i$_{corr}$) and corrosion potential (E$_{corr}$), the anodic and cathodic Tafel slopes were calculated and are presented in Table 1. From the polarization parameters, it can be seen that the current density had the highest value for the substrate. Moreover, the results indicate that the current density values recorded for the coatings decreased with increase in titanium addition, up to the addition of 10 wt% Ti. On the other hand, for the laser-cladded coating with the addition of 12.5 wt% Ti, a sudden increase in the current density values was found.

Table 1. Polarization parameters for laser-coated samples in the test solution, before the durability test.

Samples	i_{corr} [$\mu A \cdot cm^{-2}$]	E_{corr} [mV/Ag/AgCl]	$-b_c$ [$mV \cdot dec^{-1}$]	b_a [$mV \cdot dec^{-1}$]	R_p [$k\Omega \cdot cm^2$]
Substrate	2.814	−689	243	58	7
B	0.353	−158	292	176	135
B + 1.5% Ti	0.235	−165	257	144	171
B + 3% Ti	0.116	−179	212	178	362
B + 5% Ti	0.075	−199	198	168	526
B + 7% Ti	0.050	−221	184	152	723
B + 10% Ti	0.042	−241	203	121	784
B + 12.5% Ti	1.191	−342	176	150	30

Based on the polarization curves and Tafel parameters, the coatings showed good corrosion results up until the addition of 10 wt% Ti. The results were supported by EIS measurements conducted using the open circuit potential value (Figure 6). In the Nyquist plots (Figure 6a), all laser-cladded samples showed a single capacitive loop, with a substantial increment in the semicircle radius as the Ti content increased within the coating, up until the addition of 10 wt% Ti. Then, a sudden decrease was noticed. A low spectra diameter was obtained for the coating with the addition of 12.5 wt% Ti.

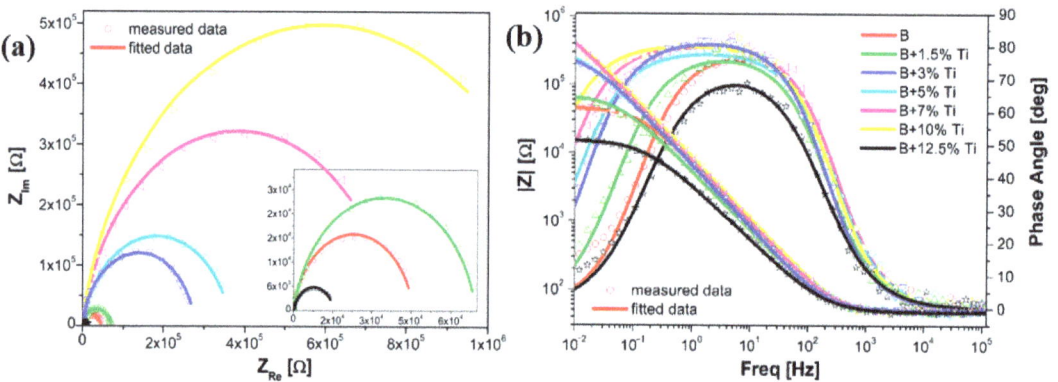

Figure 6. Nyquist (**a**) and Bode (**b**) diagrams recorded on all laser cladded samples before potentiostatic polarization.

Bode plots of the laser-cladded coatings are presented in Figure 6b. As the Ti addition increased within the coating, the Bode-|Z| diagrams showed higher impedance modulus values, while the Bode-phase diagrams showed higher phase angle plots. These results could be attributed to the development of a passive film on the surface of the coating with a growth of the corrosion resistance. A sudden decrease in the impedance modulus values and phase angle plots were noticed for laser-cladded coating with the addition of 12.5% Ti. As the sample with 12.5 wt% Ti content showed a lower corrosion resistance than the rest of the coatings, the current transients were measured only for laser-cladded samples up until the addition of 10 wt% Ti. The repeated EIS and Tafel plots were used to characterize the coatings after the durability test.

The EIS data were fitted using the equivalent electric circuit shown in Figure 7, and the calculated values of the circuit elements used for modeling the samples are presented in Table 2. In EEC, R_s stands for the solution resistance between the working and reference electrodes, R_{ct} represents the charge transfer resistance at the coating/electrolyte interface, and CPE indicates the constant phase element. From the results shown in Table 2, similar values of solution resistance were obtained for all of the samples, indicating the same ion conductivity in the test solution. As the Ti content increased within the coating, the charge transfer resistance had higher values, indicating a lower reaction rate.

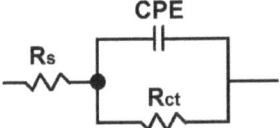

Figure 7. Electrical equivalent circuit used for the spectra fitting.

Table 2. Calculated values of the circuit elements for modeling laser-cladded samples in the test solution, after potentiostatic polarization.

Samples	R_s [$\Omega \cdot cm^{-2}$]	Y_0 [$S \cdot cm^{-2} \cdot s^n$]	n	R_{ct} [$\Omega \cdot cm^{-2}$]	Chi
B	47.56	2.93×10^{-5}	0.87	4.64×10^4	5.83×10^{-3}
B + 1.5% Ti	46.87	3.02×10^{-5}	0.88	6.92×10^4	4.77×10^{-3}
B + 3% Ti	49.50	3.33×10^{-5}	0.90	9.98×10^4	4.95×10^{-3}
B + 5% Ti	47.12	3.64×10^{-5}	0.91	2.65×10^5	6.40×10^{-3}
B + 7% Ti	47.77	3.87×10^{-5}	0.92	7.07×10^5	4.84×10^{-3}
B + 10% Ti	47.08	4.91×10^{-5}	0.92	7.95×10^5	7.12×10^{-3}
B + 12.5% Ti	47.41	1.66×10^{-4}	0.84	3.85×10^4	4.14×10^{-3}

To investigate the lower corrosion resistance of the coating manufactured with the addition of 12.5% Ti after exposure to the electrolyte, the surface was analyzed using SEM. According to the micrographs presented in Figure 8, the surface contained corrosion products and micro-cracks. The EDX spectra collected in the area marked with the red rectangle in Figure 8b show the presence of Fe and O. The increased amount of Ti led to the formation of micro-cracks within the coatings, which allowed the electrolyte to reach the substrate [38]. The micro-cracks appeared due to the residual stress caused by the thermal and chemical effects [39], attributed to the rapid cooling and to the higher amount of Ti, leading to a lower corrosion resistance.

3.4. Potentiostatic Polarization

The durability of laser-cladded coatings in simulated cathodic and anodic PEMFC environments were considered by measuring the current transients of the samples at a constant potential, as a function of time. Figure 9 shows the potentiostatic polarization curves of every coating, measured for 6 h, at a constant potential of +0.736 V and −0.493 V, which corresponded to the release of oxygen and hydrogen, respectively. Figure 9a presents the current transients measured in the cathodic PEMFC environment. It can be seen that the current density of all of the coated samples initially increased from a negative direction and then gradually stabilized at low negative values. The negative current density values could be explained because the corrosion potential of the laser-coated samples was more positive than the applied potential. Based on this, the negative values were attributed to a reduction in H$^+$ ions, which changed to H$_2$, thus providing cathodic protection to the laser-coated samples. During potentiostatic polarization, all of the samples displayed a stable corrosion current density. The cathodic current density obtained for the B coating was around −2.21 µA·cm^{-2}, while the laser-cladded coatings with Ti additions showed lower values of between −1.55 and −0.89 µA·cm^{-2}. Figure 9b presents the current transients measured in the anodic PEMFC environment. It can be seen that the current density of all of the coated samples initially decreased from a positive direction and then gradually stabilized at low positive values. This behavior could be explained by the formation of a passive film on the surface of the samples. It is clear that the current densities decreased with the increase in Ti content within the coating. During the measurement, all of the samples displayed a stable corrosion current density. The anodic current density obtained for the B coating was around 3.55 µA·cm^{-2}, while the laser-cladded coatings with Ti additions showed lower values of between 1.52 and 0.81 µA·cm^{-2}. From the presented results, the NiCr-based coating with the addition of 10 wt% Ti showed the lowest current density in both the cathodic and anodic environment.

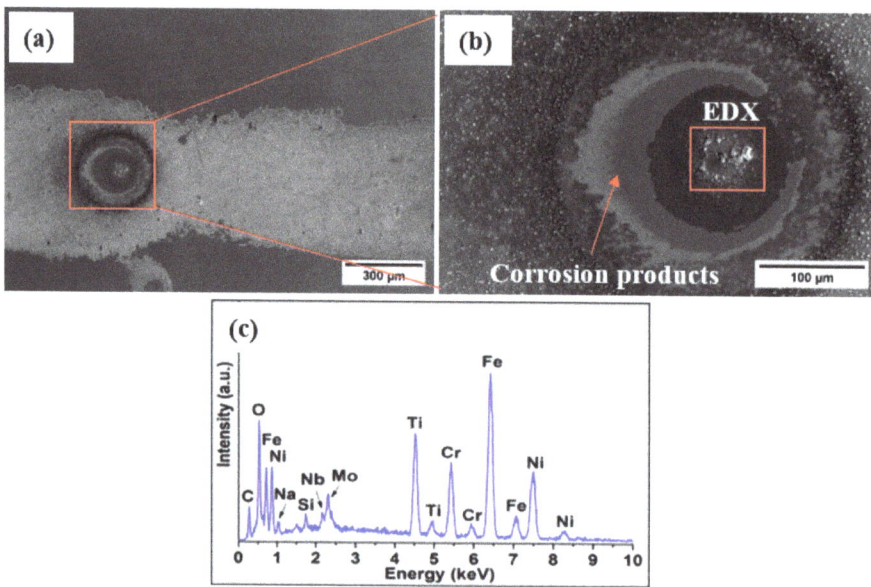

Figure 8. Surface morphology of the coating manufactured with 12.5% Ti exposed to the electrolyte: (**a**) corrosion products, (**b**) corrosion products and micro-crack, and (**c**) EDX spectra reveling the presence of Fe and O on the surface of the coating.

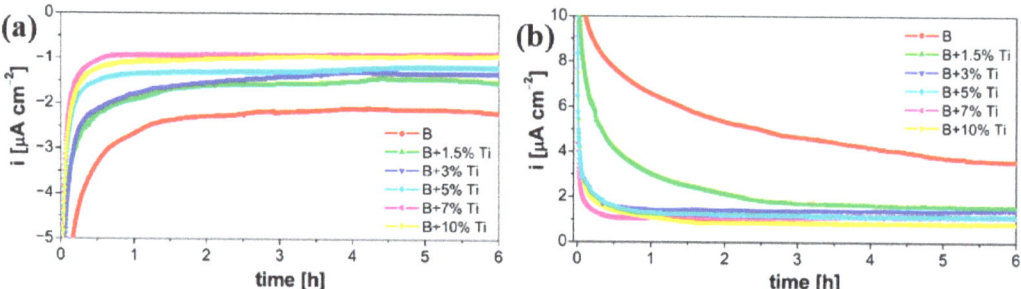

Figure 9. Potentiostatic polarization curves of laser-cladded coatings recorded for 6 h, at a constant potential of: (**a**) +0.736 V and (**b**) −0.493 V.

3.5. Electrochemical Evaluation after Durability Test

Figure 10 presents the potentiodynamic polarization curves of the laser-cladded samples with additions of up to 10 wt% Ti content, after measuring the current transients in the simulated PEMFC environment. Using the Tafel extrapolation method, the polarization parameters of the laser-cladded samples are calculated and listed in Table 3. From the polarization parameters, the corrosion current densities (i_{corr}) of the laser-cladded coatings, in both the cathodic and anodic environments, had lower values with the increase in titanium content. On the other hand, the polarization resistance (R_p) in both cases had higher values for coatings with a greater Ti content. After simulating the PEMFC cathodic environment (Figure 10a), the corrosion potential of the laser-cladded samples had a slight positive shift with the increase in Ti content. All of the laser-cladded coatings with the addition of Ti had a higher corrosion potential (E_{corr}) that the B coating. The same behavior of the corrosion potential could be observed after simulating the PEMFC anodic environment (Figure 10b).

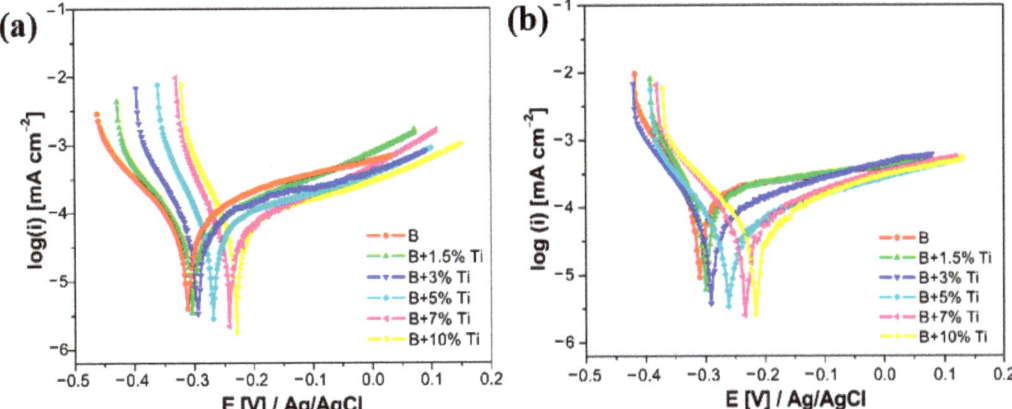

Figure 10. Potentiodynamic polarization curves recorded at a 1 mV·s^{-1} scan rate on laser-cladded samples: (**a**) after cathodic and (**b**) anodic potentiostatic polarization.

Table 3. Polarization parameters for laser-coated samples in the test solution after the durability test.

Samples	i_{corr} [µA·cm^{-2}]	E_{corr} [mV/Ag/AgCl]	$-b_c$ [mV·dec^{-1}]	b_a [mV·dec^{-1}]	R_p [kΩ·cm^2]
		Cathodic environment			
B	0.477	−315	182	134	70
B + 1.5% Ti	0.363	−308	176	122	86
B + 3% Ti	0.232	−298	183	130	142
B + 5% Ti	0.193	−276	188	119	164
B + 7% Ti	0.161	−245	186	123	200
B + 10% Ti	0.149	−236	172	120	206
		Anodic environment			
B	0.868	−320	178	122	36
B + 1.5% Ti	0.623	−312	191	120	51
B + 3% Ti	0.468	−305	179	118	65
B + 5% Ti	0.213	−268	162	115	137
B + 7% Ti	0.185	−236	174	123	169
B + 10% Ti	0.177	−227	176	117	173

Comparing the Tafel plots before and after potentiostatic polarization, a negative shift on the corrosion potential for samples up to 5% Ti content, in both cathodic and anodic environments, can be seen. On the other hand, in both environments, a small positive shift in the corrosion potential for coatings with the addition of 7 wt% and 10 wt% Ti was observed. Comparing the polarization parameters before and after potentiostatic polarization, an increase in the corrosion current density for all of the laser-cladded samples was noticed, but the values were all lower than the DOE's target for 2020, which is 1 µA·cm^{-2}.

Figure 11 shows the Nyquist and Bode plots for the coated samples after measuring the current transients. From the Nyquist plots, in both the cathodic and anodic environments (Figure 11a,c), all of the laser-cladded samples showed a single capacitive loop, with a substantial increment in the semicircle radius as the Ti content increased within the coating. In both cases, a low spectra diameter was obtained for coatings without the addition of Ti and the highest spectra diameter was recorded for coatings with the addition of 10 wt% Ti. The Bode plots in both the cathodic and anodic environments are presented in Figure 11b,d, respectively. In both cases, the Bode-|Z| diagrams showed higher impedance modulus values and higher Bode-phase angle plots as the Ti content increased within the coating.

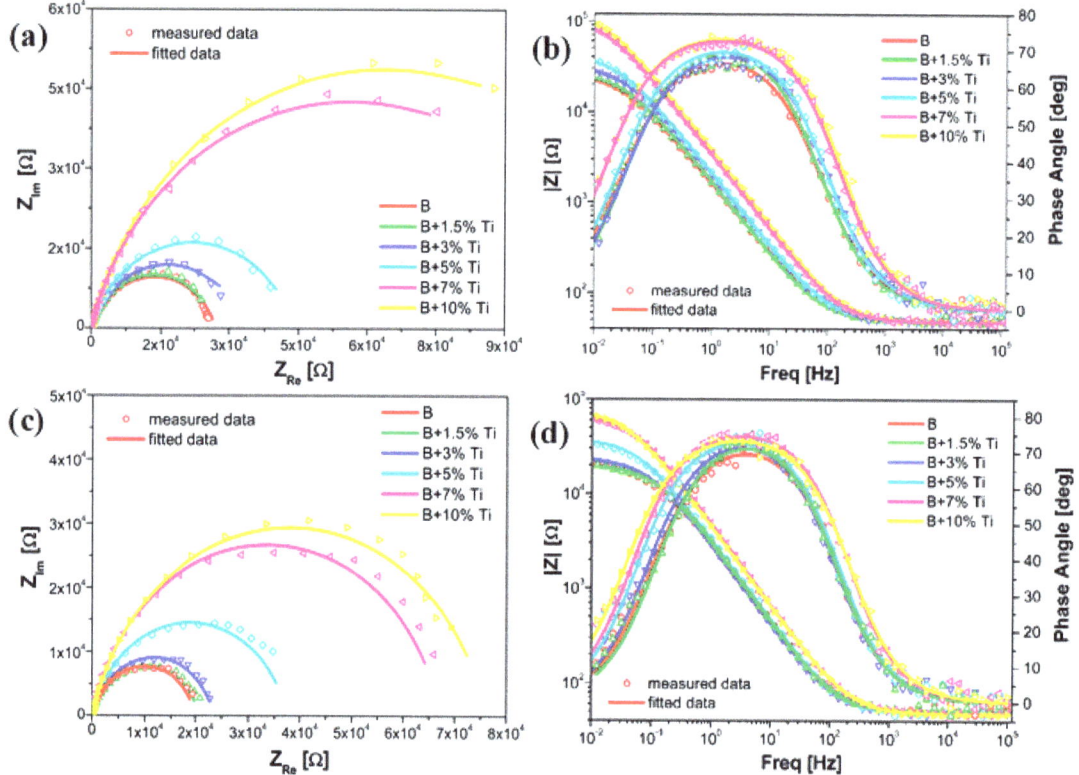

Figure 11. EIS measurements recorded on laser-cladded samples: (**a**,**b**) after cathodic and (**c**,**d**) anodic potentiostatic polarization.

Comparing the Nyquist plots before and after potentiostatic polarization, in both environments, the diameter of the semicircle decreased substantially after the durability test. Furthermore, after potentiostatic polarization, the Bode-|Z| and Bode-phase angle diagrams showed lower impedance modulus values and phase angle plots, respectively. The EIS data, after potentiostatic polarization, were fitted using the same equivalent circuit model, as presented in Figure 7. From the calculated values of the circuit elements after potentiostatic polarization, as shown in Table 4, similar values for the solution resistance were obtained for all of the samples, in both environments. Furthermore, as the Ti content increased within the coating, the charge transfer resistance had higher values. On the other hand, when comparing the results from the EEC before and after potentiostatic polarization, in both environments, a decrease in the charge transfer resistance values was noticed.

Table 4. Calculated values of the circuit elements for modeling the laser-coated samples in the test solution after potentiostatic polarization.

Samples	R_s [$\Omega \cdot cm^{-2}$]	Y_0 [$S \cdot cm^{-2} \cdot s^n$]	n	R_{ct} [$\Omega \cdot cm^{-2}$]	Chi
Cathodic environment					
B	55.61	6.17×10^{-4}	0.85	2.18×10^4	1.02×10^{-3}
B + 1.5% Ti	53.42	6.55×10^{-4}	0.85	2.71×10^4	2.73×10^{-3}
B + 3% Ti	48.13	7.39×10^{-4}	0.87	3.29×10^4	1.91×10^{-3}
B + 5% Ti	47.92	7.75×10^{-4}	0.88	1.35×10^5	2.09×10^{-3}
B + 7% Ti	47.65	8.36×10^{-4}	0.89	1.87×10^5	1.78×10^{-3}
B + 10% Ti	47.35	9.47×10^{-4}	0.90	2.55×10^5	1.53×10^{-3}
Anodic environment					
B	46.93	2.68×10^{-4}	0.77	1.16×10^4	2.42×10^{-3}
B + 1.5% Ti	49.87	2.87×10^{-4}	0.82	1.41×10^4	2.46×10^{-3}
B + 3% Ti	48.51	2.96×10^{-4}	0.87	1.86×10^4	4.32×10^{-3}
B + 5% Ti	47.43	3.11×10^{-4}	0.89	1.98×10^4	3.05×10^{-3}
B + 7% Ti	47.58	3.14×10^{-4}	0.90	2.04×10^4	2.91×10^{-3}
B + 10% Ti	46.62	3.53×10^{-4}	0.91	2.33×10^4	3.73×10^{-3}

4. Conclusions

Potential metallic bipolar plate materials were successfully manufactured using laser cladding NiCr-based alloys with different Ti additions on a low carbon steel substrate. The titanium content within the coating was added, from 1.5 to 12.5 wt% Ti, and the variation in titanium was studied in terms of the microstructure and corrosion resistance. The study focused on electrochemically testing the coatings in a mild solution, such as 0.1 M Na_2SO_4 (acidulated with H_2SO_4 at pH = 5) with the addition of 0.1 ppm F^-. The results led to the following conclusions:

- The microstructural investigation revealed a dendritic structure for all of the laser cladded coatings. Furthermore, as the Ti content increased within the coating, the size of the dendrites increased. The formation of precipitates and secondary phases was detected by EDX analysis and it was observed that the number of hard phases and the TiC particle size increased with the increase in Ti content within the coatings. The number of hard phases had a direct influence on the hardness measurement; thus, the coating containing 12.5% Ti presented the highest values.
- From the EIS measurements and potentiodynamic polarization, before measuring the current transients, it was noticed that a high content of Ti within the NiCr-base coating could provide a better corrosion resistance, up to the addition of 10% wt Ti. The corrosion current density of all coatings up to the addition of 10% wt Ti was less than DOE's 2020 target of 1 $\mu A \cdot cm^{-2}$.
- It was noticed that the coating with the addition of 12.5% Ti addition presented a lower corrosion resistance due to the micro-cracks formed due to residual stress, which might be attributed to the higher amount of Ti present in the coating.
- Potentiostatic polarization measurements revealed that the coating with 10 wt% Ti addition had the lowest and most stable current density in the PEMFC anodic and cathodic environments, respectively.
- After measuring the current transients of the laser-cladded samples, the EIS measurements and potentiodynamic polarization were repeated. The results revealed that even if the corrosion resistance decreased after the durability test, the corrosion current density for all of the coatings in both environments was less than DOE's 2020 target of 1 $\mu A \cdot cm^{-2}$. In conclusion, from the above results, we can say that the coating with the addition of 10 wt% Ti had great potential as a BP material for PEMFC applications.

Author Contributions: Conceptualization, D.N.A. and C.M.D.; methodology, D.N.A. and E.M.S.; software, I.H. and M.L.D.; validation, J.C.M.-R. and C.M.D.; formal analysis, A.P. and E.M.S.; investigation, A.P., D.N.A., and M.L.D.; resources, I.H. and A.P.; data curation, writing—original draft preparation, D.N.A.; writing—review and editing, I.H. and J.C.M.-R.; visualization, E.M.S. and C.M.D.; supervision, C.M.D. and J.C.M.-R. All authors have read and agreed to the published version of the manuscript.

Funding: This research received no external funding.

Institutional Review Board Statement: Not applicable.

Informed Consent Statement: Not applicable.

Data Availability Statement: Not applicable.

Conflicts of Interest: The authors declare no conflict of interest.

References

1. Shilling, F. *Greenhouse Gases*; Facts on File: New York, NY, USA, 1995; Volume 375, ISBN 9780816072644.
2. Dalal, R.C.; Allen, D.E. Turner Review No. 18. Greenhouse Gas Fluxes from Natural Ecosystems. *Aust. J. Bot.* **2008**, *56*, 369–407. [CrossRef]
3. Ghommem, M.; Hajj, M.R.; Puri, I.K. Influence of Natural and Anthropogenic Carbon Dioxide Sequestration on Global Warming. *Ecol. Modell.* **2012**, *235–236*, 1–7. [CrossRef]
4. Crosthwaite, J. Just Transition to a Sustainable Future without 'Fossil' Gas. In *Ecological Economics: Solutions for the Future*; Washington, H., Ed.; 2020; p. 240. ISBN 9798662828902.
5. Cao, X.; Dai, X.; Liu, J. Building Energy-Consumption Status Worldwide and the State-of-the-Art Technologies for Zero-Energy Buildings during the Past Decade. *Energy Build.* **2016**, *128*, 198–213. [CrossRef]
6. Parry, M.; Canziani, O.; Palutikof, J.; van der Linden, C.H.P. *Climate Change 2007: Impacts, Adaptation and Vulnerability*; Cambridge University Press: Cambridge, UK, 2007; ISBN 978 0 521 88010-7.
7. Sharma, P.; Pandey, O.P. Proton Exchange Membrane Fuel Cells: Fundamentals, Advanced Technologies, and Practical Applications. *PEM Fuel Cells Fundam. Adv. Technol. Pract. Appl.* **2021**, 1–24. [CrossRef]
8. Ma, S.; Lin, M.; Lin, T.E.; Lan, T.; Liao, X.; Maréchal, F.; Van herle, J.; Yang, Y.; Dong, C.; Wang, L. Fuel Cell-Battery Hybrid Systems for Mobility and off-Grid Applications: A Review. *Renew. Sustain. Energy Rev.* **2021**, *135*, 110119. [CrossRef]
9. Sapkota, P.; Boyer, C.; Dutta, R.; Cazorla, C.; Aguey-Zinsou, K.F. Planar Polymer Electrolyte Membrane Fuel Cells: Powering Portable Devices from Hydrogen. *Sustain. Energy Fuels* **2020**, *4*, 439–468. [CrossRef]
10. Jemeï, S. *Hybridization, Diagnostic and Prognostic of Proton Exchange Membrane Fuel Cells*; Wiley: Hoboken, NJ, USA, 2018; ISBN 978-1-78630-167-3.
11. Manso, A.P.; Marzo, F.F.; Garicano, X.; Alegre, C.; Lozano, A.; Barreras, F. Corrosion Behavior of Tantalum Coatings on AISI 316L Stainless Steel Substrate for Bipolar Plates of PEM Fuel Cells. *Int. J. Hydrogen Energy* **2020**, *45*, 20679–20691. [CrossRef]
12. Khosravi, H.S.; Abbas, Q.; Reichmann, K. Electrochemical Aspects of Interconnect Materials in PEMFCs. *Int. J. Hydrogen Energy* **2021**, *46*, 35420–35447. [CrossRef]
13. Asri, N.F.; Husaini, T.; Sulong, A.B.; Majlan, E.H.; Daud, W.R.W. Coating of Stainless Steel and Titanium Bipolar Plates for Anticorrosion in PEMFC: A Review. *Int. J. Hydrogen Energy* **2017**, *42*, 9135–9148. [CrossRef]
14. Wang, L.; Sun, J.; Kang, B.; Li, S.; Ji, S.; Wen, Z.; Wang, X. Electrochemical Behaviour and Surface Conductivity of Niobium Carbide-Modified Austenitic Stainless Steel Bipolar Plate. *J. Power Sources* **2014**, *246*, 775–782. [CrossRef]
15. Joseph, S.; McClure, J.C.; Sebastian, P.J.; Moreira, J.; Valenzuela, E. Polyaniline and Polypyrrole Coatings on Aluminum for PEM Fuel Cell Bipolar Plates. *J. Power Sources* **2008**, *177*, 161–166. [CrossRef]
16. Bai, C.Y.; Wen, T.M.; Hou, K.H.; Ger, M. Der The Bipolar Plate of AISI 1045 Steel with Chromized Coatings Prepared by Low-Temperature Pack Cementation for Proton Exchange Membrane Fuel Cell. *J. Power Sources* **2010**, *195*, 779–786. [CrossRef]
17. Bai, C.Y.; Wen, T.M.; Hou, K.H.; Pu, N.W.; Ger, M. Der The Characteristics and Performance of AISI 1045 Steel Bipolar Plates with Chromized Coatings for Proton Exchange Membrane Fuel Cells. *Int. J. Hydrogen Energy* **2011**, *36*, 3975–3983. [CrossRef]
18. Chiang, T.Y.; Ay-Su; Tsai, L.C.; Lee, H.B.; Lin, C.Y.; Sheu, H.H.; Chang, C.C. Effect of Metal Bipolar Plate Channel Fabrication on Electroplating—Using Nickel Electroplating of AISI 1045 Channel Substrate as an Example. *Int. J. Electrochem. Sci.* **2015**, *10*, 1926–1939.
19. Wang, S.H.; Peng, J.; Lui, W.B. Surface Modification and Development of Titanium Bipolar Plates for PEM Fuel Cells. *J. Power Sources* **2006**, *160*, 485–489. [CrossRef]
20. Lin, C.H.; Tsai, S.Y. An Investigation of Coated Aluminium Bipolar Plates for PEMFC. *Appl. Energy* **2012**, *100*, 87–92. [CrossRef]
21. Nikam, V.V.; Reddy, R.G. Corrosion Studies of a Copper-Beryllium Alloy in a Simulated Polymer Electrolyte Membrane Fuel Cell Environment. *J. Power Sources* **2005**, *152*, 146–155. [CrossRef]
22. Wang, H.; Turner, J.A. Reviewing Metallic PEMFC Bipolar Plates. *Fuel Cells* **2010**, *10*, 510–519. [CrossRef]

23. Silva, R.F.; Franchi, D.; Leone, A.; Pilloni, L.; Masci, A.; Pozio, A. Surface Conductivity and Stability of Metallic Bipolar Plate Materials for Polymer Electrolyte Fuel Cells. *Electrochim. Acta* **2006**, *51*, 3592–3598. [CrossRef]
24. Pozio, A.; Zaza, F.; Masci, A.; Silva, R.F. Bipolar Plate Materials for PEMFCs: A Conductivity and Stability Study. *J. Power Sources* **2008**, *179*, 631–639. [CrossRef]
25. Bai, C.Y.; Der Ger, M.; Wu, M.S. Corrosion Behaviors and Contact Resistances of the Low-Carbon Steel Bipolar Plate with a Chromized Coating Containing Carbides and Nitrides. *Int. J. Hydrogen Energy* **2009**, *34*, 6778–6789. [CrossRef]
26. Avram, D.N.; Davidescu, C.M.; Dan, M.L.; Mirza-Rosca, J.C.; Hulka, I.; Pascu, A.; Stanciu, E.M. Electrochemical Evaluation of Protective Coatings with Ti Additions on Mild Steel Substrate with Potential Application for PEM Fuel Cells. *Materials* **2022**, *15*, 5364. [CrossRef] [PubMed]
27. Avram, D.N.; Davidescu, C.M.; Dan, M.L.; Mirza-Rosca, J.C.; Hulka, I.; Stanciu, E.M.; Pascu, A. Corrosion Resistance of NiCr(Ti) Coatings for Metallic Bipolar Plates. *Mater. Today Proc.* **2023**, *72*, 538–543. [CrossRef]
28. Scendo, M.; Staszewska-Samson, K.; Danielewski, H. Corrosion Behavior of Inconel 625 Coating Produced by Laser Cladding. *Coatings* **2021**, *11*, 759. [CrossRef]
29. Lædre, S.; Kongstein, O.E.; Oedegaard, A.; Seland, F.; Karoliussen, H. The Effect of PH and Halides on the Corrosion Process of Stainless Steel Bipolar Plates for Proton Exchange Membrane Fuel Cells. *Int. J. Hydrogen Energy* **2012**, *37*, 18537–18546. [CrossRef]
30. Feng, K.; Wu, G.; Li, Z.; Cai, X.; Chu, P.K. Corrosion Behavior of SS316L in Simulated and Accelerated PEMFC Environments. *Int. J. Hydrogen Energy* **2011**, *36*, 13032–13042. [CrossRef]
31. Hinds, G.; Brightman, E. Towards More Representative Test Methods for Corrosion Resistance of PEMFC Metallic Bipolar Plates. *Int. J. Hydrogen Energy* **2015**, *40*, 2785–2791. [CrossRef]
32. Material Product Data Sheet Nickel-Based Superalloy Powders for Laser Cladding and Laser-Additive Manufacturing. Available online: https://www.oerlikon.com/am/en/about-us/media/oerlikon-metco-offers-materials-newly-optimized-for-additive-manufacturing-applications/ (accessed on 24 September 2021).
33. Facts, Q.; Temperature, S.; Spray, C.; Isostatic, H. Material Product Data Sheet Pure Titanium and Titanium Alloy Powders. Available online: https://www.oerlikon.com/ecoma/files/DSM-0222.0_Ti-Ti_Alloys.pdf (accessed on 18 October 2022).
34. Materials Database. Available online: https://matmatch.com/materials/minfc37596-sae-j403-grade-1010 (accessed on 18 October 2022).
35. Illana, A.; de Miguel, M.T.; García-Martín, G.; Gonçalves, F.P.; Sousa, M.G.; Pérez, F.J. Experimental Study on Steam Oxidation Resistance at 600 °C of Inconel 625 Coatings Deposited by HVOF and Laser Cladding. *Surf. Coat. Technol.* **2022**, *451*, 129081. [CrossRef]
36. Hulka, I.; Utu, D.; Serban, V.A.; Negrea, P.; Lukáč, F.; Chráska, T. Effect of Ti Addition on Microstructure and Corrosion Properties of Laser Cladded WC-Co/NiCrBSi(Ti) Coatings. *Appl. Surf. Sci.* **2020**, *504*, 144349. [CrossRef]
37. Lei, Y.; Sun, R.; Tang, Y.; Niu, W. Numerical Simulation of Temperature Distribution and TiC Growth Kinetics for High Power Laser Clad TiC/NiCrBSiC Composite Coatings. *Opt. Laser Technol.* **2012**, *44*, 1141–1147. [CrossRef]
38. Avram, D.N.; Davidescu, C.M.; Dan, M.L.; Stanciu, E.M.; Pascu, A.; Mirza-Rosca, J.C.; Iosif, H. Influence of titanium additions on the electrochemical behaviour of Nicr/Ti laser cladded coatings. *Ann. "Dunarea Jos" Univ. Galati, Fascicle XII Weld. Equip. Technol.* **2022**, *33*, 107–111. [CrossRef]
39. Uebing, S.; Brands, D.; Scheunemann, L.; Schröder, J. Residual Stresses in Hot Bulk Formed Parts: Two-Scale Approach for Austenite-to-Martensite Phase Transformation. *Arch. Appl. Mech.* **2021**, *91*, 545–562. [CrossRef]

Disclaimer/Publisher's Note: The statements, opinions and data contained in all publications are solely those of the individual author(s) and contributor(s) and not of MDPI and/or the editor(s). MDPI and/or the editor(s) disclaim responsibility for any injury to people or property resulting from any ideas, methods, instructions or products referred to in the content.

Article

Some *Brassicaceae* Extracts as Potential Antioxidants and Green Corrosion Inhibitors

Ioana Maria Carmen Ienașcu [1,2], Adina Căta [1,*], Adriana Aurelia Chis [3,*], Mariana Nela Ştefănuţ [1], Paula Sfîrloagă [1], Gerlinde Rusu [4], Adina Frum [3], Anca Maria Arseniu [3], Claudiu Morgovan [3], Luca Liviu Rus [3] and Carmen Maximiliana Dobrea [3]

[1] National Institute of Research and Development for Electrochemistry and Condensed Matter, 144 Dr. A. P. Podeanu, 300569 Timisoara, Romania
[2] Department of Pharmaceutical Sciences, Faculty of Pharmacy, "Vasile Goldiș" Western University of Arad, 86 Liviu Rebreanu, 310045 Arad, Romania
[3] Preclinical Department, Faculty of Medicine, "Lucian Blaga" University of Sibiu, 550169 Sibiu, Romania
[4] Faculty of Industrial Chemistry and Environmental Engineering, Politehnica University of Timisoara, 6 C. Telbisz, 300001 Timisoara, Romania
* Correspondence: adina.cata@yahoo.com or adina.cata@incemc.ro (A.C.); adriana.chis@ulbsibiu.ro (A.A.C.)

Abstract: Glucosinolates-rich extracts of some *Brassicaceae* sources, such as broccoli, cabbage, black radish, rapeseed, and cauliflower, were obtained using an eco-friendly extraction method, in a microwave field, with 70% ethanol, and evaluated in order to establish their in vitro antioxidant activities and anticorrosion effects on steel material. The DPPH method and Folin-Ciocâlteu assay proved good antioxidant activity (remaining DPPH, 9.54–22.03%) and the content of total phenolics between 1008–1713 mg GAE/L for all tested extracts. The electrochemical measurements in 0.5 M H_2SO_4 showed that the extracts act as mixed-type inhibitors proving their ability to inhibit corrosion in a concentration-dependent manner, with a remarkable inhibition efficiency (92.05–98.33%) achieved for concentrated extracts of broccoli, cauliflower, and black radish. The weight loss experiments revealed that the inhibition efficiency decreased with an increase in temperature and time of exposure. The apparent activation energies, enthalpies, and entropies of the dissolution process were determined and discussed, and an inhibition mechanism was proposed. An SEM/EDX surface examination shows that the compounds from extracts may attach to the steel surface and produce a barrier layer. Meanwhile, the FT-IR spectra confirm bond formation between functional groups and the steel substrate.

Keywords: *Brassicaceae* extracts; antioxidant activity; total phenolics; green corrosion inhibitors

1. Introduction

Due to their exceptional mechanical and electrical properties, metals are commonly used in human activities [1]. The corrosion process is perhaps the most common phenomenon that causes the deterioration of metals and it is due to the electrochemical interaction of metals with the corrosive environment [2,3]. Among metals, steel is usually used in many industries due to its excellent mechanical properties. Thus, finding solutions to reduce the degradation of steel by corrosion represents a high-priority matter [4].

In order to fight against corrosion, some different strategies such as design, materials selection, electrochemical protection, coatings, and the use of inhibitors were applied. The latter is considered the easiest to apply and the most cost effective [5]. Corrosion inhibitors are chemicals that are added to metal surfaces or to the aggressive medium, reducing the rate of metals dissolution. The common corrosion inhibitors, mainly chromates and their derivatives, have proven to be dangerous substances for human life and the environment [6]. Recent approaches showed the potential of plant extracts as corrosion inhibitors [7–10], so the replacement of the traditional toxic corrosion inhibitors can be achieved.

Plant extracts contain phytochemical compounds with similar characteristics to organic corrosion inhibitors and show advantages such as low cost, wide availability, nontoxicity, biodegradability, and biocompatibility, which recommend them as an eco-friendly alternative to classic corrosion inhibitors [9,11].

Crops of cabbage, acclimatized broccoli, black radishes, rape, and cauliflower are cultivated in Romania and are very cheap raw materials. Natural compounds of different species of the *Brassicaceae* family were easily extracted [12–15]. Plants from the *Brassicaceae* species contain several phytochemical compounds such as glucosinolates, glucosides, phenolic acids, erucic acid, polyphenols and tocopherols, carotenoids, flavonoids, alkaloids, terpenoids and terpenes, phytoalexins, and phytosterols [16]. Thus, the extracts of such plants are used in the food industry, pharmaceutical industry, and alternative medicine [17–22] due to their diverse biological activities, mainly antimicrobial [19–22] and antioxidant effects [14,15,19].

Several extracts from some *Brassicaceae* species demonstrated excellent inhibition effects on steel corrosion [23–25]. Hence, an aqueous extract of *Brassica oleracea* was proven to retard Q235 steel corrosion in two harsh acid environments (0.5 M H_2SO_4 and 1 M HCl) [25]. The inhibition efficiency of *Brassica oleracea* extract (99% ethanol) on the corrosion of pipeline steel in 0.5 M H_2SO_4 has also been demonstrated [24]. On the other hand, some *Brassica campestris* extracts were capable of inhibiting Cor-Ten steel corrosion in HCl and NaCl solutions [23].

Given the composition of *Brassicaceae* extracts and their proven efficiency, they can successfully replace conventional toxic inhibitors and extend the possibility of "smart coatings" by inducing a response in the coating and/or substrate to improve the inhibition of corrosion [16].

The aim of this research was to demonstrate the high antioxidant qualities and corrosion inhibition power of five indigenous *Brassicaceae* species (cabbage, broccoli, black radish, rapeseed, and cauliflower) in an aggressive 0.5 M H_2SO_4 environment.

2. Materials and Methods

2.1. Materials

The 1,1 diphenyl-2-picrylhydrazyl (DPPH), Folin-Ciocâlteu reagent, gallic acid (GA), 99% ethanol (analytical grade), methanol (analytical grade), and 98% sulphuric acid were all purchased from Merck (Germany). Romanian vegetables, white cabbage, broccoli, black radish, and cauliflower were purchased from a supermarket and rapeseed was collected from a local farm (Timiș county) in 2022. Ultrapure water was obtained in the lab (EASYpure RoDi—Barnstead apparatus).

2.2. Extraction

Plant materials were washed, air dried, chopped/ground, and freeze-dried. Then, 1 g dried material and 10 mL 70% ethanol were subjected to extraction in a microwave field (2450 MHz). Extraction was carried out at 120 °C, 15 min., in an MSW-2 Berghof oven (1000 W) equipped with a rotor with 10 Teflon vessels DAP-60K. After extraction, the solid part was removed by filtration, and the liquid phase was concentrated under a vacuum to 10 mL (extracts A).

2.3. Antioxidant Activity and Total Phenolics

For spectrophotometric measurements, a Jasco V530 apparatus (Abl&E-Jasco, Wien, Austria) was used. The antioxidant activities were determined using the DPPH method [26]. The calibration curve used was A = 11048 · C_{DPPH} + 0.0037 (R^2 = 0.999). At 2.9 mL methanolic solution of DPPH (~9.5 × 10^{-5} mol/L), 0.1 mL ethanolic extract A was added and the change of DPPH color was from mauve to yellow, caused by the consumption of

DPPH radicals by the existing antioxidants compounds, was followed (λ = 515 nm). The remaining DPPH was calculated with the following equation:

$$\text{Remaining DPPH (\%)} = \frac{C_{DPPH(t)}}{C_{DPPH(t=0)}} \times 100 \quad (1)$$

where $C_{DPPH(t)}$ is the value of the DPPH concentration in the presence of the extract at time t and $C_{DPPH(t=0)}$ at time 0.

The content of total phenolics was determined by the Folin-Ciocâlteu method, as described by Ștefănuț et al. [26]. Gallic acid was used as the reference compound and the total phenolic content was expressed as mg GAE/L.

2.4. Electrochemical Experiments

The electrochemical tests were carried out with 0.5 M H_2SO_4, in a conventional glass three-electrode cell with a Pt counter electrode, a saturated calomel (SCE) as reference electrode, and a working electrode (WE) made from a steel disk. The WE was embedded in a Teflon jacket by screwing. The exposed area was A = 0.28 cm^2. Before use, the WE was gradually polished with emery paper (1000–1400), cleaned with detergent and water, and finally, with acetone. Then, 30 mL 0.5 M H_2SO_4 plus 1 mL of each extract were used in the tests. All the experiments were carried out at an open circuit (OCP) for 30 min in order to obtain a stable potential at room temperature. Electrochemical tests were performed with a Voltalab 80 (Radiometer, Denmark). The potentiodynamic measurements were started at −600 mV cathodic potential to anodic potential +250 mV, at a scan rate 1 mV/s. The data were registered and analyzed using VoltaMaster4 software. Parameters such as corrosion potential E_{corr}, corrosion resistance Rp, Tafel slopes (βa, βc), β corrosion intensity I_{corr}, and corrosion rate, v_{corr}, were obtained by the Tafel extrapolation method.

The inhibition efficiency was defined by Equation (2).

$$IE(\%) = \left(\frac{v^0_{corr} - v_{corr}}{v^0_{corr}}\right) \times 100 \quad (2)$$

where: v^0_{corr} and v_{corr} are corrosion rates in the absence of and in the presence of different extracts, respectively.

2.5. Weight-Loss Experiments

Similar steel disks (same composition and dimensions) as those used in the electrochemical experiments were polished with different grades of emery paper (1000–1400 mesh), washed with ultrapure water, degreased with acetone, and air dried. The specimens were immersed in the corrosion medium (15 mL 0.5 M H_2SO_4) and kept for 24 h at 20, 40, 50, and 60 °C, in the absence and the presence of 0.5 mL of extract A. Finally, the steel disks were removed, rinsed with water and acetone, dried in warm air, and stored in a desiccator. Weight loss was determined by gravimetric measurements using an analytical balance with a precision of 0.1 mg.

The corrosion rate (v_{corr}, mm/year) of steel in 0.5 M H_2SO_4 with and without extract A was calculated with Formula (3) [27]:

$$v_{corr} = \frac{87.6 \cdot \Delta W}{S \cdot t \cdot D} \quad (3)$$

where ΔW is the corrosion weight loss of the steel specimen (mg), S is the area of the steel specimen (cm^2), t is the exposure time (h), and D the density of steel (g/cm^3).

The inhibition efficiency was obtained using Equation (2), using for calculation the values of corrosion rate (v_{corr}) obtained by the gravimetric method.

The temperature effect on the corrosion rate of steel in 0.5 M H_2SO_4 was studied. These tests were executed in the absence and presence of 0.5 mL extracts A for 24 h, at 20,

40, 50, and 60 °C. The relationship between the corrosion rate (v_{corr}) of steel in an acidic media and temperature (T) is expressed by the Arrhenius equation [28]:

$$v_{corr} = A \cdot e^{-E_a/RT} \quad (4)$$

where v_{corr} is the corrosion rate, A is the Arrhenius pre-exponential factor, E_a the apparent activation energy for corrosion process, R is the universal gas constant, and T the absolute temperature.

The values of enthalpy of activation (ΔH^*) and entropy of activation (ΔS^*) were calculated using an alternative form of the Arrhenius equation [27]:

$$v_{corr(G)} = \frac{RT}{Nh} \cdot e^{\Delta S^*/R} \cdot e^{-\Delta H^*/RT} \quad (5)$$

where h is the Planck's constant, N is the Avogadro's number, T is the absolute temperature, and R is the universal gas constant.

2.6. Scanning Electron Microscopy (SEM) and EDX Studies

After the corrosion tests, the steel disks were washed and dried, and subjected to an SEM/EDX analysis. The SEM images and the atomic content were registered using the scanning microscopy method (Scanning Electron Microscope Inspect S + EDAX Genesis XM 2i—FEI, Holland), at 30 kV, in vacuum mode, at 400–6000 magnification for all the samples.

2.7. FT-IR Analysis

The FT-IR spectra were recorded using a Bruker Vertex 70 spectrometer (Bruker Optik GmbH, Rosenheim, Germany) equipped with a Platinum ATR unit, Bruker Diamond A225/Q.1., at room temperature (4000–400 cm^{-1}) with a nominal resolution of 4 cm^{-1} with 64 scans.

3. Results

Five extracts from Romanian cabbage, acclimatized broccoli, black radish, cauliflower, and rapeseed were obtained (extracts A) and were analyzed by the UV-Vis technique, in order to evaluate their antioxidant effect and total phenolics. Figure 1 shows the dependence of remaining DPPH (%) on time and permits the evaluation of antioxidant activities of the studied extracts. The values obtained for antioxidant activities and the total phenolic content of the extracts A are presented in Table 1.

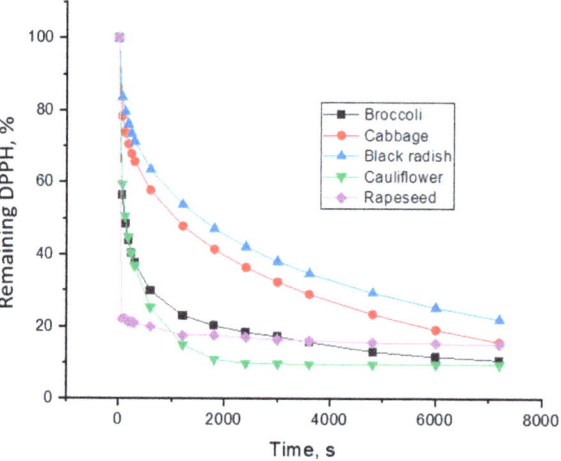

Figure 1. Antioxidant behavior of *Brassicaceae* extracts A.

Table 1. Antioxidant activities and total phenolic content of extracts A.

Extracts	Antioxidant Activity, Remaining DPPH, %	Total Phenolics, mg GAE/L
Broccoli	11.75	1623.4 ± 25.1
Cabbage	15.79	1008.8 ± 9.6
Black radish	22.03	1313.5 ± 16.4
Rapeseed	15.13	1713.0 ± 42.8
Cauliflower	9.54	1380.7 ± 19.4

For the electrochemical tests, two concentrations of the extracts were used, one corresponding to the extract A and one obtained by 20 times dilution of the extract A (extract B). The electrochemical behavior was evaluated using an experimental cell with a disk work electrode (Figure 2), the steel composition of the work electrode is presented in Table 2.

Figure 2. Electrochemical cell and disk work electrode.

Table 2. The steel composition of the working electrode.

Element	C	Si	Mn	P	S	Fe
Percentage, wt.%	0.12–0.15	0.10–0.35	0.70–1.10	0.03	0.07–0.13	98.21–98.98

The polarization curves for steel in 0.5 M H_2SO_4 (blank) with and without the two tested concentrations of broccoli, cauliflower, black radish, cabbage, rapeseed extracts are presented in Figure 3.

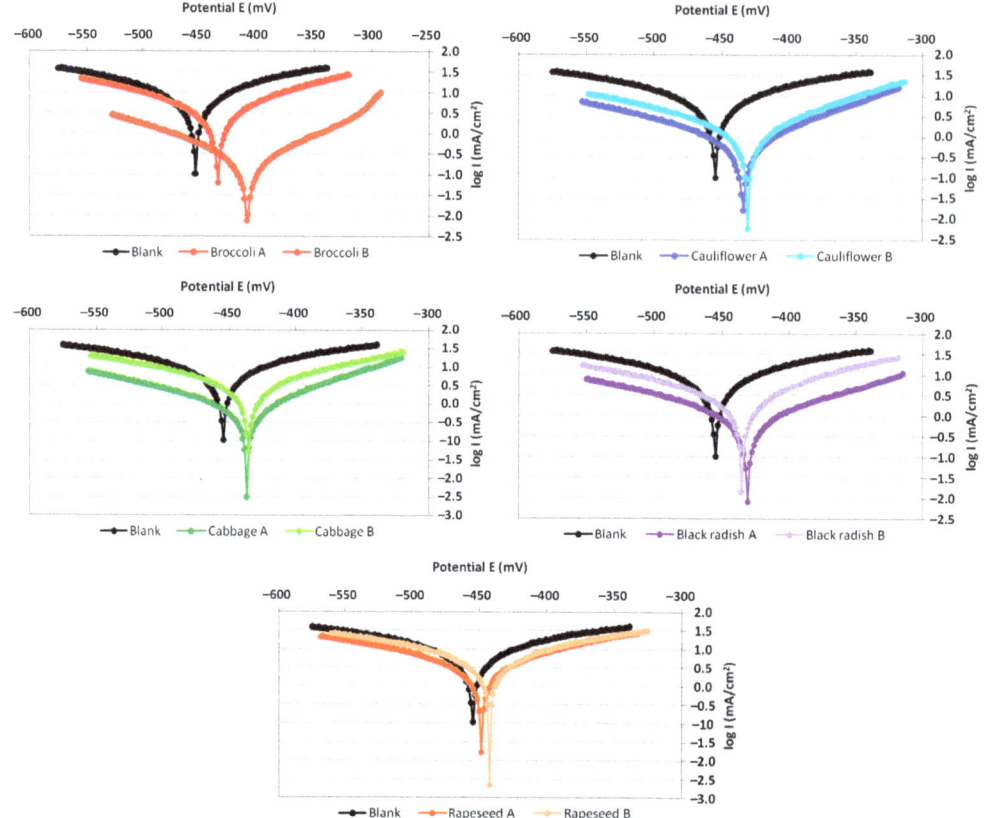

Figure 3. Potentiodynamic polarization curve for steel in 0.5 M sulfuric acid in the absence and the presence of *Brassicaceae* extracts A and B.

The electrochemical parameters, corrosion potential (E_{corr}), corrosion resistance (R_p), corrosion current density (I_{corr}), cathodic Tafel slope (βc), anodic Tafel slope (βa), corrosion rate (v_{corr}), and inhibition efficiency (IE), obtained from potentiodynamic polarization curves, are presented in Table 3.

Table 3. Potentiodynamic polarization parameters.

Sample	E_{corr}, mV	R_p, Ohm·cm²	I_{corr}, mA/cm²	βa, mV	βc, mV	v_{corr}, mm/year	IE, %
0.5 M H₂SO₄	−454.7	3.20	10.13	189.6	−202.2	118.5	-
Broccoli A	−408.9	84.60	0.17	82.6	−83.8	1.98	98.33
Broccoli B	−435.4	6.83	4.83	154.7	−187.6	56.54	52.30
Cabbage A	−436.6	24.54	1.05	95.2	−141.3	12.26	89.65
Cabbage B	−435.3	7.39	3.26	130.8	−145.0	38.14	67.82
Cauliflower A	−433.9	25.73	0.68	84.1	−108.2	7.97	93.27
Cauliflower B	−430.4	15.16	1.28	90.8	−116.4	14.99	87.35
Black radish A	−430.8	23.80	0.81	108.0	−110.8	9.43	92.05
Black radish B	−434.9	8.84	2.74	108.9	−146.5	32.02	72.98
Rapeseed A	−448.6	6.95	4.78	152.5	−184.2	55.96	52.79
Rapeseed B	−442.5	4.87	7.39	186.6	−212.3	86.38	27.13

Figure 4 shows the SEM images registered for the surface of the steel specimen before and after the electrochemical experiments, with and without extracts A and B.

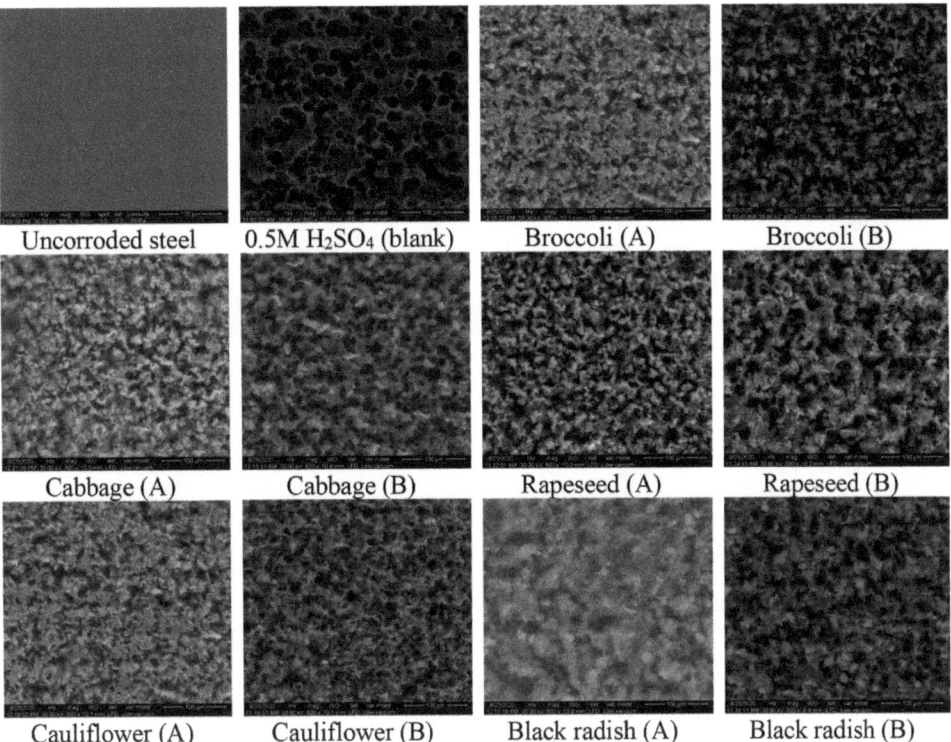

Figure 4. SEM images (800×) for steel, exposed to 0.5 M H_2SO_4 with and without *Brassicaceae* inhibitors, after electrochemical tests.

The results from the weight loss measurements for the corrosion of steel disks in 0.5 M H2SO4 in the absence and presence of extracts A, for 24 h, at four different temperatures, are given in Table 4.

Table 4. Weight loss results of steel in 0.5 M H_2SO_4, at different temperatures, in the absence and presence of *Brassicaceae* extracts A.

Sample	v_{corr} (mm/Year)				IE (%)			
	20 °C	40 °C	50 °C	60 °C	20 °C	40 °C	50 °C	60 °C
0.5 M H_2SO_4	58.66	85.70	90.41	91.69	-	-	-	-
Broccoli	0.78	33.67	56.50	79.57	98.66	60.71	37.51	13.22
Cabbage	40.98	67.41	83.12	89.57	30.13	21.35	8.06	2.32
Cauliflower	22.31	63.82	78.55	89.19	61.96	25.53	13.11	2.73
Black radish	13.55	61.59	74.13	89.31	76.90	28.14	18.00	2.60
Rapeseed	45.47	75.33	86.37	89.71	22.48	12.10	4.46	2.16

The apparent activation energies (Ea) were determined by linear regression between ln v_{corr} and 1000/T (Figure 5a) and the results are shown in Table 5. Straight lines with a regression coefficient close to unity ($R^2 > 0.9$) were plotted, from which the apparent activation energies (E_a) obtained from the slope ($-E_a/2.303R$) of the lines were determined.

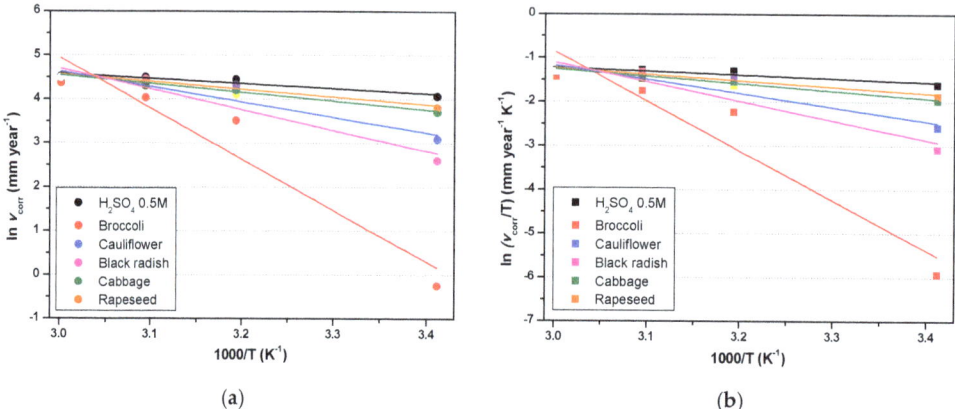

Figure 5. Arrhenius plots for steel in 0.5 M H_2SO_4 in the absence and presence of *Brassicaceae* extracts A: (**a**) $\ln v_{corr}$ vs. 1000/T; (**b**) $\ln(v_{corr}/T)$ vs. 1000/T.

Table 5. Activation parameters for steel in 0.5 M H_2SO_4 in the absence and presence of *Brassicaceae* extracts A.

Inhibitor	Linear Regression Equation (4)	R^2 Equation (4)	E_a (kJ mol^{-1})	Linear Regression Equation (5)	R^2 Equation (5)	ΔH^* (kJ mol^{-1})	ΔS^* (J mol^{-1} K^{-1})
H_2SO_4	y = −0.495x + 3.476	0.893	9.47	y = −0.828x + 1.260	0.811	6.88	−187.06
Broccoli	y = −5.055x + 17.319	0.914	96.77	y = −11.328x + 33.137	0.910	94.18	77.96
Cabbage	y = −0.861x + 4.562	0.978	16.48	y = −1.671x + 3.761	0.968	13.89	−166.27
Cauliflower	y = −1.513x + 6.555	0.942	28.97	y = −3.173x + 8.352	0.930	26.38	−128.10
Black radish	y = −2.047x + 8.186	0.921	39.19	y = −4.403x + 12.108	0.910	36.60	−96.88
Rapeseed	y = −0.752x + 4.245	0.943	14.40	y = −1.421x + 3.031	0.916	11.81	−172.34

The enthalpy of activation (ΔH^*) and entropy of activation (ΔS^*) for steel dissolution in 0.5 M H_2SO_4 with and without inhibitor extracts, were established by plotting the ln (v_{corr}/T) against 1000/T. The straight lines plotted are illustrated in Figure 5b. The values of ΔH^* and ΔS^*, calculated from the slope $-\Delta H^*/R$ and the intercept ($\ln(R/Nh) + \Delta S^*/R$) are shown in Table 5.

The SEM images recorded for the steel surface after the weight loss experiments, with and without extracts A and the EDX spectra of the surface of the steel disks before and after immersion for 24 h, at 20 °C, in 0.5 M H_2SO_4 solution, with and without extracts A, are displayed in Figure 6 and the atomic content is presented in Table 6.

Table 6. Percentage of atomic contents of elements obtained from EDX spectra.

Elements	Atomic content (%)						
	Steel Disks	H_2SO_4	Broccoli	Cabbage	Cauliflower	Black radish	Rapeseed
C	0.22	-	19.61	-	17.24	18.92	2.45
Si	0.07	0.35	2.64	0.30	1.70	0.68	1.03
P	0.02	0.59	0.44	0.30	0.30	0.37	0.34
S	0.09	1.42	0.65	11.22	0.63	0.68	2.67
Mn	1.04	0.63	-	0.34	0.78	0.68	56.46
Fe	98.56	57.35	60.99	29.63	57.00	46.47	37.05
O	-	39.52	15.67	58.07	22.00	31.55	-
Cl	-	0.15	-	0.14	0.34	0.66	-

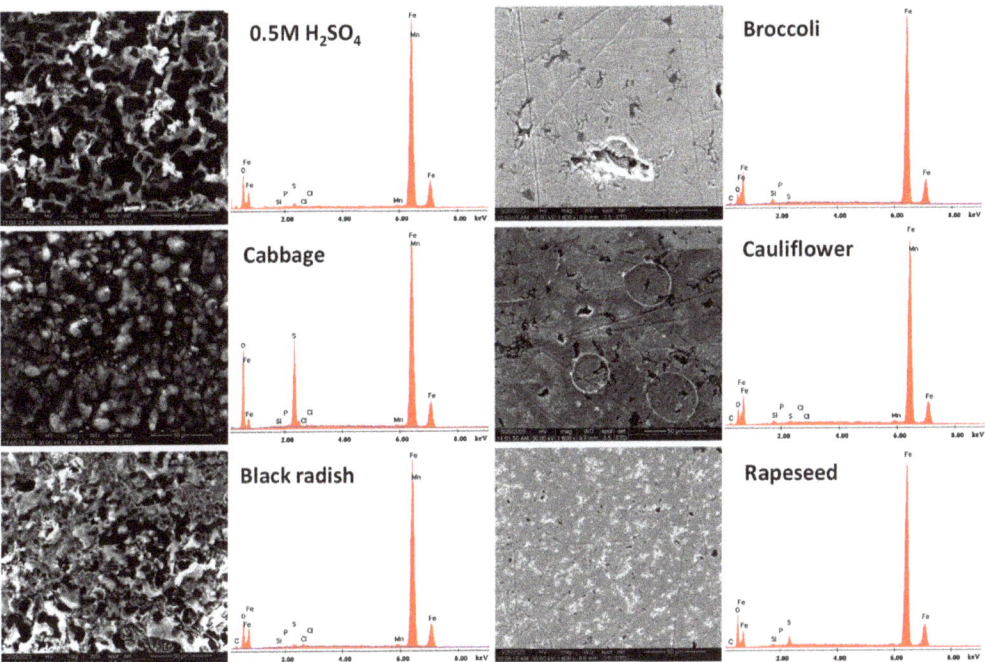

Figure 6. SEM images/EDX spectra of the steel surface after the weight-loss experiments, with and without extracts A.

FT-IR analysis (Figure 7) of the extracts and for the steel surface after 24 h of exposure to a solution of 0.5 M H_2SO_4, with and without *Brassicaceae* extracts, at 20 °C, was carried out. The changes in the FT-IR spectra of the inhibitor film compared to *Brassicaceae* extracts are presented in Table 7.

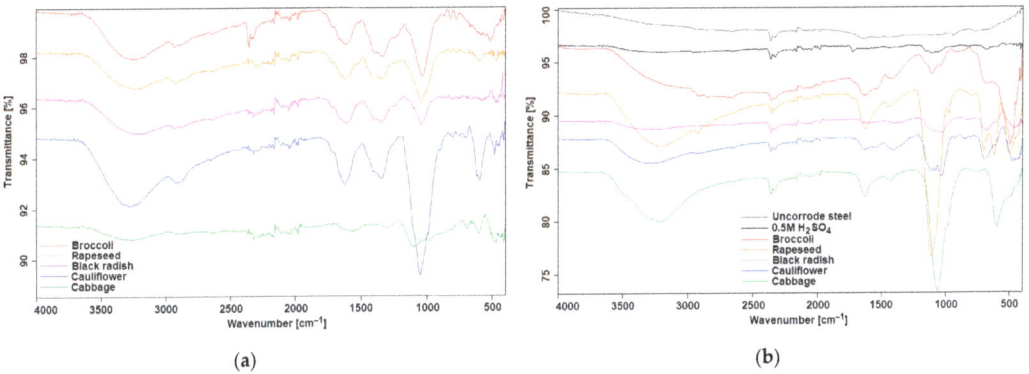

Figure 7. FT-IR analysis: (**a**) *Brassicaceae* extracts; (**b**) Steel surface after 24 h of exposure to a solution of 0.5 M H_2SO_4, with and without *Brassicaceae* extracts, at 20 °C.

Table 7. The changes in the FT-IR spectra of inhibitor film compared to Brassicaceae extracts.

Brassicaceae Extracts	Inhibitor Film
3284 cm^{-1} OH stretching, intermolecular bonded, intense, large	Decreased intensity or shifted to 3330 cm^{-1}
1661 cm^{-1} C = N stretching, medium	Decreased intensity, absent or shifted to 1644 cm^{-1}
1625 cm^{-1} C = C stretching, medium	Absent
1392, 1337 cm^{-1} OH bending, intense	Shifted to 1416 cm^{-1}
1029 cm^{-1} C-O stretching, intense	Decreased intensity, shifted to 1115 cm^{-1}

4. Discussion

According to our previous research [19], the microwave method using 70% EtOH, is a proper eco-friendly method for plant extraction. Regarding the antioxidant capacity and total phenolics of the *Brassicaceae* extracts, all analyzed extracts showed very good antioxidant activities (9.54–22.03% remaining DPPH) (Figure 1, Table 1), however, the results do not correlate with the values obtained for total phenolics (Table 1). This could be due to the presence of other compounds in the plant matrix, i.e., glucosinolates [19,29].

In a previous study, we determined the composition of the glucosinolates of the extracts using HPLC-DAD. Sinigrin was found in all extracts, being the predominant glucosinolate in the cabbage extract, meanwhile, gluconapin was identified in broccoli, cauliflower, and black radish. Neoglucobrassicin was the major glucosinolate found only in cauliflower. Methoxyglucobrassicin from broccoli, glucobrassicanapin, 4-hydroxyglucobrassicin, and glucoraphasatin from black radish completed the chromatographic profile of the extracts [19].

The structures of the glucosinolates commonly found in *Brassicaceae* plants are shown in Scheme 1.

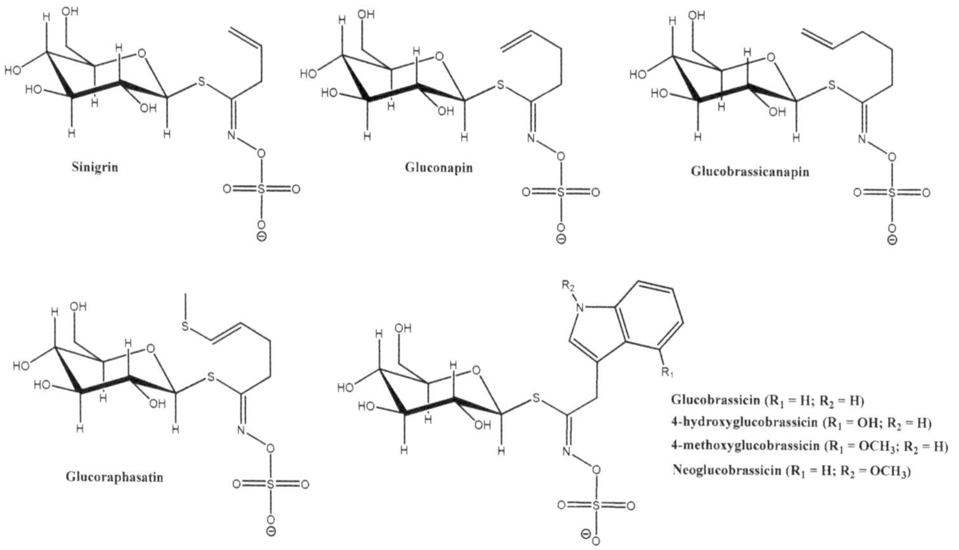

Scheme 1. The structures of some major glucosinolates found in *Brassicaceae* plants.

The highest antioxidant activity was obtained for cauliflower (9.54%) and broccoli (11.75%), while higher values of phenolic content were obtained for rapeseed (1713 mg GAE/L) and broccoli (1623 mg GAE/L).

The electrochemical behavior of the *Brassicaceae* extracts (extracts A and B) was evaluated. To achieve a corrosive environment, an aggressive support electrolyte (0.5 M H_2SO_4) was used.

Figure 3 reveals that the corrosion potential in the presence of all tested extracts shifted to noble values compared to blank. Moreover, both the cathodic and anodic current density present decreased values, meanwhile, Table 3 shows a change in E_{corr} between 6–46 mV. This trend has been reported by other researchers [30,31]. An inhibitor is classified as an anodic-type or cathodic-type inhibitor when the change in E_{corr} is greater than 85 mV [30,31]. Also, a mixed-type inhibitor produces a reduction in both anodic and cathodic current densities [32].

It can be observed from Table 3 that the addition of *Brasicaceae* extracts leads to a significant decrease in the corrosion current densities (I_{corr}) with a more pronounced drop for higher concentrations. The corrosion potential (E_{corr}) values were only slightly affected by the presence of extracts and no explicit tendency in the change of E_{corr} values depending on the extract concentration was observed. This type of behavior suggests that the extracts might act as pickling inhibitors [33]. This also indicates that the adsorption of the tested extracts on the steel surface leads to the blocking of the active sites slowing down corrosion. Considering the changes in the cathodic Tafel slope (βc) and anodic Tafel slope (βa), it can be concluded that the extracts' actions are exerted on both anodic and cathodic reactions, thus resulting in a decrease in anodic dissolution and a delay of the cathodic hydrogen reaction [27,34]. These results suggest that all tested extracts act as mixed-type corrosion inhibitors.

Furthermore, in addition to the fact that E_{corr} values have been slightly positively shifted, a more obvious decrease in the βa compared to the βc values in the presence of extracts can be noticed, which indicates a mixed-type inhibition behavior with predominant control of the anodic reaction [35].

The adsorption of organic compounds from the extracts at the active sites of the electrode surface leads to the delay of metallic dissolution and hydrogen evolution reaction. These results prove the capacity of the studied extracts to act as green corrosion inhibitors. The presence of extracts in the corrosive medium diminishes the corrosion rates concomitant with the shifting of the corrosion current density to lower values relative to the blank (Table 3).

The lowest corrosion rates, 1.98, 7.97, and 9.43 mm/year, were obtained for extracts A of broccoli, cauliflower, and black radish, respectively. This means that inhibitor efficiency increases with an increase in the inhibitor concentrations.

For plant extracts, both in anodic and cathodic domains, after a specific potential, the current-vs.-potential characteristics no longer change significantly. This behavior could be associated with the desorption of the adsorbed film of inhibitors on the surface of the electrode in acidic media. Above the desorption potential, the desorption rate of inhibitors is raised more than its adsorption [27].

The values obtained for the corrosion rates correlate with those calculated for inhibition efficiency (*IE*) and are presented in Table 3.

The values of inhibition efficiency obtained from the electrochemical measurements clearly increase with the concentration of the extracts and follow the order of broccoli > cauliflower > black radish > cabbage > rapeseed for extracts A. As for extracts B, the extract of broccoli can be found before rapeseed and the rest maintain the same trend. The inhibition efficiency for extracts A ranged between 89–99%. Only for rapeseed, an inhibition of 52.79% was obtained. All these results recommend the tested extracts as potent corrosion inhibitors.

Weight loss experiments (Table 4) were chosen to complete the results obtained by electrochemical tests for the determination of corrosion rates and inhibition efficiency. Although it is a time-consuming method, the benefit of this gravimetric method recommends its use. The advantages are based on the use of experimental conditions that more closely resemble real-life environments and the provided results that are likely to be more reliable [36].

All tested extracts proved to inhibit corrosion even if the exposure time was increased from minutes in the case of electrochemical tests to 24 h in the case of gravimetrical measurements, with inhibition efficiency values between 22.48 and 98.66%. However, except for broccoli extract, for which similar results were obtained by both tests (IE = 98–99%), the values of inhibition efficiency for the rest of the extracts, after 24 h of immersion in a corrosive media, decreased considerably. Thus, the cabbage extract lost 67% of its effectiveness, cauliflower 34%, black radish 17%, and rapeseed 58%. So, it seems that the difference in measurement times generally influences the inhibition capacity of the tested extracts. The weight loss experiments demonstrated the great potential of broccoli extract to inhibit the corrosion of the steel at 20 °C after a long period of immersion in an acidic solution.

The effect of temperature on the corrosion of steel in the presence and absence of *Brassicaceae* extracts was also monitored by weight loss measurements between 20–60 °C. The results presented in Table 4 reveal that the corrosion rate increases with the increase in the temperature, though to a lesser extent in the case of the presence of inhibitors. This behavior of the extracts in an acidic media is due to the increase in surface coverage by increasing the inhibitor concentration [28].

Temperature plays an important role in understanding the inhibitive mechanism of the corrosion process. To assess the temperature effect, the weight loss results (Table 4) were used to study the activation of the inhibition process by the means of Arrhenius Equation (4).

It is obvious that the apparent energy of activation increased in the presence of *Brassicaceae* extracts, compared to the uninhibited solution (Table 5). This increase suggests physical adsorption of the inhibitor on the steel surface. Also, the increase in the activation energy can be explained by an appreciable decrease in the adsorption of the inhibitor on the steel surface by an increase in the temperature. This decrease in adsorption leads to higher corrosion rates due to the increased exposed surface area of the steel towards the corrosive solution [28].

However, chemical adsorption of the inhibitor on a steel surface can also be involved. This kind of interaction involves charge sharing or charge transfer from the inhibitor to the atoms of the Fe in order to form a coordinate bond [28]. Moreover, the physical adsorption suggested by the trend of activation energy cannot be considered critical due to the competitive adsorption with water molecules, whose removal from the steel surface also requires some activation energy. Therefore, it can be considered that the adsorption of extracts' compounds on the steel surface occurs through both physical and chemical phenomena, concurrently [37].

The values of the enthalpy of activation (ΔH^*) and entropy of activation (ΔS^*) for steel dissolution in 0.5 M H_2SO_4 in the presence and absence of inhibitor extracts (Table 5) were calculated using Equation (5). The positive values for ΔH^* in the absence and presence of extracts reflect the endothermic nature of metal dissolution during the corrosion process [27,38,39]. In addition, the enthalpy values increase in the presence of the extracts compared to free the 0.5 M H_2SO_4 solution indicating a higher protection efficiency [28,40].

Regarding the entropy of activation (ΔS^*), it can be seen that its values have increased in the presence of the inhibitor compared to the uninhibited solution. The gain in entropy suggests an increase in the disordering on going from the reactant to the metal–solution interface and can be attributed to the increase in solvent entropy and to more positive water desorption enthalpy [27].

Positive entropy of activation was obtained in the presence of broccoli extract while negative values but higher than ΔS^* value for the free acid solution were obtained for the other tested extracts. Large and negative values of ΔS^* indicate that the activation complex in the rate-determining step represents an association rather than a dissociation step, meaning that a decrease in disordering takes place on going from reactants to the activated complex [37,41].

The surface morphology of both steel electrodes from potentiodynamic polarization tests (Figure 4) and steel disks from weight-loss experiments (Figure 6) was examined using the SEM technique. The inhibition effect of the extracts can be clearly observed from the SEM images, especially for more concentrated extracts. The SEM images reveal that the surface was intensely injured in the absence of extracts. The surface damage was reduced in the presence of inhibitors, probably due to the protective film adsorbed on the steel surface that is responsible for the corrosion inhibition. The protective effect increased when using higher concentrated extracts. The aspect of the steel surface proved the presence of the shielding film adsorbed on it and is in agreement with the IE values presented in Tables 3 and 4.

The results of the EDX analysis on the steel surface before and after the weight loss tests conducted at 20 °C are displayed in Figure 6 and Table 6. The EDX spectra of uncorroded steel shows the characteristics' peaks of the elements constituting the steel sample (C, Si, P, S, Mn, Fe). The EDX spectra of the uninhibited steel disk (blank) show the normal peaks ascribed to general corrosion in sulfuric acid. As for inhibited solutions, the EDX spectra showed additional peaks characteristic of the existence of O and Cl, and differences in the weight percentage of the elements. These elements can originate from the compounds contained in the extracts. For example, we have demonstrated that the studied *Brassicaceae* extracts are rich in glucosinolates. These are sulfur- and nitrogen-containing glycosides, with a 2-hydroxymethyl-tetrahydro-pyran-3,4,5-triol moiety and a sulfide group, among other structures, which can contribute to the change in the weight percentage of elements such as C, O, and S on the steel surface. These results indicate that the inhibitor extracts adsorbed on the steel surface with different degrees of surface coverage, which can be correlated to their capacity to inhibit corrosion.

In order to elucidate the nature of the protective layer formed at the steel surface, the FT-IR spectra of the extracts and that of the steel surface after immersion in the inhibited and uninhibited solution of 0.5 M H_2SO_4 for 24h, at 20 °C, were evaluated. Generally, the FT-IR spectra of the films look almost similar to that of the corresponding extract. However, the intensity of some absorption bands of the steel surface decreased or their vibrations were shifted.

As shown in Table 7, the deviations are observed for functional groups from glucosinolates, so we can conclude that the glucosinolates were adsorbed on the steel surface as a protective anticorrosion film. A [Fe-extract functional groups]$^{2+}$ complex is formed by covalent or coordinate bonds between nonbonding electrons in N, O, or C = C and vacant Fe d-orbital. It should be mentioned that plant extracts are complex matrices, so such complexes can be stable or soluble and consequently can act through a corrosion retarding or accelerating mechanism, concurrently. This can explain the extracts' behaviors when increasing the temperature, i.e., the stable complex formation dominates at an increasing concentration until a critical concentration, where the formation of a soluble complex dominates [24].

5. Conclusions

This paper has demonstrated the anticorrosion properties of *Brassicaceae* extracts on steel materials in an acidic environment. Inhibition efficiency values increased with the increase in inhibitor concentration and decreased with an increase in temperature. The tested extracts act as mixed-type corrosion inhibitors, proving anticorrosion effects even at a low concentration. No correlation between antioxidant activities, total phenolic, and inhibition efficacy was observed. The mechanisms of corrosion inhibition consist of physical and chemical adsorption of glucosinolates on a steel surface concomitant with the formation of [Fe-extract functional groups]$^{2+}$ complexes. Among the five *Brassicaceae* ethanolic extracts, broccoli extract was the best inhibitor for the corrosion of steel in a 0.5 M H_2SO_4 solution. Given the behaviors of the *Brassicaceae* extracts, these can successfully substitute for the conventional toxic inhibitors and could be used as green corrosion inhibitors.

Author Contributions: Conceptualization, I.M.C.I., A.C., M.N.Ş., P.S. and C.M.D.; methodology, I.M.C.I., A.C., M.N.Ş., P.S. and A.A.C.; software, I.M.C.I., A.C., A.M.A., L.L.R., A.F. and C.M.; validation, I.M.C.I., A.C., M.N.Ş., P.S. and C.M.D.; formal analysis, I.M.C.I., A.C., L.L.R., G.R. and C.M.D.; investigation I.M.C.I., A.C., M.N.Ş., P.S., G.R. and A.F.; resources, I.M.C.I., A.C. and C.M.D.; writing—original draft preparation, I.M.C.I., A.C., A.A.C., A.M.A. and C.M; writing—review and editing, I.M.C.I., A.C., M.N.Ş., P.S. and C.M.D.; visualization, I.M.C.I., A.C., M.N.Ş., P.S., A.A.C.,

A.M.A., A.F., L.L.R., C.M. G.R. and C.M.D.; supervision, I.M.C.I., A.C., M.N.Ş., P.S. and C.M.D. All authors have read and agreed to the published version of the manuscript.

Funding: This work is part of the project PN 19 22 03 01/2019–2022 "Supramolecular inclusion complexes of some natural and synthetic compounds with health applications", carried out under the NUCLEU Program funded by the Romanian Ministry of Research, Innovation and Digitization. Project financed by Lucian Blaga University of Sibiu through the research grant LBUS-IRG-2022-08/No. 2905/18.07.2022.

Institutional Review Board Statement: Not applicable.

Informed Consent Statement: Not applicable.

Data Availability Statement: Not applicable.

Acknowledgments: This work is part of the project PN 19 22 03 01/2019-2022 "Supramolecular inclusion complexes of some natural and synthetic compounds with health applications", carried out under the NUCLEU Program funded by the Romanian Ministry of Research, Innovation and Digitization. Project financed by Lucian Blaga University of Sibiu through the research grant LBUS-IRG-2022-08/ No. 2905/18.07.2022.

Conflicts of Interest: The authors declare no conflict of interest.

References

1. Verma, C.; Ebenso, E.E.; Bahadur, I.; Quraishi, M.A. An Overview on Plant Extracts as Environmental Sustainable and Green Corrosion Inhibitors for Metals and Alloys in Aggressive Corrosive Media. *J. Mol. Liq.* **2018**, *266*, 577–590. [CrossRef]
2. Zhu, Y.; Wang, L.; Behnamian, Y.; Song, S.; Wang, R.; Gao, Z.; Hu, W.; Xia, D.H. Metal Pitting Corrosion Characterized by Scanning Acoustic Microscopy and Binary Image Processing. *Corros. Sci.* **2020**, *170*, 108685. [CrossRef]
3. Singh, P.; Srivastava, V.; Quraishi, M.A. Novel Quinoline Derivatives as Green Corrosion Inhibitors for Mild Steel in Acidic Medium: Electrochemical, SEM, AFM, and XPS Studies. *J. Mol. Liq.* **2016**, *216*, 164–173. [CrossRef]
4. Ladan, M.; Basirun, W.J.; Kazi, S.N.; Rahman, F.A. Corrosion Protection of AISI 1018 Steel Using Co-Doped TiO_2/Polypyrrole Nanocomposites in 3.5% NaCl Solution. *Mater. Chem. Phys.* **2017**, *192*, 361–373. [CrossRef]
5. Umoren, S.A.; Solomon, M.M.; Obot, I.B.; Suleiman, R.K. A Critical Review on the Recent Studies on Plant Biomaterials as Corrosion Inhibitors for Industrial Metals. *J. Ind. Eng. Chem.* **2019**, *76*, 91–115. [CrossRef]
6. Jiang, S.; Chai, F.; Su, H.; Yang, C. Influence of Chromium on the Flow-Accelerated Corrosion Behavior of Low Alloy Steels in 3.5% NaCl Solution. *Corros. Sci.* **2017**, *123*, 217–227. [CrossRef]
7. Pradeep Kumar, C.B.; Mohana, K.N. Phytochemical Screening and Corrosion Inhibitive Behavior of Pterolobium Hexapetalum and Celosia Argentea Plant Extracts on Mild Steel in Industrial Water Medium. *Egypt. J. Pet.* **2014**, *23*, 201–211. [CrossRef]
8. Soltani, N.; Tavakkoli, N.; Khayat Kashani, M.; Mosavizadeh, A.; Oguzie, E.E.; Jalali, M.R. Silybum Marianum Extract as a Natural Source Inhibitor for 304 Stainless Steel Corrosion in 1.0 M HCl. *J. Ind. Eng. Chem.* **2014**, *20*, 3217–3227. [CrossRef]
9. Sedik, A.; Lerari, D.; Salci, A.; Athmani, S.; Bachari, K.; Gecibesler, H.; Solmaz, R. Dardagan Fruit Extract as Eco-Friendly Corrosion Inhibitor for Mild Steel in 1 M HCl: Electrochemical and Surface Morphological Studies. *J. Taiwan Inst. Chem. Eng.* **2020**, *107*, 189–200. [CrossRef]
10. Faiz, M.; Zahari, A.; Awang, K.; Hussin, H. Corrosion Inhibition on Mild Steel in 1 M HCl Solution by Cryptocarya Nigra Extracts and Three of Its Constituents (Alkaloids). *RSC Adv.* **2020**, *10*, 6547–6562. [CrossRef]
11. Dehghani, A.; Bahlakeh, G.; Ramezanzadeh, B.; Ramezanzadeh, M. Potential of Borage Flower Aqueous Extract as an Environmentally Sustainable Corrosion Inhibitor for Acid Corrosion of Mild Steel: Electrochemical and Theoretical Studies. *J. Mol. Liq.* **2019**, *277*, 895–911. [CrossRef]
12. Doheny-Adams, T.; Redeker, K.; Kittipol, V.; Bancroft, I.; Hartley, S.E. Development of an Efficient Glucosinolate Extraction Method. *Plant Methods* **2017**, *13*, 17. [CrossRef] [PubMed]
13. Crocoll, C.; Halkier, B.A.; Burow, M. Analysis and Quantification of Glucosinolates. *Curr. Protoc. Plant Biol.* **2016**, *1*, 385–409. [CrossRef] [PubMed]
14. Grosser, K.; van Dam, N.M. A Straightforward Method for Glucosinolate Extraction and Analysis with High-Pressure Liquid Chromatography (HPLC). *J. Vis. Exp.* **2017**, *2017*, 55425. [CrossRef]
15. Moreno, D.A.; Carvajal, M.; López-Berenguer, C.; García-Viguera, C. Chemical and Biological Characterisation of Nutraceutical Compounds of Broccoli. *J. Pharm. Biomed. Anal.* **2006**, *41*, 1508–1522. [CrossRef]
16. Ungureanu, C.; Fierascu, I.; Fierascu, R.C. Sustainable Use of Cruciferous Wastes in Nanotechnological Applications. *Coatings* **2022**, *12*, 769. [CrossRef]
17. Mazumder, A.; Dwivedi, A.; Plessis, J. Du Sinigrin and Its Therapeutic Benefits. *Molecules* **2016**, *21*, 416. [CrossRef]
18. Kumar, V.; Thakur, A.K.; Barothia, N.D.; Chatterjee, S.S. Therapeutic Potentials of Brassica Juncea: An Overview. *CellMed* **2011**, *1*, 2.1–2.16. [CrossRef]

19. Muntean, D.; Ştefănuţ, M.N.; Căta, A.; Buda, V.; Danciu, C.; Bănică, R.; Pop, R.; Licker, M.; Ienaşcu, I.M.C. Symmetrical Antioxidant and Antibacterial Properties of Four Romanian Cruciferous Extracts. *Symmetry* 2021, *13*, 893. [CrossRef]
20. Njumbe Ediage, E.; Diana Di Mavungu, J.; Scippo, M.L.; Schneider, Y.J.; Larondelle, Y.; Callebaut, A.; Robbens, J.; Van Peteghem, C.; De Saeger, S. Screening, Identification and Quantification of Glucosinolates in Black Radish (*Raphanus sativus* L. Niger) Based Dietary Supplements Using Liquid Chromatography Coupled with a Photodiode Array and Liquid Chromatography-Mass Spectrometry. *J. Chromatogr. A* 2011, *1218*, 4395–4405. [CrossRef]
21. Vale, A.P.; Santos, J.; Melia, N.; Peixoto, V.; Brito, N.V.; Oliveira, M.B.P.P. Phytochemical Composition and Antimicrobial Properties of Four Varieties of Brassica Oleracea Sprouts. *Food Control.* 2015, *55*, 248–256. [CrossRef]
22. Hu, S.H.; Wang, J.C.; Kung, H.F.; Wang, J.T.; Lee, W.L.; Yang, Y.H. Antimicrobial Effect of Extracts of Cruciferous Vegetables. *Kaohsiung J. Med. Sci.* 2004, *20*, 591–599. [CrossRef]
23. Casaletto, M.P.; Figà, V.; Privitera, A.; Bruno, M.; Napolitano, A.; Piacente, S. Inhibition of Cor-Ten Steel Corrosion by "Green" Extracts of Brassica Campestris. *Corros. Sci.* 2018, *136*, 91–105. [CrossRef]
24. Ngobiri, N.C.; Oguzie, E.E.; Li, Y.; Liu, L.; Oforka, N.C.; Akaranta, O. Eco-Friendly Corrosion Inhibition of Pipeline Steel Using Brassica Oleracea. *Int. J. Corros.* 2015, *2015*, 404139. [CrossRef]
25. Li, H.; Qiang, Y.; Zhao, W.; Zhang, S. A Green Brassica Oleracea L Extract as a Novel Corrosion Inhibitor for Q235 Steel in Two Typical Acid Media. *Colloids Surfaces A Physicochem. Eng. Asp.* 2021, *616*, 126077. [CrossRef]
26. Ştefănuţ, M.N.; Căta, A.; Pop, R.; Tănasie, C.; Boc, D.; Ienaşcu, I.; Ordodi, V. Anti-Hyperglycemic Effect of Bilberry, Blackberry and Mulberry Ultrasonic Extracts on Diabetic Rats. *Plant Foods Hum. Nutr.* 2013, *68*, 378–384. [CrossRef]
27. Ahamad, I.; Prasad, R.; Quraishi, M.A. Thermodynamic, Electrochemical and Quantum Chemical Investigation of Some Schiff Bases as Corrosion Inhibitors for Mild Steel in Hydrochloric Acid Solutions. *Corros. Sci.* 2010, *52*, 933–942. [CrossRef]
28. Shukla, S.K.; Ebenso, E.E. Corrosion Inhibition, Adsorption Behavior and Thermodynamic Properties of Streptomycin on Mild Steel in Hydrochloric Acid Medium. *Int. J. Electrochem. Sci.* 2011, *6*, 3277–3291.
29. Vicas, S.I.; Teusdea, A.C.; Carbunar, M.; Socaci, S.A.; Socaciu, C. Glucosinolates Profile and Antioxidant Capacity of Romanian Brassica Vegetables Obtained by Organic and Conventional Agricultural Practices. *Plant Foods Hum. Nutr.* 2013, *68*, 313–321. [CrossRef]
30. Huang, J.; Cang, H.; Liu, Q.; Shao, J. Environment Friendly Inhibitor for Mild Steel by Artemisia Halodendron. *Int. J. Electrochem. Sci.* 2013, *8*, 8592–8602.
31. Li, W.H.; He, Q.; Zhang, S.T.; Pei, C.L.; Hou, B.R. Some New Triazole Derivatives as Inhibitors for Mild Steel Corrosion in Acidic Medium. *J. Appl. Electrochem.* 2008, *38*, 289–295. [CrossRef]
32. Umoren, S.A.; Solomon, M.M.; Obot, I.B.; Suleiman, R.K. Comparative Studies on the Corrosion Inhibition Efficacy of Ethanolic Extracts of Date Palm Leaves and Seeds on Carbon Steel Corrosion in 15% HCl Solution. *J. Adhes. Sci. Technol.* 2018, *32*, 1934–1951. [CrossRef]
33. Abdel-Gaber, A.M.; Abd-El-Nabey, B.A.; Sidahmed, I.M.; El-Zayady, A.M.; Saadawy, M. Inhibitive Action of Some Plant Extracts on the Corrosion of Steel in Acidic Media. *Corros. Sci.* 2006, *48*, 2765–2779. [CrossRef]
34. Hussin, M.H.; Jain Kassim, M.; Razali, N.N.; Dahon, N.H.; Nasshorudin, D. The Effect of Tinospora Crispa Extracts as a Natural Mild Steel Corrosion Inhibitor in 1 M HCl Solution. *Arab. J. Chem.* 2016, *9*, S616–S624. [CrossRef]
35. Wang, H.; Gao, M.; Guo, Y.; Yang, Y.; Hu, R. A Natural Extract of Tobacco Rob as Scale and Corrosion Inhibitor in Artificial Seawater. *Desalination* 2016, *398*, 198–207. [CrossRef]
36. de Souza, F.S.; Spinelli, A. Caffeic Acid as a Green Corrosion Inhibitor for Mild Steel. *Corros. Sci.* 2009, *51*, 642–649. [CrossRef]
37. Hamani, H.; Daoud, D.; Benabid, S.; Douadi, T.; Al-Noaimi, M. Investigation on Corrosion Inhibition and Adsorption Mechanism of Azomethine Derivatives at Mild Steel/0.5 M H2SO4 Solution Interface: Gravimetric, Electrochemical, SEM and EDX Studies. *J. Indian Chem. Soc.* 2022, *99*, 100330. [CrossRef]
38. Chaouiki, A.; Chafiq, M.; Lgaz, H.; Al-Hadeethi, M.R.; Ali, I.H.; Masroor, S.; Chung, I.M. Green Corrosion Inhibition of Mild Steel by Hydrazone Derivatives in 1.0 M HCl. *Coatings* 2020, *10*, 640. [CrossRef]
39. Aziz, I.A.A.; Abdulkareem, M.H.; Annon, I.A.; Hanoon, M.M.; Al-Kaabi, M.H.H.; Shaker, L.M.; Alamiery, A.A.; Isahak, W.N.R.W.; Takriff, M.S. Weight Loss, Thermodynamics, SEM, and Electrochemical Studies on N-2-Methylbenzylidene-4-Antipyreamine as an Inhibitor for Mild Steel Corrosion in Hydrochloric Acid. *Lubricants* 2022, *10*, 23. [CrossRef]
40. Abdul Rahiman, A.F.S.; Sethumanickam, S. Corrosion Inhibition, Adsorption and Thermodynamic Properties of Poly(Vinyl Alcohol-Cysteine) in Molar HCl. *Arab. J. Chem.* 2017, *10*, S3358–S3366. [CrossRef]
41. Hamdy, A.; El-Gendy, N.S. Thermodynamic, Adsorption and Electrochemical Studies for Corrosion Inhibition of Carbon Steel by Henna Extract in Acid Medium. *Egypt. J. Pet.* 2013, *22*, 17–25. [CrossRef]

Disclaimer/Publisher's Note: The statements, opinions and data contained in all publications are solely those of the individual author(s) and contributor(s) and not of MDPI and/or the editor(s). MDPI and/or the editor(s) disclaim responsibility for any injury to people or property resulting from any ideas, methods, instructions or products referred to in the content.

Article

Electrochemical Impedance Analysis for Corrosion Rate Monitoring of Sol–Gel Protective Coatings in Contact with Nitrate Molten Salts for CSP Applications

V. Encinas-Sánchez [1], A. Macías-García [2], M. T. de Miguel [1], F. J. Pérez [1] and J. M. Rodríguez-Rego [2,*]

[1] Surface Engineering and Nanostructured Materials Research Group, Complutense University of Madrid, Complutense Avenue s/n, 28040 Madrid, Spain
[2] Department of Mechanical, Energetic and Materials Engineering, School of Industrial Engineering, University of Extremadura, Avda. de Elvas, s/n, 06006 Badajoz, Spain
* Correspondence: jesusrodriguezrego@unex.es

Abstract: The protective behaviour of ZrO_2-3%molY_2O_3 sol–gel coatings, deposited with an immersion coating technique on 9Cr-1Mo P91 steel, was evaluated with corrosion monitoring sensors using the electrochemical impedance spectroscopy technique. The tests were carried out in contact with solar salt at 500 °C for a maximum of 2000 h. The results showed the highly protective behaviour of the coating, with the corrosion process in the coated system being controlled by the diffusion of charged particles through the protective layer. The coating acts by limiting the transport of ions and slowing down the corrosive process. The system allowed a reduction in the corrosion rate of uncoated P91 steel. The estimated corrosion rate of 22.62 µm·year^{-1} is lower than that accepted for in-service operations. The proposed ZrO_2-3%molY_2O_3 sol–gel coatings are an option to mitigate the corrosion processes caused by the molten salts in concentrated solar power plants.

Keywords: coating; sol–gel; solar salt; corrosion; corrosion monitoring

1. Introduction

Interest in Concentrated Solar Power (CSP) plants has increased in recent years [1]. This interest has grown steadily due to its high potential for improvements in efficiency and dispatchability when compared with other renewable energy technologies [2]. However, better dispatchability remains a crucial issue for increasing its competitiveness [3], and thermal energy storage could provide a realistic solution [4]. Although many studies are being performed with other salt mixtures [5–7], the most industrially used compound is 60 wt.% $NaNO_3$/40 wt.% KNO_3 (Solar Salt®) [8].

The degradation of materials in contact with molten salts in CSP plants has been studied by different researchers [9–15]. These studies recommend the use of carbon steel at temperatures below 300 °C, stainless steels between 300 °C and 550 °C and Ni-based alloys for temperatures above 550 °C [16]. CSP technologies are expensive and require the use of efficient and cheaper materials [17]. The replacement of stainless steels and Ni-based alloys (known for their high costs [18]) with low-alloy steels could be a solution [19]. However, these steels in contact with molten nitrate salts do not offer good corrosion behaviour [20]. Other alloys with low Cr content, such as P91 and X20CrMoV, also did not show good corrosion behaviour in studies carried out during 2500 h at 600 °C [18,21]. P91 steel in contact with molten nitrates for 1000 h at 580 °C showed less pronounced corrosion [21]. Other authors explained this behaviour by the formation of two layers, a superficial Fe_2O_3 layer and a protective interior layer, rich in chromium oxide, for a 1000 h test. At longer times, the protective chromium oxide layer decreases [22]. Therefore, the development of protective coatings for these alloys could be an economical alternative against corrosion for industrial applications in contact with molten salts [21]. Sol–gel coatings seem to

be an interesting option because of their numerous advantages [23], including their low processing temperature and the ability to be deposited on complex shapes.

The sol–gel process uses a precursor solution as a protective coating for a certain substrate. This solution is transformed into a gel layer by the evaporation of the solvent and/or the chemical reactions that take place [24]. The use of yttria-stabilized zirconia (YSZ) sol–gel solutions is based on the properties provided by ZrO_2 with good thermal shock resistance, mechanical and chemical properties on the one hand, and Y_2O_3 with thermal stability and anti-aging performance on the other hand [25,26]. Both components allow for good high-temperature stability of YSZ [27]. Previous studies have been promising [28]. P91 steel was dip-coated using a sol–gel ZrO_2-3%molY_2O_3, and the results were comparable to those obtained with uncoated 304 steel. SEM micrographs corresponding to the above sol–gel solution gave a compact coating with a thickness ranging from 1 to 1.4 µm. Additionally, YSZ has been used as a protective coating in molten vanadates and sulphate media [29,30]). These media are well known for being highly corrosive, which suggests the great potential of YSZ as a protective material in molten Solar Salt environments.

Most molten salt corrosion analysis studies have been performed using techniques such as optical microscopy, scanning electron microscopy energy/scatter X-ray spectroscopy and X-ray diffraction [8,31,32]. However, these techniques used are laborious and complex [33]. Techniques that allow the register of corrosion rate and mechanism in real-time can help in gaining a better understanding of the corrosion processes. Electrochemical techniques are a good solution for the corrosion monitoring of materials, especially at high temperatures in the presence of molten salts [34].

Electrochemical Impedance Spectroscopy (EIS) is a technique that allows the recording of experimental data in real-time [33,35]. The main advantage of this technique is the low intensity of the excitation signal required and a reduction of the error rate associated with the measurement process [36,37]. This technique allows us to evaluate the corrosion process and to determine the corrosion rate [33]. For this purpose, the steel in contact with molten salt can be represented by circuits composed of resistance, capacitance and inductance elements under an alternating current [38].

The EIS technique has been previously used in many studies for evaluating the corrosion behaviour of different substrates in contact with molten salts (such as nitrates/nitrites [39,40], chlorides [41] and vanadates [42]). Corrosion investigations by EIS on different materials, such as 316 stainless steels in molten HITEC salts [39], Inconel 718 superalloy in molten Na_2SO_4, $80V_2O_5$-$20Na_2SO_4$, $NaVO_3$ [42], ferritic–martensitic steel with molten $NaNO_3/KNO_3$ [33] and 9Cr-1Mo steel in molten LiCl-KCl salt [34] showed that the corrosion processes exhibited different mechanisms. Zhu et al. [39] found that the corrosion of 316 stainless steels in molten salt HITEC was controlled by the outward diffusion of metal ions, while Jagadeeswara-Rao et al. [34] registered the formation of intermittent oxide films in 9Cr-1Mo steels in molten salt LiCl-KCl. Thus, the main purpose of this study is to assess the corrosion resistance of ZrO_2-Y_2O_3 sol–gel coatings on 9Cr-1Mo P91 ferritic–martensitic steel in contact with Solar Salt at 500 °C for up to 2000 h by employing corrosion monitoring sensors that are based on the EIS technique.

2. Methodology

2.1. Materials

2.1.1. Preparation of the Nitrate Salt Mixture and Steel Samples

The nitrate mixture, 60 wt.% $NaNO_3$/40 wt.% KNO_3, was prepared using $NaNO_3$ from BASF (Ludwigshafen, Germany) with 99% purity and KNO_3 from Haifa (Madrid, Spain) with 98% purity. The required quantity of each compound was weighed and mixed in an alumina crucible. The impurity level present in both nitrates is gathered in Table 1.

Table 1. Impurities composition in the chemicals used.

Chemicals	Cl$^-$	SO$_4^{2-}$	CO$_3^{2-}$
NaNO$_3$	0.02	0.005	0.02
KNO$_3$	0.015	<0.0005	<0.02

The substrate used as the base material for CSP applications was 9Cr-1Mo P91 steel, (weight composition of 0.12% C, 0.21% Si, 0.49% Mn, 0.014% P, 0.002% S, 0.01% Al, 8.70% Cr, 0.85% Mo, 0.02% Ni, 0.18% V, 0.06% Nb and 0.053% N). The substrates were machined to a size of 20 × 10 × 2 mm^3 and subsequently sanded and polished using 240, 600 and 800 grit sandpaper and 9 µm, 6 µm and 3 µm polishing cloths.

2.1.2. Coating Preparation and Deposition

The procedure of the sol–gel solution has been previously described in [27,43].

The coatings on the P91 substrates with the above solution were carried out using the immersion technique [27,43] at an extraction velocity of 25 mm-min^{-1} and subsequent heat treatment at 500 °C in a Hobersal® Furnace (Barcelona, Spain) for 2 h, at a heating/cooling velocity of 3 °C-min^{-1}.

In order to reduce stresses in the coating, a drying process was performed on the coated samples before being heat-treated, thus minimizing the organic residue content in the coating [44]. This initial drying phase was carried out at 100 °C for 60 min, applying the previously mentioned heating/cooling rate. In the sintering process of the coatings, the formation of cracks occurs due to the existing tensions in the crystallisation process and/or thermal expansion; to inhibit them, slow heating/cooling ramps are carried out [45].

The structural quality and morphology of the coatings as deposited have been deeply analysed in previous research and can be consulted in [43].

2.2. Electrochemical Impedance Corrosion Study

Electrochemical impedance spectroscopy (EIS) allows observing the corrosion of the coated P91 samples in contact with the molten binary salt during 2000 h. For this purpose, points of the coated P91 sample were electrically connected by welding to a Kanthal wire. This joint was protected by the application of a ceramic slurry mixture. The wire was introduced in an alumina tube sealed with the same ceramic mixture avoiding contact between the sample–wire connection and the molten salts (Figure 1). The EIS uses the working electrode (WE), the auxiliary electrode (AE) and the reference electrode (RE) (patent reference code WO2017046427).

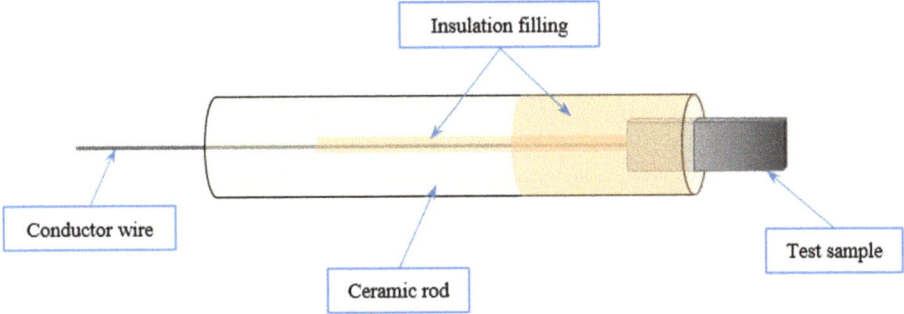

Figure 1. Sketch of an electrode used in the electrochemical sensor.

The crucible containing the molten salt mixture was placed in an electric camera furnace (Carbolite, Hope Valley, UK) with the electrodes immersed.

The surface of each electrode was in contact with the liquid salt (60 wt.% NaNO$_3$/40 wt.% KNO$_3$) at 500 °C, and to a depth of about 3.5 cm. EIS measurements were taken at 0, 24, 72, 500, 1000, 1500 and 2000 h.

It has been reported that this salt mixture starts to degrade at 535 °C [46]. The temperature selected for this study was 500 °C in order to assure the stability of the salt during the experiments. In addition, the parabolic shape of the reflecting mirrors used in the parabolic troughs makes the heat flux in the inferior part of the absorption tube greater than in the superior part, giving a non-uniform distribution of the temperature and thermal stresses that are accentuated above 500 °C. The presence of thermal stresses can deform the absorber tube and deviate the focal line, which affects its optical properties [47].

The EIS measurements were taken by means of VoltaLab 80 equipment (Radiometer Analytical SAS, Villeurbanne, France). The amplitude of the voltage perturbation was fixed at 10 mV with a sweep frequency from 50 kHz to 10 MHz. Impedance data fitting and simulations were performed with the software Zview (version 2018).

3. Results and Discussion

3.1. Electrochemical Impedance Corrosion Test

The 3% yttria-doped zirconia sol–gel was used to coat P91 steel by dip-coating and subsequent sintering heat treatment. Coated and uncoated samples are shown in Figure 2.

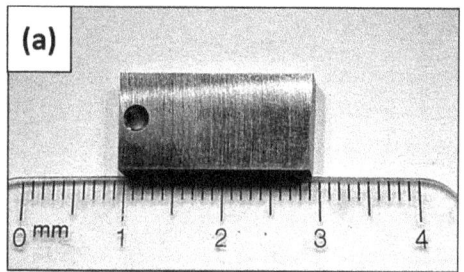 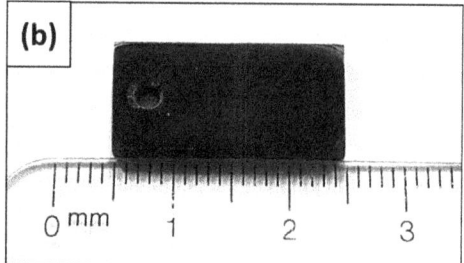

Figure 2. Surface appearance macrographs of (**a**) uncoated P91; (**b**) coated P91.

As expected, and in line with the previously reported study [48], coated samples showed good uniformity given the high homogeneity in colour, which was a methodology proposed by Morrow et al. [49].

Methods used to date for analysing the protective behaviour of the proposed coatings against molten salt corrosion are based on conventional techniques. This typical procedure is tedious and time-consuming, which makes it unsuitable for monitoring corrosion in real-time. EIS measurements represent an interesting solution for achieving rigorous and controlled degradation monitoring of the coated samples in molten salts. The corrosion monitoring system allowed the recording of information on the corrosive process, allowing, inter alia, the estimation of the corrosion rate.

Figure 3 shows the impedance spectra and the best-fitted equivalent electrical circuit of the coated P91 samples in molten binary salt at various times. The electrochemical data registered were adjusted to different models in order to assess the corrosion mechanism taking place. The results showed that the system is consistent with the theoretical model of the protective layer at all measuring times. The best fit was determined by a goodness-of-fit test, analysing the chi-square and relative error values. In line with this model, the impedance spectra show two semicircles, with the one at the high frequency being smaller. This loop at high frequencies refers to the charges that are circulating in the interface between the material and the electrolyte. Here, R_e represents the molten-salt (electrolyte) resistance, and R_t is the electrochemical transfer resistance.

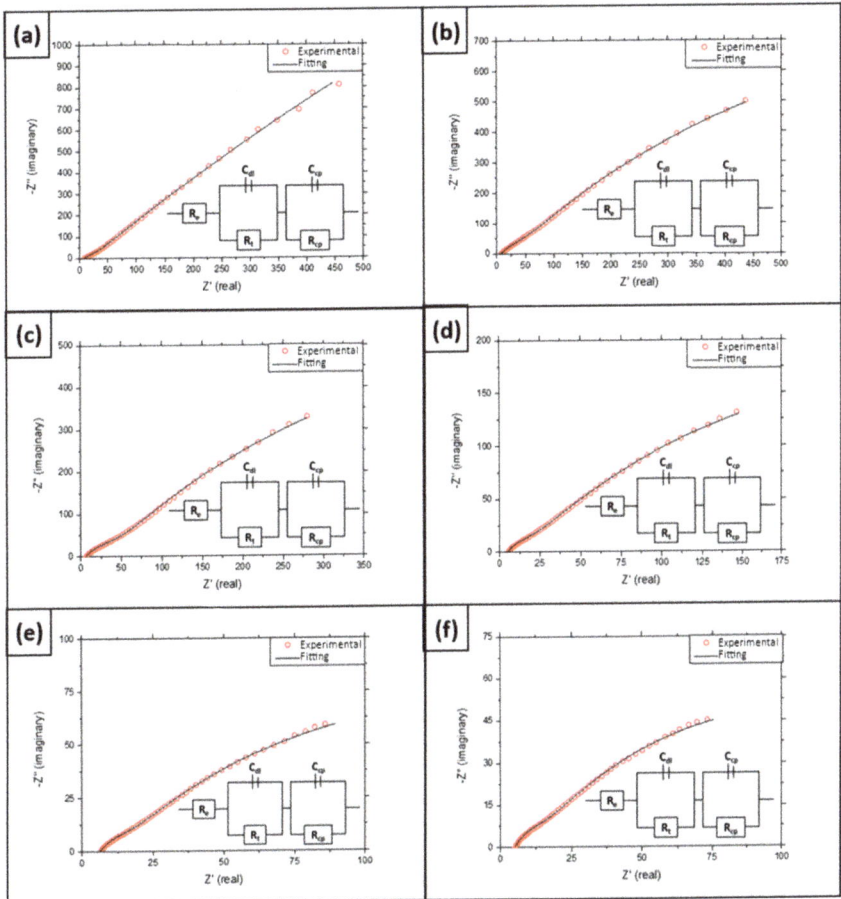

Figure 3. Impedance spectra of the coated P91 samples tested at 500 °C in Solar Salt at (**a**) 24 h; (**b**) 72 h; (**c**) 500 h; (**d**) 1000 h; (**e**) 1500 h; and (**f**) 2000 h.

This transfer resistance is related by the simplified Butler–Volmer equation (Equation (1)), which reveals that it has an inverse relationship with the current density exchanged.

$$R_t = \frac{R \cdot T}{n \cdot F \cdot i_o} \quad (1)$$

where R is the gas constant, T is temperature, n is the number of electrons involved, F is the Faraday constant and i_o is the exchange current density. C_{dl} is the capacitance of the double layer that appears at the coating/salt interface, and n_{dl}/n_{cp} represent, respectively, the constant phase element coefficient of the first and second capacitance loops. It is common for the EIS experiments that capacitors do not behave ideally. There are several theories regarding the cause of this deviation, e.g., surface roughness, non-uniform current distribution or varying thickness or composition of the oxide scale [39,40]. To avoid this non-ideal behaviour, without dependence on its origin, constant phase elements (CPE) are used in the equivalent circuits instead of pure capacitors. In Table 2, C_{dl} and C_{cp} are the modulus values of the two constant phase elements (CPE) and n_{dl} and n_{cp} are their respective indices.

Table 2. Values of each of the equivalent circuit elements in compliance for each testing time with the protective layer model.

Time, h	R_e, Ω	R_t, Ω	C_{dl}, $\Omega^{-1} \cdot s^n$	n_{dl}	R_{cp}, Ω	C_{cp}, $\Omega^{-1} \cdot s^n$	n_{cp}
24	7.557	13.87	$1.506 \cdot 10^{-3}$	0.777	14488	$1.483 \cdot 10^{-3}$	0.731
72	7.095	29.09	$1.973 \cdot 10^{-3}$	0.748	2526	$1.884 \cdot 10^{-3}$	0.711
500	6.171	33.99	$1.632 \cdot 10^{-3}$	0.804	1657	$3.060 \cdot 10^{-3}$	0.738
1000	5.065	11.81	$3.014 \cdot 10^{-3}$	0.794	595	$5.982 \cdot 10^{-3}$	0.669
1500	6.327	6.64	$4.350 \cdot 10^{-3}$	0.808	261	$1.021 \cdot 10^{-2}$	0.615
2000	5.072	6.65	$4.440 \cdot 10^{-3}$	0.809	188	$1.195 \cdot 10^{-2}$	0.610

The protective layer model indicates that ion transport in the layer is the stage that limits the process and slows down the corrosion process. From a physicochemical point of view, the layer resembles a capacitor in series with the double-layer capacitance (see Figure 4). Thus, C_{cp} and R_{cp} respectively represent the protective layer capacitance and its resistance to the charged particles transfer. By way of example, Figure 5 shows the adjustment made after 2000 h of testing, including the models for a protective layer, localized corrosion and porous layer.

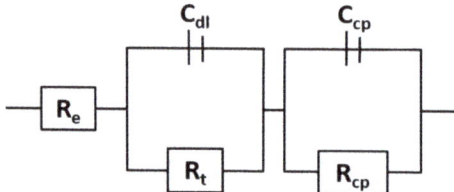

Figure 4. Equivalent circuit to which experimental data are fitted.

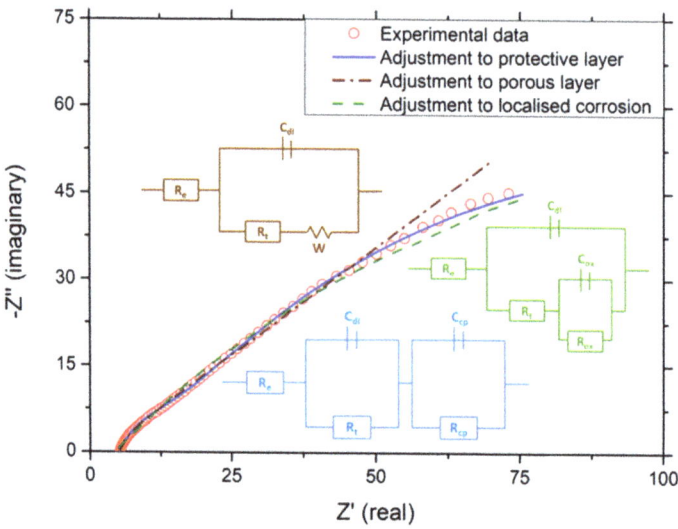

Figure 5. Adjustment of experimental data obtained after 2000 h.

Therefore, given the good adjustment of the experimental data to the protective layer model, the good behaviour of the protective coatings throughout the test can be confirmed.

However, it is worth noting that the behaviour of the system is unstable in the early stages of the test, with the formation of the two semicircles becoming clearer after 72 h of testing (see Figure 3b). This initial instability may be related to the weight loss that this type of coating suffers during the initial stages of the test due to the excess coating deposited [28], which implies that the system becomes more protective after this initial stage. Additionally, as previously reported in [33], a corrosion-monitoring system usually requires a few hours for stabilisation, which may lead to instability in the measurements taken during the initial stages.

Table 2 contains the values of each of the equivalent circuit elements in compliance with each testing time with the protective layer model. According to the obtained results and as shown in Figure 6, the resistance of the electrolyte (R_e) (i.e., the electrical resistance of the molten salt) remains roughly constant in a range of between 5.051 Ω and 7.557 Ω. However, the resistance peak of 7.557 Ω obtained after 24 h of the test is attributed to the instability of the system during the initial stages. This instability is also visible when observing the remaining parameters. Thus, without considering the values obtained during the stability period (first 72 h of testing), it may be affirmed that the molten salt resistance has an average value of 5.914 ± 0.820 Ω. This value differs from that obtained in other studies [33,50,51], which may be due to the coating material that was detached during the initial stages of the test [28]. ZrO_2 is a semiconductor material and leads to an increase in molten salt electrical resistance [50]. Additionally, the presence of coating components in the salt and the formation of volatiles throughout the test led to a reduction in ionic species concentration and, therefore, a reduction in electrolyte conductivity. The effect of the system on the R_e value seems to be evident according to the dissimilar values found in those works cited [33,50,51].

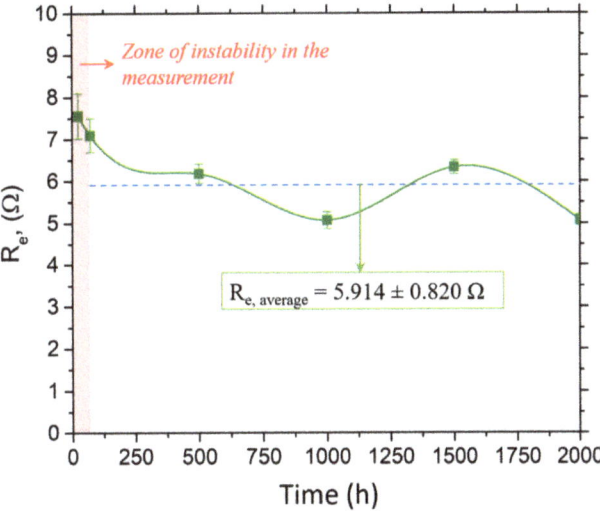

Figure 6. Variation in the resistance of the electrolyte (R_e) during testing.

By comparing charge transfer resistance (R_t) and protective layer resistance (R_{cp}), it is observed that R_{cp} is higher than R_t at every testing time (see Figure 7).

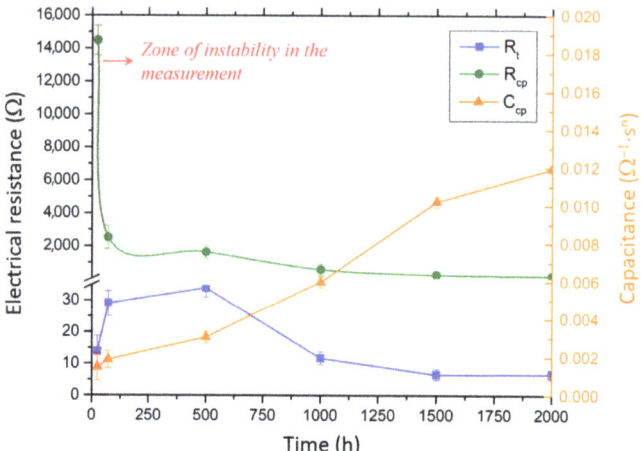

Figure 7. Variation in charge transfer resistance (R_t), protective layer resistance (R_{cp}) and protective layer capacitance (C_{cp}) during testing.

This difference implies that the corrosion process is controlled by the transport of charged particles moving through the protective layer since it is the slower process [50]. This fact is indicative of the good behaviour of the coating. In addition, it is important to highlight the significant reduction in time in both parameters. The decrease in both resistances reveals that the transference of charges and charged particles is eased so the kinetic of the degradation processes can be accelerated. This, together with the increase in C_{cp} (see Figure 7), may be due to the degradation of the coating over time. However, after 1000 h of testing, the system seems to be more stable, leading to slower degradation.

Finally, with regard to the exponents n_{dl} and n_{cp}, they remained constant during the entire test and remained below 1 (see Figure 8). These two exponents show a value of 0.79 ± 0.02 and 0.68 ± 0.06, respectively. According to Omar et al. [52], an *n* value equal to 1 indicates that the system acts as an ideal capacitor, which would be the closest to an ideal coating.

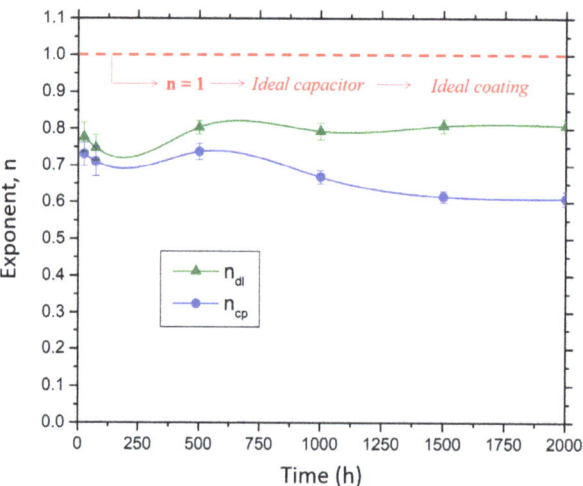

Figure 8. Variation in n_{dl} and n_{cp} exponents during testing.

Corrosion Rate Estimation

As explained above, electrochemical impedance monitoring allows not only the evaluation of the controlling corrosion mechanism but also an estimation of the corrosion rate along the experiment. To this end, following the ASTM-G102 Standard Practice [53], Equation (2) was used for calculating the corrosion rate:

$$v_{corr} = K \cdot \frac{i_{corr} \cdot EW}{\rho} \qquad (2)$$

where v_{corr} is the corrosion rate in $\mu m \cdot cm^{-2}$, K is a system-type dependent constant, i_{corr} is the corrosion current density, EW is the equivalent weight of the material and ρ is its density. For P91 steel, the parameters K, EW and ρ are, respectively, $3.27 \cdot 10^{-3}$ $\mu g \cdot \mu A^{-2} \cdot cm^{-1} \cdot year^{-1}$, 25.3, and 7.76 $g \cdot cm^{-3}$ [33]. The current density is given by the Stern–Geary equation [54]:

$$i_{corr} = K \cdot \frac{B}{R_p} \qquad (3)$$

where B is the Stern–Geary constant (26 mV) and R_p is the polarisation resistance obtained from the experimental values of R_e and Z_{real}:

$$Z_{real} = R_e + R_p \qquad (4)$$

Thus, Figure 9 summarises the corrosion rate determined from the EIS test results during the entire experiment. It is important to highlight that the method used considers generalised corrosion.

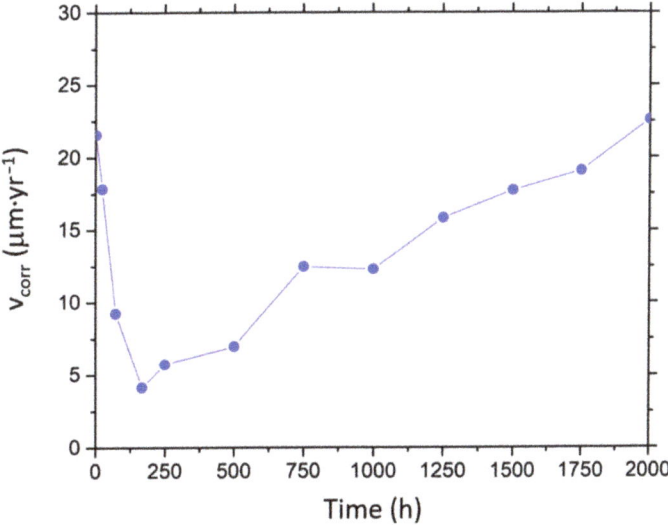

Figure 9. Corrosion rates estimated by EIS in coated P91.

Several results can be highlighted. Firstly, the corrosion rate of 22.62 $\mu m \cdot year^{-1}$ of the coated steel, estimated after 2000 h of testing, falls well below those estimated in other studies for uncoated P91 steel (118 $\mu m \cdot year^{-1}$ [16] and 300 $\mu m \cdot year^{-1}$ [33]). Additionally, according to the guide for corrosion rates used in the industry, where recommendations concerning the use of materials in molten salts are considered (Table 3 [55]), it can be observed that the estimated corrosion rate is closer to the upper limit set for materials recommended for long-term services (0.4–13 $\mu m \cdot year^{-1}$) [55]. This fact suggests that the proposed coated system could be an interesting option from an industrial point of view.

Table 3. Guide for corrosion rates used in the industry [55].

Corrosion Rate, mm·yr^{-1}	Recommendation
>1275	Completely destroyed within days
127–1274	Not recommended for service greater than 1 month
64–126	Not recommended for service greater than 1 year
14–63	Caution recommended, based on the specific application
0.4–13	Recommended for long-term service
<0.3	Recommended for long-term service; no corrosion, other than as a result of surface cleaning, was evidenced

Furthermore, and considering the corrosion rates estimated at each testing time (see Figure 9), the initial stabilisation period is clearly observed (0–72 h), as well as the good behaviour of the coating during the intermediate periods (72–1000 h). Within the intermediate times, the corrosion rate remains within the range recommended by the industry for this material in long-term tests (see Table 3 [55]). However, the increase in the corrosion rate after 1000 h of testing is worth noting. Even though the protective properties of the coating are still suitable after 2000 h of testing (according to the EIS spectra), the coating is steadily degraded.

4. Conclusions

The protective behaviour of sol–gel ZrO_2–3%molY_2O_3 coatings deposited using a dip-coating technique on 9Cr-1Mo P91 steel was isothermally assessed at 500 °C for up to 2000 h in contact with Solar Salt by employing corrosion monitoring sensors that use the electrochemical impedance spectroscopy technique.

The obtained results showed the highly protective character of the proposed coating, which acts by limiting ion transport, thus slowing down the corrosive process. According to the values obtained in terms of charge transfer resistance and coating layer resistance, the corrosive process in the coated system is controlled by the transport of charged particles moving through the protective layer, which indicates that this is the stage that limits the process, confirming the protective behaviour. The results made it possible to estimate the corrosion rate of the coated system in contact with Solar Salt at 500 °C, this being 22.62 µm·year^{-1}. The estimated corrosion rate falls well below those estimated for uncoated P91 steel and those accepted for in-service operations over 1 year, being even closer to the upper limit established for materials recommended for long-term services.

The results suggest that the proposed coating system could be an interesting option for industrial concentrated solar power plants as a potential solution to the severe corrosion issues that currently affect industrial tanks and pipes operating in contact with Solar Salt.

Author Contributions: V.E.-S.: Conceptualization, Investigation, Validation, Writing—Original Draft; A.M.-G.: Methodology, Visualization, Writing—Review & Editing; M.T.d.M.: Conceptualization, Validation, Investigation, Writing—Review & Editing; F.J.P.: Conceptualization, Supervision, Funding acquisition; J.M.R.-R.: Visualization, Writing—Review & Editing. All authors have read and agreed to the published version of the manuscript.

Funding: This work received funding from the Agencia Estatal de Investigation in the frame of the "Proyectos de I+D+I «Programación Conjunta Internacional» 2020" program and under project reference number PCI2020-120703-2-EUROPATMOS.

Institutional Review Board Statement: Not applicable.

Informed Consent Statement: Not applicable.

Data Availability Statement: The data presented in this study are available on request from the corresponding author.

Conflicts of Interest: The authors declare no conflict of interest.

References

1. Zhang, H.; Baeyens, J.; Degrève, J.; Cáceres, G. Concentrated solar power plants: Review and design methodology. *Renew. Sustain. Energy Rev.* **2013**, *22*, 466–481. [CrossRef]
2. Crespi, F.; Toscani, A.; Zani, P.; Sánchez, D.; Manzolini, G. Effect of passing clouds on the dynamic performance of a CSP tower receiver with molten salt heat storage. *Appl. Energy* **2018**, *229*, 224–235. [CrossRef]
3. Grosu, Y.; Bondarchuk, O.; Faik, A. The effect of humidity, impurities and initial state on the corrosion of carbon and stainless steels in molten HitecXL salt for CSP application. *Sol. Energy Mater. Sol. Cells* **2017**, *174*, 34–41. [CrossRef]
4. de Miguel, M.; Encinas-Sánchez, V.; Lasanta, M.; García-Martín, G.; Pérez, F. Corrosion resistance of HR3C to a carbonate molten salt for energy storage applications in CSP plants. *Sol. Energy Mater. Sol. Cells* **2016**, *157*, 966–972. [CrossRef]
5. Grosu, Y.; Udayashankar, N.; Bondarchuk, O.; González-Fernández, L.; Faik, A. Unexpected effect of nanoparticles doping on the corrosivity of molten nitrate salt for thermal energy storage. *Sol. Energy Mater. Sol. Cells* **2018**, *178*, 91–97. [CrossRef]
6. Encinas-Sánchez, V.; de Miguel, M.; García-Martín, G.; Lasanta, M.; Pérez, F. Corrosion resistance of Cr/Ni alloy to a molten carbonate salt at various temperatures for the next generation high-temperature CSP plants. *Sol. Energy Mater.* **2018**, *171*, 286–292. [CrossRef]
7. Villada, C.; Bonk, A.; Bauer, T.; Bolívar, F. High-temperature stability of nitrate/nitrite molten salt mixtures under different atmospheres. *Appl. Energy* **2018**, *226*, 107–115. [CrossRef]
8. García-Martín, G.; Lasanta, M.; Encinas-Sánchez, V.; de Miguel, M.; Pérez, F. Evaluation of corrosion resistance of A516 Steel in a molten nitrate salt mixture using a pilot plant facility for application in CSP plants. *Sol. Energy Mater. Sol. Cells* **2017**, *161*, 226–231. [CrossRef]
9. Fernández, A.; Cortes, M.; Fuentealba, E.; Pérez, F. Corrosion properties of a ternary nitrate/nitrite molten salt in concentrated solar technology. *Renew. Energy* **2015**, *80*, 177–183. [CrossRef]
10. Slusser, J.W.; Titcomb, J.B.; Heffelfinger, M.T.; Dunbobbin, B.R. Corrosion in Molten Nitrate-Nitrite Salts. *J. Met.* **1985**, *37*, 24–27. [CrossRef]
11. Kong, Z.; Jin, Y.; Hossen, G.M.S.; Hong, S.; Wang, Y.; Vu, Q.-V.; Truong, V.-H.; Tao, Q.; Kim, S.-E. Experimental and theoretical study on mechanical properties of mild steel after corrosion. *Ocean Eng.* **2022**, *246*, 110652. [CrossRef]
12. de Miguel, M.; Lasanta, M.; García-Martín, G.; Díaz, R.; Pérez, F. Temperature effect and alloying elements impact on the corrosion behaviour of the alloys exposed to molten carbonate environments for CSP application. *Corros. Sci.* **2022**, *201*, 110274. [CrossRef]
13. García-Martin, G.; Lasanta, M.I.; de Miguel, M.T.; Sánchez, A.I.; Pérez-Trujillo, F.J. Corrosion Behavior of VM12-SHC Steel in Contact with Solar Salt and Ternary Molten Salt in Accelerated Fluid Conditions. *Energies* **2021**, *14*, 5903. [CrossRef]
14. Goods, S.H.; Bradshaw, R.W. Corrosion of Stainless Steels and Carbon Steel by Molten Mixtures of Commercial Nitrate Salts. *J. Mater. Eng. Perform.* **2004**, *13*, 78–87. [CrossRef]
15. Turchi, C.; Vidal, J.; Bauer, M. Molten salt power towers operating at 600–650 °C: Salt selection and cost benefits. *Sol. Energy* **2018**, *164*, 38–46. [CrossRef]
16. Walczak, M.; Pineda, F.; Fernández, G.; Mata-Torres, C.; Escobar, R.A. Materials corrosion for thermal energy storage systems in concentrated solar power plants. *Renew. Sustain. Energy Rev.* **2018**, *86*, 22–44. [CrossRef]
17. Pizzolato, A.; Donato, F.; Verda, V.; Santarelli, M. CFD-based reduced model for the simulation of thermocline thermal energy storage systems. *Appl. Therm. Eng.* **2015**, *76*, 391–399. [CrossRef]
18. Dorcheh, A.S.; Durham, R.N.; Galetz, M.C. Corrosion behavior of stainless and low-chromium steels and IN625 in molten nitrate salts at 600 °C. *Sol. Energy Mater. Sol. Cells* **2016**, *144*, 109–116. [CrossRef]
19. Agüero, A.; Audigié, P.; Rodríguez, S.; Encinas-Sánchez, V.; de Miguel, M.T.; Pérez, F.J. Protective coatings for high temperature molten salt heat storage systems in solar concentration power plants. *AIP Conf. Proc.* **2018**, *2033*, 090001. [CrossRef]
20. Dorcheh, A.S.; Galetz, M. Slurry aluminizing: A solution for molten nitrate salt corrosion in concentrated solar power plants. *Sol. Energy Mater. Sol. Cells* **2016**, *146*, 8–15. [CrossRef]
21. Audigié, P.; Encinas-Sánchez, V.; Juez-Lorenzo, M.; Rodríguez, S.; Gutiérrez, M.; Pérez, F.; Agüero, A. High temperature molten salt corrosion behavior of aluminide and nickel-aluminide coatings for heat storage in concentrated solar power plants. *Surf. Coat. Technol.* **2018**, *349*, 1148–1157. [CrossRef]
22. Fähsing, D.; Oskay, C.; Meißner, T.; Galetz, M. Corrosion testing of diffusion-coated steel in molten salt for concentrated solar power tower systems. *Surf. Coat. Technol.* **2018**, *354*, 46–55. [CrossRef]
23. Wang, D.; Bierwagen, G.P. Sol–gel coatings on metals for corrosion protection. *Prog. Org. Coat.* **2009**, *64*, 327–338. [CrossRef]
24. Berlin, I.J.; Lekshmy, S.S.; Ganesan, V.; Thomas, P.; Joy, K. Effect of Mn doping on the structural and optical properties of ZrO_2 thin films prepared by sol-gel method. *Thin Solid Films* **2014**, *550*, 199–205. [CrossRef]
25. Díaz-Parralejo, A.; Ortiz, A.; Caruso, R. Effect of sintering temperature on the microstructure and mechanical properties of ZrO_2-3mol%Y_2O_3 sol–gel films. *Ceram. Int.* **2010**, *36*, 2281–2286. [CrossRef]
26. Encinas-Sánchez, V.; Macías-García, A.; Pérez, F. Effect of withdrawal rate on the evolution of optical properties of dip-coated yttria-doped zirconia thin films. *Ceram. Int.* **2017**, *43*, 13094–13100. [CrossRef]
27. Kirubaharan, A.K.; Kuppusami, P.; Priya, R.; Divakar, R.; Gupta, M.; Pandit, D.; Ningshen, S. Synthesis, microstructure and corrosion behavior of compositionally graded Ni-YSZ diffusion barrier coatings on inconel-690 for applications in high temperature environments. *Corros. Sci.* **2018**, *135*, 243–254. [CrossRef]

28. Encinas-Sánchez, V.; Batuecas, E.; Garcia, A.M.; Mayo, C.; Díaz, R.; Perez-Trujillo, F.J. Corrosion resistance of protective coatings against molten nitrate salts for thermal energy storage and their environmental impact in CSP technology. *Sol. Energy* **2018**, *176*, 688–697. [CrossRef]
29. Loghman-Estarki, M.; Razavi, R.S.; Jamali, H. Effect of molten V_2O_5 salt on the corrosion behavior of micro- and nano-structured thermal sprayed SYSZ and YSZ coatings. *Ceram. Int.* **2016**, *42*, 12825–12837. [CrossRef]
30. Hajizadeh-Oghaz, M.; Razavi, R.S.; Ghasemi, A.; Valefi, Z. Na_2SO_4 and V_2O_5 molten salts corrosion resistance of plasma-sprayed nanostructured ceria and yttria co-stabilized zirconia thermal barrier coatings. *Ceram. Int.* **2016**, *42*, 5433–5446. [CrossRef]
31. Liu, M.; Bell, S.; Segarra, M.; Tay, N.S.; Will, G.; Saman, W.; Bruno, F. A eutectic salt high temperature phase change material: Thermal stability and corrosion of SS316 with respect to thermal cycling. *Sol. Energy Mater. Sol. Cells* **2017**, *170*, 1–7. [CrossRef]
32. Sarvghad, M.; Will, G.; Steinberg, T.A. Corrosion of steel alloys in molten $NaCl+Na_2SO_4$ at 700 °C for thermal energy storage. *Sol. Energy Mater. Sol. Cells* **2018**, *179*, 207–216. [CrossRef]
33. Encinas-Sánchez, V.; de Miguel, M.T.; Lasanta, M.I.; García-Martín, G.; Pérez, F.J. Electrochemical impedance spectroscopy (EIS): An efficient technique for monitoring corrosion processes in molten salt environments in CSP applications. *Sol. Energy Mater. Sol. Cells* **2019**, *191*, 157–163. [CrossRef]
34. Jagadeeswara-Rao, C.; Venkatesh, P.; Ningshen, S. Corrosion assessment of 9Cr-1Mo steel in molten LiCl-KCl eutectic salt by electrochemical methods. *J. Nucl. Mater.* **2019**, *514*, 114–122.
35. Han, Y.; Wang, J.; Zhang, H.; Zhao, S.; Ma, Q.; Wang, Z. Electrochemical impedance spectroscopy (EIS): An efficiency method to monitor resin curing processes. *Sens. Actuators A Phys.* **2016**, *250*, 78–86. [CrossRef]
36. Macdonald, J.R. Impedance spectroscopy: Old problems and new developments. *Electrochim. Acta* **1990**, *35*, 1483–1492. [CrossRef]
37. Macdonald, D.D. Reflections on the history of electrochemical impedance spectroscopy. *Electrochim. Acta* **2006**, *51*, 1376–1388. [CrossRef]
38. Montemor, M.; Simões, A.; Ferreira, M. Chloride-induced corrosion on reinforcing steel: From the fundamentals to the monitoring techniques. *Cem. Concr. Compos.* **2003**, *25*, 491–502. [CrossRef]
39. Zhu, M.; Zeng, S.; Zhang, H.; Li, J.; Cao, B. Electrochemical study on the corrosion behaviors of 316 SS in HITEC molten salt at different temperatures. *Sol. Energy Mater. Sol. Cells* **2018**, *186*, 200–207. [CrossRef]
40. Encinas-Sánchez, V.; Lasanta, M.; de Miguel, M.; García-Martín, G.; Pérez, F. Corrosion monitoring of 321H in contact with a quaternary molten salt for parabolic trough CSP plants. *Corros. Sci.* **2021**, *178*, 109070. [CrossRef]
41. Salinas-Solano, G.; Porcayo-Calderon, J.; Gonzalez-Rodriguez, J.; Salinas-Bravo, V.; Ascencio-Gutierrez, J.; Martinez-Gomez, L. High temperature corrosion of Inconel 600 in NaCl-KCl molten salts. *Adv. Mater. Sci. Eng.* **2014**, *2014*, 8. [CrossRef]
42. Trinstancho-Reyes, J.; Sandoval-Jabalera, R.; Orozco-Carmona, V.; Almeraya-Calderón, F.; Chacón-Nava, J.; Gonzalez-Rodriguez, J.; Martínez-Villafañe, A. Electrochemical impedance spectroscopy investigation of alloy Inconel 718 in molten salts at high temperature. *Int. J. Electrochem. Sci.* **2011**, *6*, 419–431.
43. Encinas-Sánchez, V.; Macías-García, A.; Díaz-Díez, M.; Brito, P.; Cardoso, D. Influence of the quality and uniformity of ceramic coatings on corrosion resistance. *Ceram. Int.* **2015**, *41*, 5138–5146. [CrossRef]
44. Croll, S.G. The origin of residual internal stress in solvent-cast thermoplastic coatings. *J. Appl. Polym. Sci.* **1979**, *23*, 847–858. [CrossRef]
45. Hamden, Z.; Boufi, S.; Conceição, D.; Ferraria, A.; Rego, A.B.D.; Ferreira, D.; Ferreira, L.V.; Bouattour, S. Li–N doped and codoped TiO_2 thin films deposited by dip-coating: Characterization and photocatalytic activity under halogen lamp. *Appl. Surf. Sci.* **2014**, *314*, 910–918. [CrossRef]
46. Fernández, A.; Lasanta, M.; Pérez, F. Molten salt corrosion of stainless steels and low-Cr steel in CSP plants. *Oxid. Met.* **2012**, *78*, 329–348. [CrossRef]
47. Fuqiang, W.; Ziming, C.; Jianyu, T.; Yuan, Y.; Yong, S.; Linhua, L. Progress in concentrated solar power technology with parabolic trough collector system: A comprehensive review. *Renew. Sustain. Energy Rev.* **2017**, *79*, 1314–1328. [CrossRef]
48. Encinas-Sánchez, V.; Macías-García, A.; Díaz-Díez, M.; Díaz-Parralejo, A. Characterization of sol–gel coatings deposited on a mechanically treated stainless steel by using a simple non-destructive electrical method. *J. Ceram. Soc. Jpn.* **2016**, *124*, 185–191. [CrossRef]
49. Morrow, J.; Tejedor-Anderson, M.I.; Anderson, M.A.; Ruotolo, L.A.; Pfefferkorn, F.E. Sol-gel synthesis of ZrO_2 coatings on micro end mills with sol-gel processing. *J. Micro Nano-Manuf.* **2014**, *2*, 041002–041012. [CrossRef]
50. Ni, C.; Lu, L.; Zeng, C.; Niu, Y. Evaluation of corrosion resistance of aluminium coating with and without annealing against molten carbonate using electrochemical impedance spectroscopy. *J. Power Sources* **2014**, *261*, 162–169. [CrossRef]
51. Fernandez, A.; Rey, A.; Lasanta, I.; Mato, S.; Brady, M.; Pérez, F.J. Corrosion of alumina-forming austenitic steel in molten nitrate salts by gravimetric analysis and impedance spectroscopy. *Mater. Corros.* **2014**, *65*, 267–275. [CrossRef]
52. Omar, S.; Repp, F.; Desimone, P.M.; Weinkamer, R.; Wagermaier, W.; Ceré, S.; Ballarre, J. Sol–gel hybrid coatings with strontium-doped 45S5 glass particles for enhancing the performance of stainless steel implants: Electrochemical, bioactive and in vivo response. *J. Non-Cryst. Solids* **2015**, *425*, 1–10. [CrossRef]
53. *ASTM G102*; Standard Practice for Calculation of Corrosion Rates and Related Information from Electrochemical Measurements. ASTM: West Conshohocken, PA, USA, 1994.

54. Stern, M.; Geary, A.L. Electrochemical Polarization: I. A Theoretical Analysis of the Shape of Polarization Curves. *J. Electrochem. Soc.* **1957**, *104*, 56–63. [CrossRef]
55. Ruiz-Cabañas, F.J.; Prieto, C.; Osuna, R.; Madina, V.; Fernández, A.I.; Cabeza, L.F. Corrosion testing device for in-situ corrosion characterization in operational molten salts storage tanks: A516 Gr70 carbon steel performance under molten salts exposure. *Sol. Energy Mater. Sol. Cells* **2016**, *157*, 383–392. [CrossRef]

Disclaimer/Publisher's Note: The statements, opinions and data contained in all publications are solely those of the individual author(s) and contributor(s) and not of MDPI and/or the editor(s). MDPI and/or the editor(s) disclaim responsibility for any injury to people or property resulting from any ideas, methods, instructions or products referred to in the content.

Article

Comparing the Corrosion Resistance of 5083 Al and Al₂O₃3D/5083 Al Composite in a Chloride Environment

Liang Yu [1,2,3], Chen Zhang [1,2,3], Yuan Liu [1,2,3], Yulong Yan [1,2,3], Pianpian Xu [1,2,3], Yanli Jiang [1,2,3,*] and Xiuling Cao [4,5,*]

[1] Key Laboratory of New Processing Technology for Nonferrous Metals & Materials, Guilin University of Technology, Guilin 541004, China
[2] Collaborative Innovation Center for Exploration of Nonferrous Metal Deposits and Efficient Utilization of Resources, Guilin University of Technology, Guilin 541004, China
[3] Guangxi Modern Industry College of Innovative Development in Nonferrous Metal Material, Guilin 541004, China
[4] Hebei Technology Innovation Center for Intelligent Development and Control of Underground Built Environment, Shijiazhuang 050031, China
[5] School of Urban Geology and Engineering, Hebei GEO University, Shijiazhuang 050031, China
* Correspondence: 2010043@glut.edu.cn (Y.J.); caoxlhbdz@163.com (X.C.);
Tel.: +86-138-7834-0301 (Y.J.); +86-135-1331-0032 (X.C.)

Citation: Yu, L.; Zhang, C.; Liu, Y.; Yan, Y.; Xu, P.; Jiang, Y.; Cao, X. Comparing the Corrosion Resistance of 5083 Al and Al₂O₃3D/5083 Al Composite in a Chloride Environment. *Materials* **2023**, *16*, 86. https://doi.org/10.3390/ma16010086

Academic Editors: Jose M. Bastidas, Costica Bejinariu and Nicanor Cimpoesu

Received: 12 November 2022
Revised: 4 December 2022
Accepted: 15 December 2022
Published: 22 December 2022

Copyright: © 2022 by the authors. Licensee MDPI, Basel, Switzerland. This article is an open access article distributed under the terms and conditions of the Creative Commons Attribution (CC BY) license (https://creativecommons.org/licenses/by/4.0/).

Abstract: In this study, an Al₂O₃3D/5083 Al composite was fabricated by infiltrating a molten 5083 Al alloy into a three-dimensional alumina reticulated porosity ceramics skeleton preform (Al₂O₃3D) using a pressureless infiltration method. The corrosion resistance of 5083 Al alloy and Al₂O₃3D/5083 Al in NaCl solution were compared via electrochemical impedance spectroscopy (EIS), dynamic polarization potential (PDP), and neutral salt spray (NSS) tests. The microstructure of the two materials were investigated by 3D X-ray microscope and scanning electron microscopy aiming at understanding the corrosion mechanisms. Results show that an Al₂O₃3D/5083 Al composite consists of interpenetrating structure of 3D-continuous matrices of continuous networks 5083 Al alloy and Al₂O₃3D phase. A large area of strong interfaces of 5083 Al and Al₂O₃3D exist in the Al₂O₃3D/5083 Al composite. The corrosion development process can be divided into the initial period, the development period, and the stability period. Al₂O₃3D used as reinforcement in Al₂O₃3D/5083 Al composite improves the corrosion resistance of Al₂O₃3D/5083 Al composite via electrochemistry tests. Thus, the corrosion resistance of Al₂O₃3D/5083 Al is higher than that of 5083 Al alloy. The NSS test results indicate that the corrosion resistance of Al₂O₃3D/5083 Al was lower than that of 5083 Al alloy during the initial period, higher than that of 5083 Al alloy during the development period, and there was no obvious difference in corrosion resistance during the stability period. It is considered that the elements in 5083 Al alloy infiltrated into the Al₂O₃3D/5083 Al composite are segregated, and the uniform distribution of the segregated elements leads to galvanic corrosion during the corrosion initial period. The perfect combination of interfaces of Al₂O₃3D and the 5083 Al alloy matrix promotes excellent corrosion resistance during the stability period.

Keywords: Al₂O₃3D/5083 Al; corrosion mechanism; electrochemistry; neutral salt spray; interface; interpenetrating structure

1. Introduction

The interpenetrating phase composites (IPCs) with percolating metallic and ceramic phases offer manifold benefits, such as a good combination of strength, toughness, and stiffness, very good thermal properties, excellent wear resistance, as well as the flexibility of microstructure and processing route selection, etc. [1]. The interconnectedness of the phases provides some promising benefits. Each phase contributes to the final composite's properties, with the metal part increasing strength and fracture toughness compared to monolithic ceramics and the ceramic part increasing dimensional and mechanical stability at high temperatures compared to pure metal [2].

The fabrication of metal/ceramic IPCs typically involves two steps: (1) Processing of a reticulated porosity ceramic preform; (2) Infiltration of metallic melt in the pores of reticulated porosity ceramic preform to fabricate the IPC [3].

$Al_2O_3$3D/Al alloy IPC is one of metal/ceramic IPCs, which consists of 3D-continuous matrices of continuous networks Al alloy and $Al_2O_3$3D reticulated porosity ceramics phase [4]. $Al_2O_3$3D can inhibit the nucleation and growth of columnar crystals and reduce the area of exposed Al alloy matrix. $Al_2O_3$3D are used as reinforcement in $Al_2O_3$3D/Al alloy composite to achieve the high temperature mechanical and the wear resistance properties [5]. In the $Al_2O_3$3D/Al alloy IPCs, much of the driving force for investigating interpenetrating microstructures has been the toughening of $Al_2O_3$3D ceramic preforms by the addition of Al alloy metal phase [6]. $Al_2O_3$3D/Al alloy IPCs have a random, usually isotropic, spatial distribution of phases. Compared with particle-reinforced Al_2O_3P/Al, $Al_2O_3$3D/Al exhibits the advantages of high hardness, low density, high corrosion resistance, less corrosion spalling, and fewer defects. Due to their light weight quality, high mechanical properties, and excellent wear, $Al_2O_3$3D/Al alloy IPC is especially preferred for high temperature applications in many areas such as the automotive, space, and aviation industries [7–9].

Although Al alloys and aluminum metal matrix composites (AMCs) can improve the physical/mechanical properties, the presence of alloying elements and reinforcements can also increase the susceptibility to the more severe forms of localized corrosion: intergranular, exfoliation, and stress corrosion cracking [10]. $Al_2O_3$3D/Al alloy IPC corrosion often begins at the interface between the Al matrix and the composite reinforcements. Organic coatings with advantages of effectiveness and convenience have been widely applied to mitigate corrosion of AMCs, which can provide a robust physical barrier against the permeation of corrosive media [11]. The current protective methods to improve the corrosion resistance of the AMCs mainly include anodic oxidation, chemical conversion coating, surface facial mask layer, reinforcement surface coating and matrix alloying, etc. [12]. The heat treatment process, as well as the adjustment of reinforcement content and distribution, will also improve the corrosion resistance of AMCs [13].

The 5083 Al alloy is a high-magnesium alloy that exhibits good strength, corrosion resistance, good weldability, and machinability among non-heat treatable alloys [14]. In addition, 5083 Al alloy is widely used in maritime applications, automotive and aircraft weldments, and subway light rails for its excellent corrosion resistance [15]. Shuiqing Liu investigated the corrosion resistance of 5083 aluminum alloy after refining with nano-CeB6/Al, and found that nano-CeB6/Al inoculant showed a significant grain-refining effect on 5083 Al alloy, and the corrosion resistance of 5083 Al alloy was improved as well [16]. Roseline investigated the corrosion behaviour of heat treated Al metal matrix composites reinforced with fused zirconia alumina, and found the corrosion current density of the composites decreased with an increase in volume % of the heat-treated composite, comparatively more than the specimens that were not heat-treated [17].

Our previous research shows that, as a wear-resistant brake material, reticulated porosity SiC3D ceramic skeleton reinforced 6061 Al alloy metal composite (SiC3D/6061Al alloy IPCs) has excellent friction and wear performance, which can meet the requirements of high-speed train brake discs under emergency braking conditions [18] due to the SiC3D skeleton as a support to enhance the wear resistance of the material. However, obvious corrosion can be observed in the NSS corrosion experiment due to the weak interface bonding between SiC and Al [19]. It is worth improving corrosion resistance of IPC materials.

In this work, reticulated porosity $Al_2O_3$3D ceramic skeleton and 5083 Al alloy were used to improve the corrosion resistance of IPCs materials. The $Al_2O_3$3D skeleton can reduce the area of Al matrix exposed to air and reduce the occurrence of corrosion reactions. $Al_2O_3$3D is tightly bonded to the Al matrix to reduce the defects of composites. In addition, the Al_2O_3 film generated by the oxidation of Al matrix adheres to the $Al_2O_3$3D skeleton. The research on corrosion mechanism of $Al_2O_3$3D/5083Al alloy IPC will promote industrialization of the high-speed train brake materials.

2. Materials and Methods

2.1. Experimental Materials

The reticulated porosity $Al_2O_3$3D ceramic skeleton ($Al_2O_3$3D) was prepared using a polymer replication technique. Polyurethane sponge (Shenzhen Green-tron Environmental Protection Filter Material Co., Ltd., Shenzhen, China) with 10 PPI porosity was used as a template to impregnate the Al_2O_3 slurry. Then, the excess Al_2O_3 slurry on the polyurethane sponge was removed. To improve the surface hanging slurry of the polyurethane sponge, the polyurethane sponge was dipped into 75 °C and 25 wt.% NaOH solution for 2 h to increase surface roughness. A round polyurethane sponge with dimensions of Φ = 500 mm and H = 100 mm was impregnated with Al_2O_3 covering slurry. The sponge was dried in a microwave oven for 15 min to obtain a green Al_2O_3 reticulated body with a good structure, cured at room temperature for 24 h, dried at 120 °C for 6 h. The $Al_2O_3$3D was produced in a graphite-resistance furnace (Jinzhou Santai Electric Furnace Factory, China) with argon gas as the sintering atmosphere. The sintering temperature was increased from 25 °C to 1600 °C with 2 °C/min, held at 1600 °C for 3 h, and cooled at room temperature to produce $Al_2O_3$3D.

Figure 1 shows the preliminary preparation and corrosion direction. The pressureless infiltration method was used to create the $Al_2O_3$3D/5083 Al alloy composite. The volume ratio of 5083Al to $Al_2O_3$3D is about 8:2. 5083 Al alloy and $Al_2O_3$3D were put in two corundum crucibles, respectively, and both were heated to temperatures ranging from 25 °C to 800 °C. The liquid 5083 Al alloy in the crucible was manually agitated, and after being agitated, the aluminum solution was put back into the crucible which containing $Al_2O_3$3D. The $Al_2O_3$3D was gradually positioned on the aluminum solution and held there for 30 min. The aluminum solution was allowed to slowly infiltrate into the $Al_2O_3$3D and then cooled to obtain the $Al_2O_3$3D/5083 Al composite.

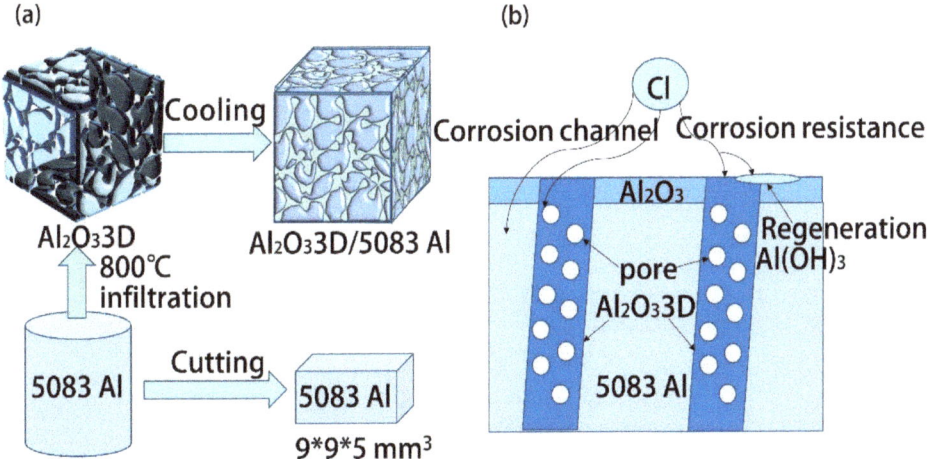

Figure 1. Preliminary preparation and corrosion direction: (a) preparation flow chart; (b) corrosion direction.

A metallographic cutting machine (Laizhou Weiyi Test Machinery Manufacturing Co., Ltd., Laizhou, China) was used to cut the $Al_2O_3$3D/5083 Al material into small squares that measured $9 \times 9 \times 5$ mm^3. A comparative study of corrosion performance was conducted with 5083 Al alloy. The composition of 5083 Al alloy is shown in Table 1.

Table 1. Composition of 5083 Al alloy (mass fraction).

Elements	Si	Cu	Mg	Zn	Mn	Ti	Cr	Fe	Al
Wt.%	0.14	0.03	3.90	0.02	0.60	0.15	0.07	0.42	Balance

2.2. Characterization

Platinum sheet as the auxiliary electrode, and a sample as the working electrode, potentio-dynamic polarization (PDP) experiments and electrochemical impedance spectroscopy (EIS) were carried out on an electrochemical workstation (CHI 790E, Shang-hai Chenhua Instruments Co., Ltd., Shanghai, China). Hot-melt adhesive was used to bind copper wires (Jiangsu Jinzi Xuan Metal Technology Co., Ltd., Wuxi, China) to the back of each sample before it was put into a mold. The electrochemical test samples were made using proportional metallographic cold-mounting fluid (Shanghai Dental Materials Co., Ltd., Shanghai, China). The test sample's working area was 0.81 cm^2. The sample was polished with SiC sandpaper (Eagle) from 800 mesh to 1500 mesh, rinsed with deionized water, polished with 2.5 purpose metallographic polish, rinsed with deionized water, wiped with alcohol, and dried with a hair dryer in cold air.

Tests for open circuit potential (OCP), PDP, and EIS were performed on a sample of a polished surface. For the OCP testing and PDP investigations, the test samples were placed in a glass cell with 3.5 wt.% NaCl solution (Dongguan Xunye Chemical Reagent Co., Ltd., Dongguan, China) at room temperature for 6 min. The OCP testing hold was 1000 s with a starting potential of −1.3 V and a scan direction from cathode to anode, with a scanning interval of roughly 600 mV in relation to the self-corrosive potential, the scanning speed from cathode to anode direction was 0.25 mv/s. Samples of polished surface were measured at OCP stabilization by using a 10 mV perturbed potential sine wave for EIS, and samples were measured within the frequency range of 10^{-2}–10^6 Hz. Samples of polished surface were observed using a high-resolution 3D X-ray Microscope (Zeiss Xradia 510 Versa, Carl Zeiss) for the microstructure characterization of the samples.

A laboratory-prepared Keller solution (95 mL deionized water + 2.5 mL HNO$_3$ + 1.5 mL HCl + 1.0 mL HF) was used to etch the specimens for the intergranular corrosion test. The samples were etched with Keller's solution for 10–20 s, rinsed with deionized water, and wiped with alcohol. Sample tissues were observed using an inverted metallographic microscope (MS600 Hangzhou Jingke Testing Instruments Co., Ltd., Hangzhou, China). A neutral salt spray (NSS) corrosion test was performed using a fully automatic salt spray tester (ZK-60K, Dongguan Zhenke Testing Equipment Co., Ltd., Dongguan, China). The experiments were performed with 5% mass fraction of corrosive liquid and neutral NaCl solution with pH 6.5–7.2. The experimental times were 24, 72, 144, 240 and 360 h at (35 ± 1) °C. The experimental samples were observed using Zeiss GeminiSEM 300 field-emission scanning electron microscope (Oberkochen, Germany) equipped with an energy-dispersive X-ray spectrometer (Oberkochen, Germany) for the microstructure characterization of the samples and corrosion products.

3. Results

3.1. Sample of Polished Surface of the Two Materials

Figure 2a shows the burned round disks shaped Al$_2$O$_3$3D. The overall skeleton is white, round, and exhibits certain hardness and strength. The Vickers hardness tester measures compressive strength at 3.9 MPa. The flexural strength of sintered Al$_2$O$_3$ foams was determined from three-point bending. The loaded surfaces were covered with a thin sponge layer to obtain uniform load distribution throughout the faces. In all mechanical determinations, results were based on an average of five samples. The measured flexural resistance was 2.7 MPa. The pores of Al$_2$O$_3$3D are approximate round holes, and measured at 2–3 mm.

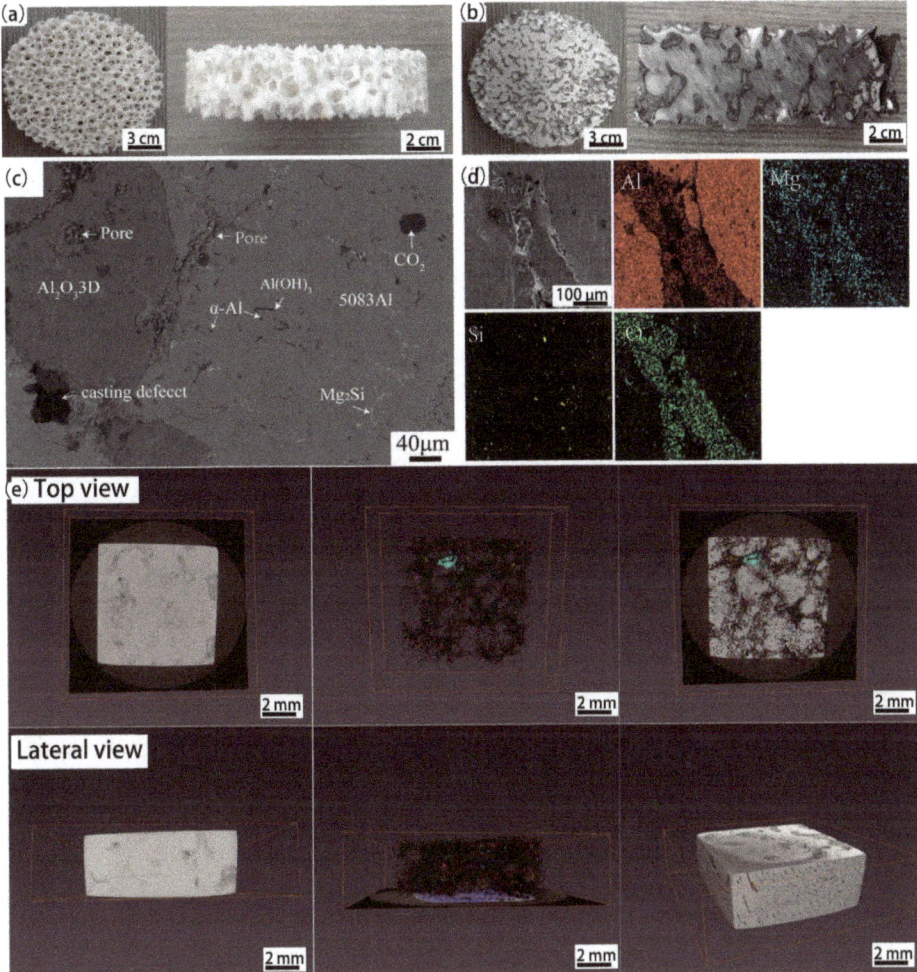

Figure 2. (**a**) Al$_2$O$_3$3D; (**b**) sampling point of Al$_2$O$_3$3D/5083Al; (**c**) SEM (Scanning Electron Microscope) of the Al$_2$O$_3$3D/5083Al; (**d**) EDS of the Al$_2$O$_3$3D/5083Al; (**e**) 3D XRM (High resolution 3D X-ray Microscope) of Al$_2$O$_3$3D/5083Al.

Figure 2b shows the prepared Al$_2$O$_3$3D/5083 Al composite. The prepared sample by a pressureless infiltration method shows typical IPC structural characteristics and exhibits a light silver metallic luster. The sample is compact in structure, without obvious pores on the outer surface, which implies it has high strength and tightness.

Figure 2c shows the scanning electron microscopy (SEM) image of the Al$_2$O$_3$3D/5083 Al composite without corrosion. The brighter part is the 5083 Al matrix, while the darker part is the Al$_2$O$_3$3D. The second phase was observed in the 5083 Al matrix. The round dot is α-Al$_2$O$_3$, and the thin strip is Al$_6$ (iron, manganese), Mg$_2$Si. Because there is a large amount of Al$_2$O$_3$ powder in the green Al$_2$O$_3$ reticulated body, it is difficult to form a sintering neck during the sintering process. Consequently, Al$_2$O$_3$3D is not completely dense, leaving defects such as pores. In addition, in the process of molten 5083 Al liquid infiltrating into the Al$_2$O$_3$3D, the thermal stress of Al$_2$O$_3$3D is transferred and released in the direction of the Al$_2$O$_3$3D, causing damage to the Al$_2$O$_3$3D. The Al$_2$O$_3$3D has cracks and dark pits on the surface and inside, exerting a negative effect on the performance of the Al$_2$O$_3$3D/5083 Al composite [20].

Figure 2d shows the energy-dispersive X-ray spectroscopy (EDS) diagram of the uncorroded Al$_2$O$_3$3D/5083 Al composite. No delamination occurred between the 5083 Al matrix and the Al$_2$O$_3$3D, and the Mg appeared enriched in the Al$_2$O$_3$3D. During the preparation, the molten 5083 Al liquid released energy when it cooled down. The diffusion ability of the Mg in the molten 5083 Al liquid was considerably enhanced, and it entered through the Al$_2$O$_3$3D. The Al$_2$O$_3$ particles in the Al$_2$O$_3$3D exhibited an adsorption effect, showing the enrichment of the Mg in the Al$_2$O$_3$3D. During the cooling process of the 5083 Al liquid, the second phase precipitated. The EDS indicates a point-like enrichment of the Si, which exerts minimal effect on the performance of the Al$_2$O$_3$3D/5083 Al. The O demonstrated enrichment in the Al$_2$O$_3$3D [21].

Figure 2e shows the test diagram of the 3D X-ray microscopy (XRM) of uncorroded Al$_2$O$_3$3D/5083 Al. The pores are concentrated at the Al$_2$O$_3$3D. A few pores were observed on the surface of the Al matrix. .A large area of strong interfaces of 5083 Al and Al$_2$O$_3$3D was observed in the Al$_2$O$_3$3D/5083 Al composite. No evident delamination is found between the two phases.

Figure 3a shows the optical microscopy (OM) image of the Al$_2$O$_3$3D/5083 Al composite. The eroded out metallographic phase by the etching solution can be seen between the Al$_2$O$_3$3D and the 5083 Al matrix. The boundary has a thicker layer because, when the material is compounded, the aluminum liquid touches the Al$_2$O$_3$3D skeleton, resulting in faster cooling of the parts in contact. The aluminum liquid will preferentially solidify at the Al$_2$O$_3$3D skeleton, resulting in the uneven local solidification of the aluminum liquid. The Al$_2$O$_3$3D skeleton will have more solidified parts. During the preparation of Al$_2$O$_3$3D/5083 Al composites, second phases were precipitated, mostly Al$_3$Mg$_2$, α-Al, and Mg$_2$Si [22].

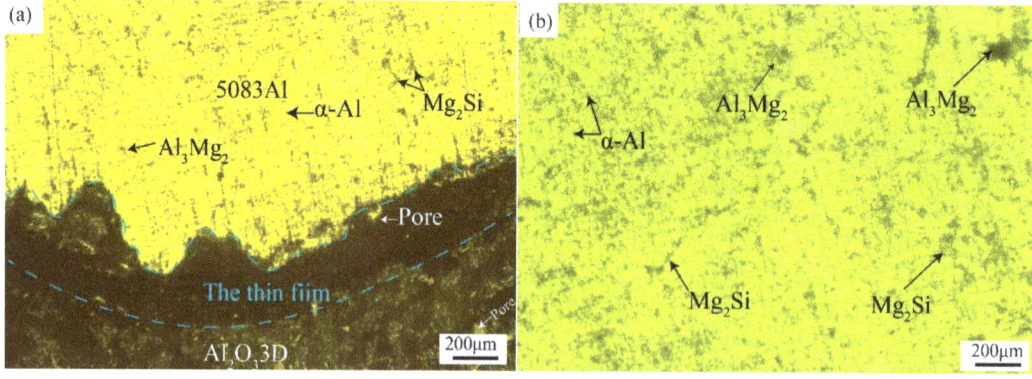

Figure 3. Optical micrograph (OM) image of two materials: (**a**) Al$_2$O$_3$3D/5083 Al; (**b**) 5083 Al.

Figure 3b shows the optical micrograph of the 5083 Al alloy. The Al also precipitates Al$_3$Mg$_2$, α-Al, and Mg$_2$Si during solidification. The Al$_2$O$_3$3D in the composite can inhibit interfacial reaction, exerting a positive effect on material properties [23].

3.2. Polarization Curve

The electrochemical corrosion of the Al$_2$O$_3$3D/5083 Al composite includes interfacial and intergranular corrosion. The OCP and PDP of the Al$_2$O$_3$3D/5083 Al composite and 5083 Al are depicted in Figure 4a,b. The OCP, corrosion potential (E$_{corr}$), and corrosion current density (I$_{corr}$) values obtained from the curves in Figure 4 are given in Table 2. The OCP voltage of the Al$_2$O$_3$3D/5083 Al composite is more negative than that of 5083 Al due to certain defects in the sample preparation.

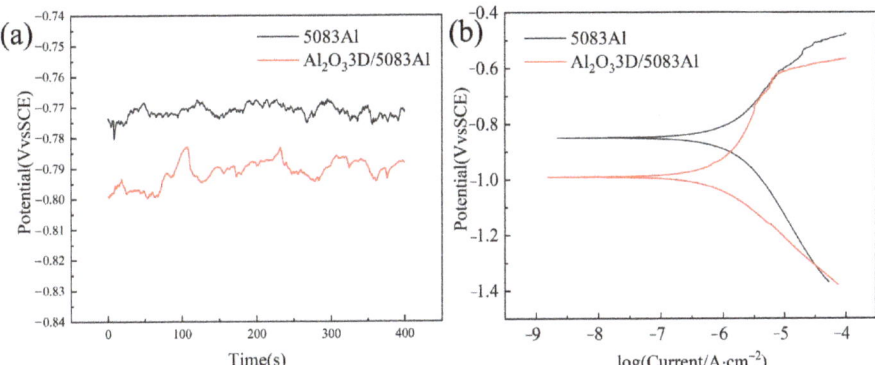

Figure 4. (**a**) Open circuit potential (OCP) curve; (**b**) potentio-dynamic polarization (PDP) curve.

Table 2. Corrosion potential and corrosion current.

Sample	Open Circuit Potential OPC (mV)	E_{corr} (mV)	I_{corr} ($\mu A \cdot cm^2$)
$Al_2O_3 3D/5083Al$	−791.6	−1069	6.410
5083Al	−773.9	−849	9.879

In the polarization curve, the E_{corr} value of the $Al_2O_3 3D/5083$ Al composite is smaller than that of 5083 Al, while its I_{corr} value is smaller than that of 5083 Al, indicating that the corrosion tendency of the $Al_2O_3 3D/5083$ Al composite is higher than that of 5083 Al. The corrosion current is the determining factor of the corrosion resistance of the materials. The I_{corr} value of the $Al_2O_3 3D/5083$ Al composite is smaller than that of 5083 Al, indicating that the corrosion resistance of the $Al_2O_3 3D/5083$ Al composite is better than that of 5083 Al.

Figure 5 shows the optical micrographs after electrochemical corrosion. More Mg_2Si particles precipitated in the $Al_2O_3 3D/5083$ Al composite in Figure 5d than in the 5083 Al alloy in Figure 5f. The 5083 Al matrix in the $Al_2O_3 3D/5083$ Al composite exhibits a higher tendency to corrode and is less sensitive to early pitting microporous nucleation due to the small potential difference between Mg_2Si particles and the 5083 Al matrix.

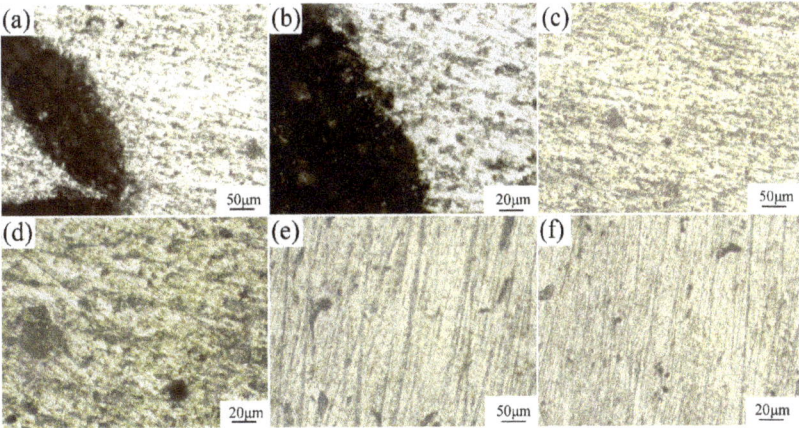

Figure 5. Optical micrographs of two materials after Tafel corrosion: (**a**–**d**) $Al_2O_3 3D/5083Al$; (**e**,**f**) 5083 Al.

During the corrosion process, Mg_2Si particles hindered the continuity of the matrix and inhibited corrosion [24]. The precipitation of the second phase reduces the corrosion sensitivity of the material, inhibits corrosion tendency, and reduces corrosion current density. Notably, the $Al_2O_3$3D/5083 Al composite not only exhibits the advantages of large interfacial composite, less interfacial concentration, small specific surface area, anticorrosive, and antioxidant IPC structure, but the $Al_2O_3$3D also simultaneously enhances the corrosion resistance of the $Al_2O_3$3D/5083 Al composite. In the preparation of the $Al_2O_3$3D/5083 Al composite, the $Al_2O_3$3D demonstrates strongly bonding properties with the 5083 Al to reduce the generation of voids. When the 5083 Al matrix in the $Al_2O_3$3D/5083 Al composite was oxidized, the Al_2O_3 film generated by oxidation combined with $Al_2O_3$3D to fill the voids of the composite, making the IPC structure denser.

3.3. EIS of Polished Surface Materials

Figure 6 shows the EIS plots of $Al_2O_3$3D/5083 Al and 5083 Al in the absence of salt spray corrosion. The Nyquist plots of two materials are capacitive reactance plots in Figure 6a. The impedance spectra are capacitive reactance arcs in the high-frequency region, reflecting the electrochemical reaction of corrosion on the electrode surface. The two materials show a similar EIS curve at high frequencies ranging from 1.0×10^5–5.0×10^5 Hz, and $Al_2O_3$3D/5083 Al is considerably larger than 5083 Al at low frequencies ranging from 0.01–10 Hz, indicating greater resistance and corrosion resistance of $Al_2O_3$3D/5083 Al [25]. The Nyquist plot in Figure 6a is a semicircle. Therefore, the control step of the electrode process is determined by the electrochemical reaction step (charge transfer process). The impedance caused by the diffusion process can be disregarded. Figure 6b,c show that the Bode plot has two time constants. The EIS results of $Al_2O_3$3D/5083 Al and 5083 Al are consistent with the PDP results presented in Figure 4.

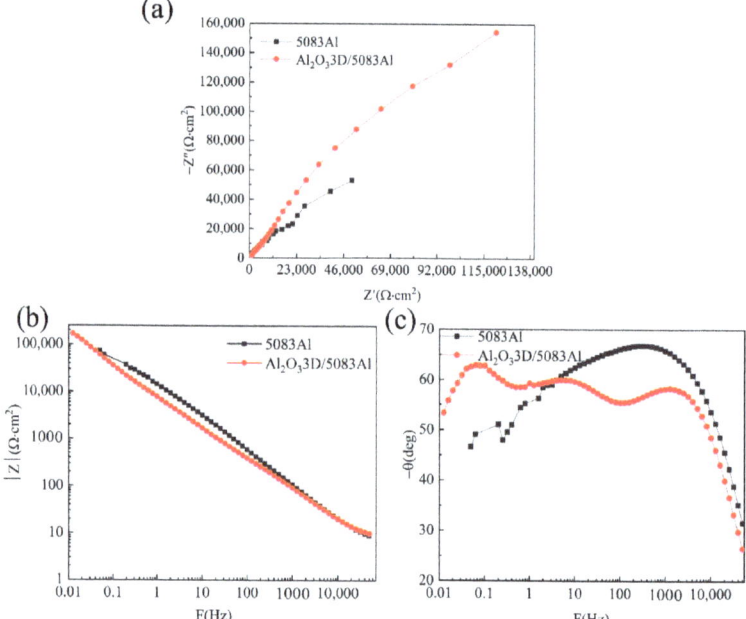

Figure 6. Uncorroded electrochemical alternating current impedance spectroscopy (EIS): (**a**) Nyquist diagram; (**b**) Bode diagram (|Z|-F); (**c**) Bode diagram (−θ-F).

3.4. Corrosion Morphology Analysis

The NSS tests show that the corrosion process of $Al_2O_3 3D/5083$ Al and 5083 Al consists of pitting, intergranular corrosion, and spalling corrosion [26]. The surface morphology of the NSS-corroded $Al_2O_3 3D/5083$ Al and 5083 Al specimens is shown in Figure 7, the red and blue circles in the figure are a partial enlargement of the original Figure 7. To observe the microstructure of the 5083 Al alloy and the degree of corrosion of 5083 Al matrix in an $Al_2O_3 3D/5083$ Al composite, the metal phases of both materials were amplified and processed.

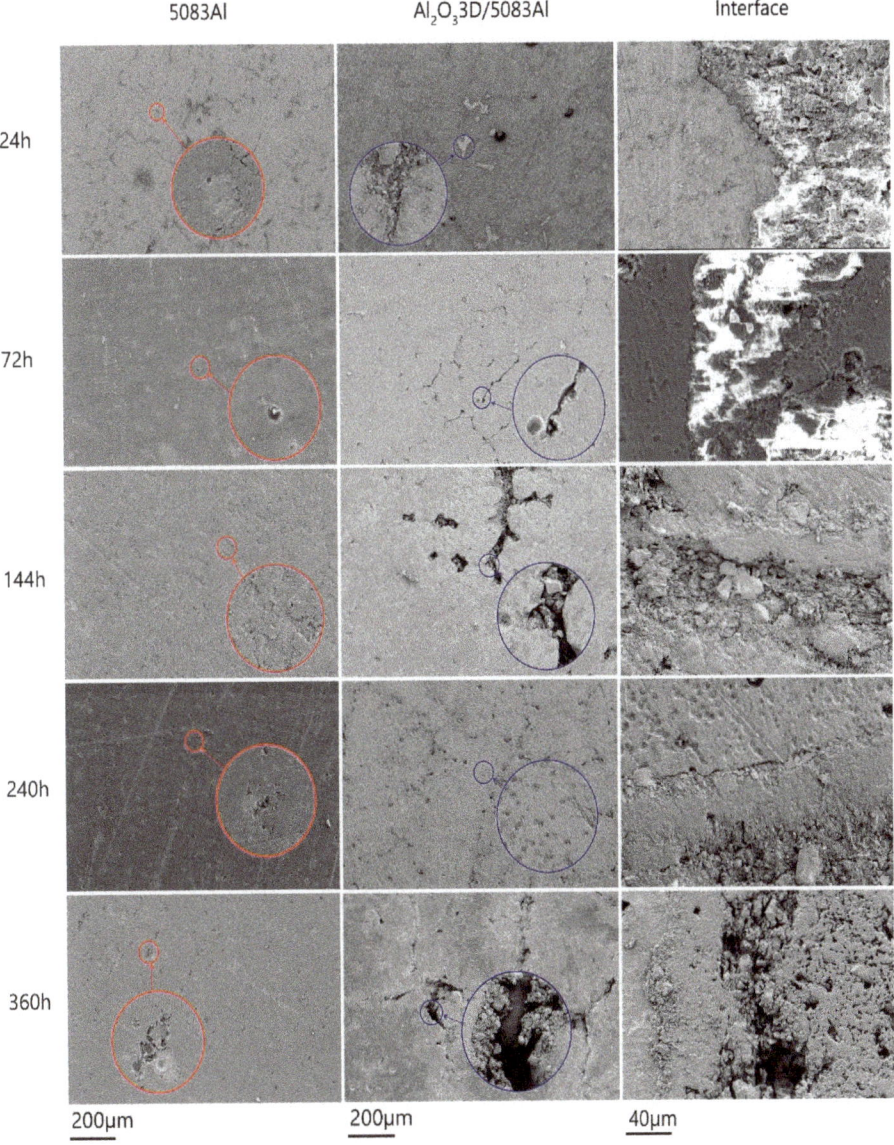

Figure 7. Corrosion morphology of $Al_2O_3 3D/5083Al$ and 5083Al under different times after NSS (neutral salt spray) corrosion.

After 24 h, pitting appeared in 5083 Al. The size of pitting was about 0.5 µm, and the overall structure was well maintained. During the initial period, $Al_2O_3$3D/5083 Al has fine corrosion pits in the Al matrix. These pits defects were formed in the process of pressureless infiltration of $Al_2O_3$3D/5083 Al. At the bond of the interface, no evident corrosion pits and no evident damage of $Al_2O_3$3D are found [27].

After 72 h, a gradual increase in the pitting of 5083 Al was observed. During the initial period, pitting occurred around the precipitates, which was driven by the galvanic coupling effect. The pitting rapidly extended horizontally with shallow circular structures caused by the deposition around corrosion pits. $Al_2O_3$3D/5083 Al exhibited an increase in corrosion pits in the Al matrix, the oxide film on the surface was destroyed, no significant corrosion change occurred in the aluminum matrix at the boundary, and $Al_2O_3$3D presented corrosion pits.

After 144 h, 5083 Al pitting developed substantially, small pieces of spalling corrosion appeared, and the Al_2O_3 film broke down. During the development period, $Al_2O_3$3D/5083 Al demonstrated substantial development of corrosion pits in the Al matrix. The substrate Al was exposed, pitting deepened, with corrosion and evident reaction on a small area. No evident corrosion occurred on a large area, with pitting in the Al matrix at the interface and spalling of Al_2O_3 particles in $Al_2O_3$3D.

After 240 h, 5083 Al spalling corrosion increased, accompanied by the development of pitting and cracking. During the development period, the structure of 5083 Al was not significantly damaged. The generation of Al_2O_3 appeared on the surface, and the self-healing of Al began to occur. $Al_2O_3$3D/5083 Al presented a large area of pitting in the Al matrix, accompanied by the deepening of local corrosion pits. The Al_2O_3 film on the surface was destroyed, and a large area of pits were observed at the interface. Corrosion developed downward along the pores of $Al_2O_3$3D.

After 360 h, 5083 Al spalling corrosion was enhanced, and large corrosion pits appeared in the spalled Al matrix and developed downward in depth. During the stability period, $Al_2O_3$3D/5083 Al corrosion pit depth was elevated in the Al matrix. Corrosion pits became larger, and pitting corrosion appeared on the interface between the $Al_2O_3$3D and 5083 Al two phases. Depth development of pitting corrosion occurred at the Al matrix, and the surface of $Al_2O_3$3D located at the interface was destroyed. Deepening and expansion of corrosion pits were demonstrated on $Al_2O_3$3D.

Therefore, 5083 Al present pitting after corrosion testing, milder than a composite metal matrix surface. $Al_2O_3$3D/5083 Al has a part of the corrosion enrichment of the Al matrix, and evident damage was observed. However, it exerted a minimal effect on the overall corrosion resistance, and IPC structure before and after corrosion is maintained well.

Figure 8 shows the morphology of the corrosion products of NSS-corroded $Al_2O_3$3D/5083 Al and 5083 Al.

After 24 h, 5083 Al showed fine pitting with more pitting divisions accompanied by small pieces of Al (OH)$_3$ adsorbed onto the surface. During the initial period, $Al_2O_3$3D/5083 Al showed pitting that was larger compared with that of 5083 Al but less in number [28]. The substrate Al was maintained well.

After 72 h, 5083 Al showed deepening of pitting, and visible Al (OH)$_3$ was generated on the corrosion pits. During the initial period, Pitting Al (OH)$_3$ appeared around the corrosion pits. $Al_2O_3$3D/5083 Al exhibited partial deepening of pitting pits. Large pieces of Al (OH)$_3$ appeared around the pitting pits, and the IPC structure was maintained well.

After 144 h, 5083 Al showed corrosion cracks. Corrosion led to the formation of the second phase, where a rupture of the oxide film occurred. During the development period, Al exuded from the crack and was subjected to oxidation to generate pitting Al_2O_3. A small amount of Al (OH)$_3$ enriched at the crack played a role in repairing the crack, and $Al_2O_3$3D/5083 Al showed a small area of pitting. Metal elements exuded at the pitting pits on the substrate Al, and a small amount of Al (OH)$_3$ agglomerated at the pitting pits, increasing roughness of the surface.

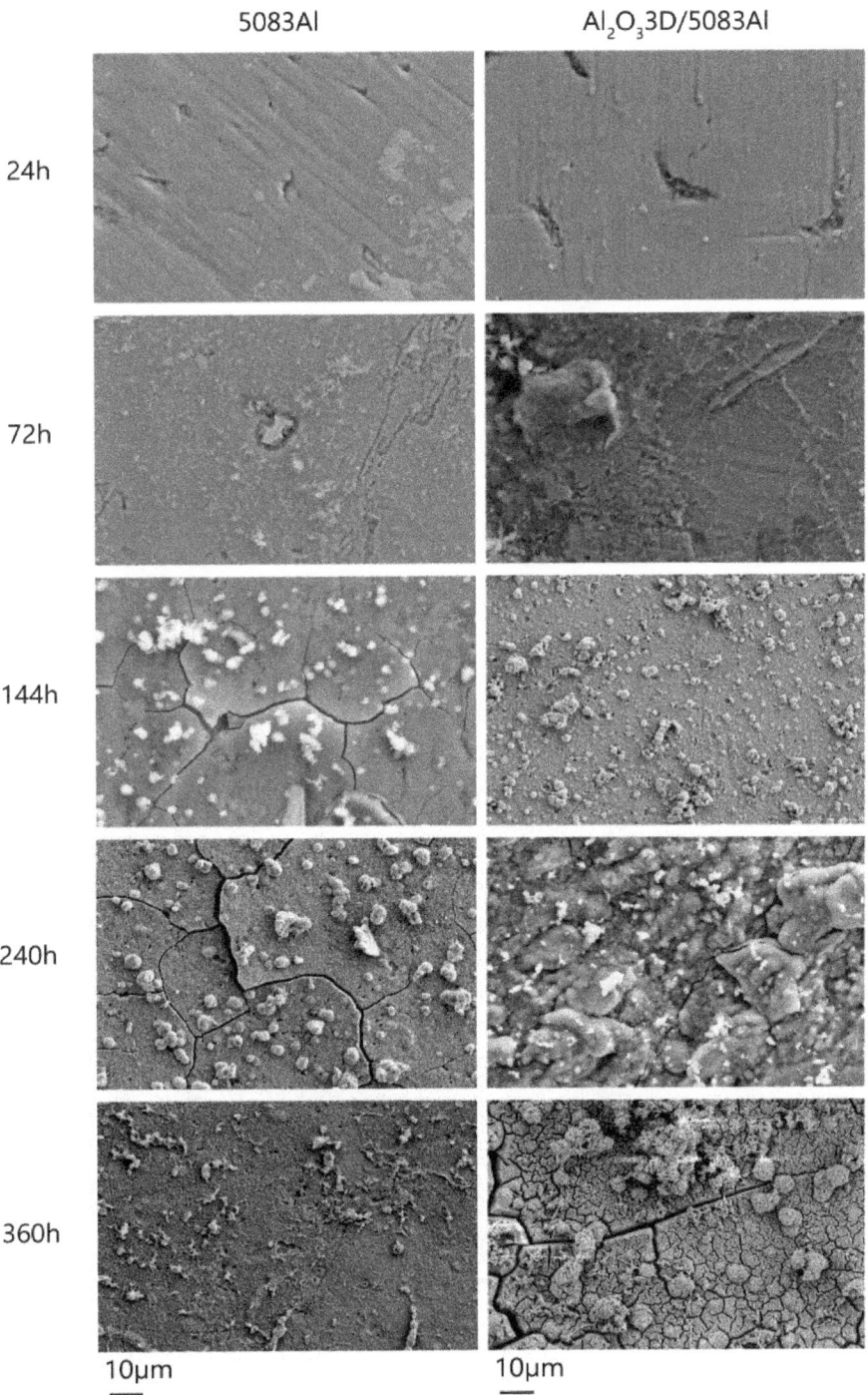

Figure 8. Corrosion product morphology under different times after NSS corrosion.

After 240 h, the pitting deepened on the surface of 5083 Al, causing a large expansion of cracks. During the development period, the passivation film was destroyed. The surface of $Al_2O_3$3D/5083 Al showed a deepening of pitting. A large number of elements exuded from the Al matrix and $Al_2O_3$3D. The Al element was oxidized on the surface, forming a new Al_2O_3 film with enhanced corrosion resistance.

After 360 h, the surface of 5083 Al formed a porous oxide film layer. During the stability period, small agglomerates of Al (OH)$_3$ appeared on the Al_2O_3 film of $Al_2O_3$3D/5083 Al. Meanwhile, repaired cracks reappeared. Spherical oxide particles were distributed around for the self-repair of the oxide film.

Figure 9 shows the EDS of corrosion products at different times after NSS corrosion. The scale bar used for the elements is the same as that used for the corresponding microstructure.

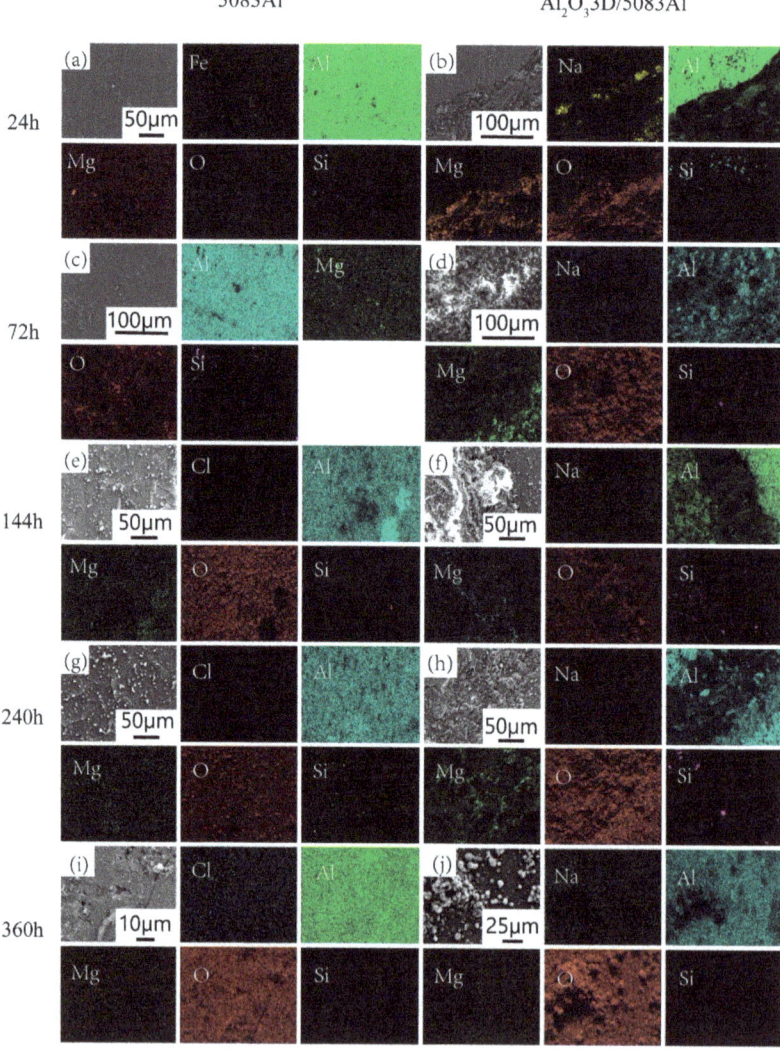

Figure 9. Energy dispersive X-ray spectrometer (EDS) of corrosion products with different times after NSS corrosion: (**a,c,e,g,i**) 5083Al; (**b,d,f,h,j**) $Al_2O_3$3D/5083Al.

After 24 h, the EDS of 5083 Al showed no significant changes on the surface. During the initial period, Al and Fe atoms diffused out of the pitting and were oxidized. The EDS of $Al_2O_3$3D/5083 Al showed that the Mg was enriched toward the $Al_2O_3$3D during corrosion.

After 72 h, a small amount of Si precipitated on the surface of the Al matrix in a punctuated division. During the initial period, an enrichment of Na occurred at the interface, and an aggregation of elements on the skeleton of $Al_2O_3$3D was observed. The Na on the surface of Al originated from the atmosphere, and Na was involved in the generation of corrosion products $NaAlCO_3(OH)_2$. The content of $NaAlCO_3(OH)_2$ decreased as the depth increased following an exponential power function.

After 144 h, the EDS of 5083 Al indicated that a small amount of Cl^- was uniformly divided on the surface. During the development period, Al atoms were gathered at the corrosion pits, and the oxide film was broken. Spalling corrosion exposed the aluminum substrate to air. The EDS showed that a large number of Mg atoms were precipitated at the corrosion pits. Corrosion was relatively fast, and Mg atoms acting as the anode were oxidized. The corrosion of the 5083 Al began as intergranular corrosion and galvanic corrosion. Si atoms were also gathered at the corrosion pits in the form of spots, with the active state of the damaged Al_2O_3 film acting as the anode and the passive state of the undamaged film acting as the cathode, constituting an activation–passivation cell, wherein the redox reaction caused the metal to dissolve in the pore. The pore was maintained as electrically neutral, with Cl^- migrating into the pore. The Al atoms were activated by the action of Cl^-, forming a pore activation (internal)–passivation (external) corrosion cell with increased migration of Cl^-. The EDS of $Al_2O_3$3D/5083 Al showed a small amount of uniform precipitation of Na atoms. Corrosion products of Al gathered on the $Al_2O_3$3D. Corrosion products of Mg atoms were distributed on both sides of the interface and enriched in the two-phase interface. O atoms were found on the surface of the uniform division, indicating that the oxide film structure was maintained well. Corrosion products of Si atoms were distributed on both sides of the interface. The performance of the division state was a small amount of aggregation at the interface, and a small amount of corrosion products on the surface of the Al matrix. The second phase generation was enriched at the interface, enhancing the corrosion resistance of the interface.

After 240 h, the EDS of 5083 Al presented a small amount of Cl^- that was uniformly distributed on the surface. During the development period, Al_2O_3 was generated again, with no vacancy or agglomeration of Al and Mg divisions, indicating that corrosion pits were covered, and corrosion resistance was enhanced. A small amount of Si was distributed on the surface, showing a dotted division. The precipitated second phase mixed with oxidation-generated Al_2O_3 encapsulated the metal surface. Corrosion products of Mg precipitated on the $Al_2O_3$3D. O divisions indicated that the $Al_2O_3$3D was bonded well with the interface of 5083 Al, and the oxide film was not damaged. Si divisions were concentrated at the interface.

After 360 h, the EDS of 5083 Al presented an increase in brightness of the residual Cl^- with an increase in corrosion time and an increase in the residual amount of Cl^-. During the stability period, a small amount of Mg gathered around the corrosion pits, and O indicated the destruction of the oxide film. The corrosion products of Si decreased, and the secondary phase was consumed. The corrosion map of $Al_2O_3$3D/5083 Al showed $Al(OH)_3$ encapsulation at the corrosion pits. O were distributed to agglomerate at the corrosion pits and covered the corrosion pits.

3.5. EIS of Corrosion Product

The EIS of corrosion products was conducted to study the influence of corrosion product layer on the material surface of the corrosion process. As shown in Figure 10a,d, the Nyquist diagram of 5083 Al, the high- and medium-frequency capacitance resistance arc after corrosion, the capacitance resistance in the high-frequency region was larger at 72 h, and it gradually decreased with time. The $Al_2O_3$3D/5083 Al Nyquist plot consisted of a semicircular inductive resistance arc. The high- and medium-frequency capacitive–resistance arcs corresponded to the corrosion products on the surface of $Al_2O_3$3D/5083 Al. The low-frequency capacitance–resistance arc corresponded to the electrochemical corrosion reaction on the electrode surface. With an increase in salt spray time, the capacitance resistance of $Al_2O_3$3D/5083 Al in the high-frequency area decreased rapidly from 24 h to 144 h, and corrosion products were slowly formed on the surface after 240 h. The capacitance resistance in the high-frequency area increased slowly. The impedance modulus of $Al_2O_3$3D/5083 Al exhibited a rise, then a fall, and then a rise again at high frequency. This pattern was the same as the phase angle, and corrosion resistance was enhanced at 240 h.

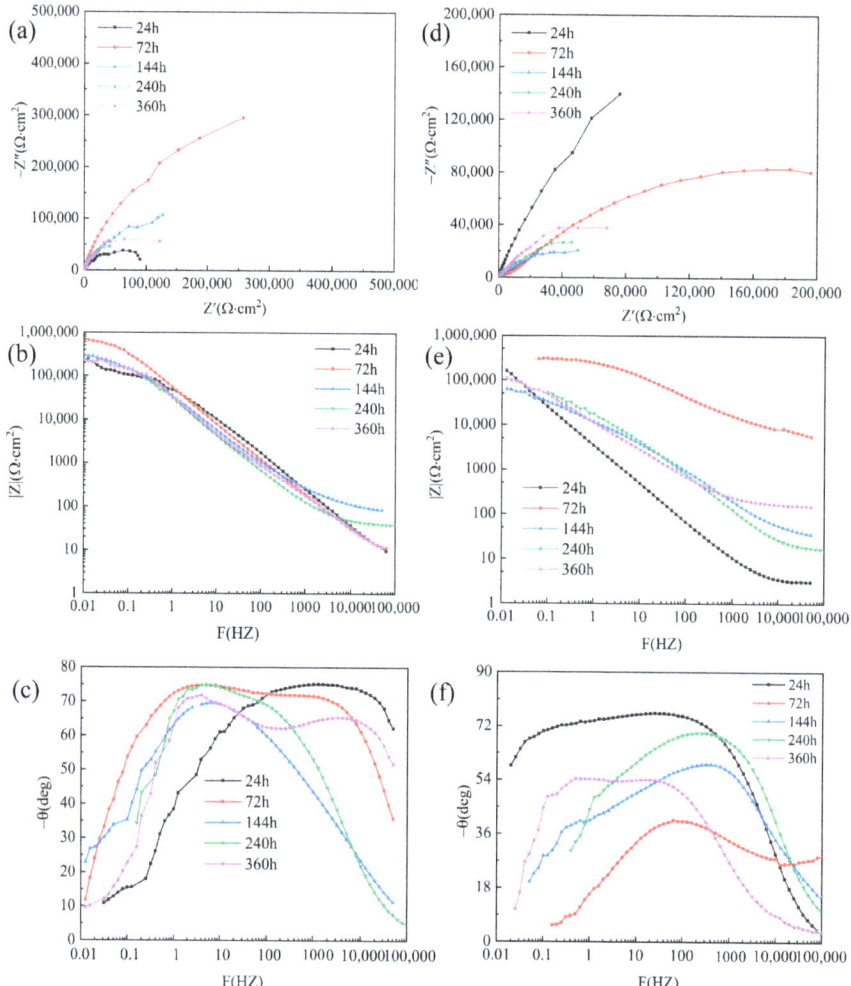

Figure 10. EIS of $Al_2O_3$3D/5083 Al (**d**–**f**) and EIS of 5083Al alloy (**a**–**c**) after NSS at different times: (**a**,**d**) Nyquist diagram; (**b**,**e**) Bode diagram (|Z|-F); (**c**,**f**) Bode diagram (−θ-F).

3.6. Tafel of Corrosion Product

The Tafel of corrosion products was performed to study the effect of corrosion products on corrosion performance. Galvanic coupling corrosion is a key issue in the localized corrosion of Al alloys. It is caused by the nonhomogeneous microstructure of these alloys. Pitting corrosion caused by galvanic coupling reactions may be the initiation point of cracks, leading to failure. The current interactions depend on their electrochemical properties, which are closely related to the surrounding environment and are the key to understanding corrosion formation. The electro-couple interaction between the two phases can lead to crater formation through the dissolution of the particles or the corrosion of the substrate adjacent to the particles.

As shown in Figure 11a, the corrosion potential of 5083 Al exhibited a successive decline followed by a rise. The corrosion anode demonstrates evident fluctuations. The corrosion reaction was more obvious.

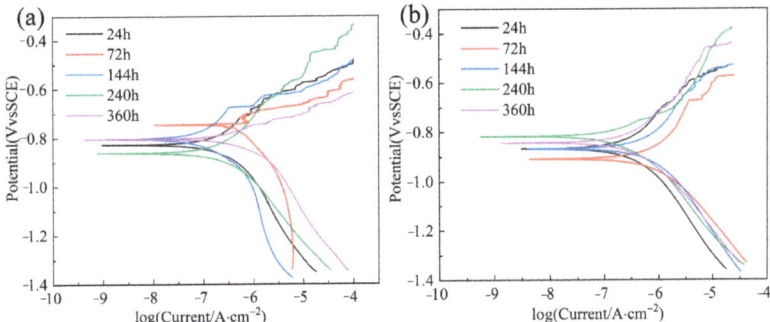

Figure 11. (a) Tafel of 5083Al after NSS at different times; (b) Tafel of $Al_2O_3 3D/5083Al$ after NSS at different times.

At 72 h, the metal passivation phenomenon appeared, corrosion potential increased, corrosion tendency decreased, corrosion current density was reduced, and corrosion resistance improved. Corrosion resistance was maintained better.

Figure 11b shows the corrosion potential of $Al_2O_3 3D/5083$ Al at different times. Corrosion current density was larger, and then smaller, and then larger again. The phenomenon of anodic passivation appeared after 72 h, but the addition of $Al_2O_3 3D$ reduced the exposed area of the metal, weakening the passivation effect of the substrate Al. Corrosion potential decreased, and corrosion current density rose, with a decrease in corrosion resistance.

Ten samples were tested using electrochemistry, and the results listed in Tables 3 and 4. Corrosion current density of the 5083 Al after different times that NSS was 2.115 µA·cm², 1.024 µA·cm², 1.086 µA·cm², 2.114 µA·cm², and 2.580 µA·cm², respectively, for the 24 h, 72 h, 144 h, 240 h, and 360 h of NSS, as shown in Table 3. Corrosion current density of the $Al_2O_3 3D/5083$ Al after different time NSS was 2.808 µA·cm², 7.048 µA·cm², 5.343 µA·cm², 1.094 µA·cm², and 2.823 µA·cm², respectively, for the 24 h, 72 h, 144 h, 240 h, and 360 h of NSS, as shown in Table 4.

Table 3. Tafel of 5083 Al after NSS at different times.

	Tafel of 5083 Al after NSS at Different Times				
Time/h	24	72	144	240	360
Ecorr/mV	−827	−720	−803	−860	−803
Icorr/(µA·cm²)	2.115	1.024	1.086	2.114	2.580

Table 4. Tafel of $Al_2O_3 3D/5083$ Al after NSS at different time.

	Tafel of $Al_2O_3 3D/5083$ Al after NSS at Different Times				
Time/h	24	72	144	240	360
Ecorr/mV	−867	−908	−866	−816	−842
Icorr/(µA·cm²)	2.808	7.048	5.343	1.094	2.823

Tables 3 and 4 exhibited the phenomenon of passivation of 5083 Al metal at 72 h with enhanced corrosion resistance and weakened corrosion resistance of the composite material at 72 h. In addition, 5083 Al with weakened corrosion resistance at 240 h and a decrease in corrosion rate of Al$_2$O$_3$3D/5083 Al at 240 h with enhanced corrosion resistance. The residual salt crystal accumulated at 24–72 h of the Al$_2$O$_3$3D/5083 Al, Al matrix corrosion resistance was weakened. A large number of corrosion products were generated, and corrosion products had uneven distribution. Al atoms produced by corrosion were oxidized into Al (OH)$_3$ with O$_2$, which coated the Al matrix surface and hindered the corrosion reaction. After 240 h of the NSS experiment, the corrosion products of 5083 Al matrix in Al$_2$O$_3$3D/5083 Al are covered on the surface of Al$_2$O$_3$3D/5083 Al, which improves the corrosion resistance of the Al$_2$O$_3$3D/5083 Al. After the 144–240 h NSS experiment, the number of Mg$_2$Si had increased. Mg$_2$Si are one of the second phases in the Al matrix, which can be used as a corrosion anode. The increase of Mg$_2$Si will enhance the corrosion resistance of Al$_2$O$_3$3D/5083 Al. After 360 h of the NSS experiment, corrosion current density increased and the corrosion resistance of Al$_2$O$_3$3D/5083 Al decreased.

3.7. Analysis of Corrosion Rate (CR)

CR is calculated by weight loss in accordance with ASTM-G31-72:

$$CR = (K \times W)/(A \times T \times D), \quad (1)$$

where the constant K = 8.76 × 10^3, W is the weight loss, A is the area of the sample exposed to the NaCl solution, T is the exposure time, and D is the standard density of the material under test. In this experiment, the dimensions of the parallel hexahedral sample are 9 × 9 × 5 mm^3. The exposed surface of the sample is 81 mm^2. The density of Al$_2$O$_3$3D/5083 Al is 2.19 g/cm^3, while the density of the 5083 Al alloy is 2.71 g/cm^3.

The variation of CR with NSS time for both materials in Figure 12 shows that the weight loss of the 5083 Al alloy is less than that of the Al$_2$O$_3$3D/5083 Al composite. It illustrates the corrosion development process can be divided into the initial period, the development period, and the stability period. Compared with the 5083 Al alloy, (1) the skeleton of Al$_2$O$_3$3D is also capable of corrosion, and corrosion will lead to loose and porous bone; and (2) the skeleton structure of Al$_2$O$_3$3D is not sufficiently dense, pores occur, and after a long period of corrosion, the Al$_2$O$_3$3D skeleton appears to shed Al$_2$O$_3$ particles.

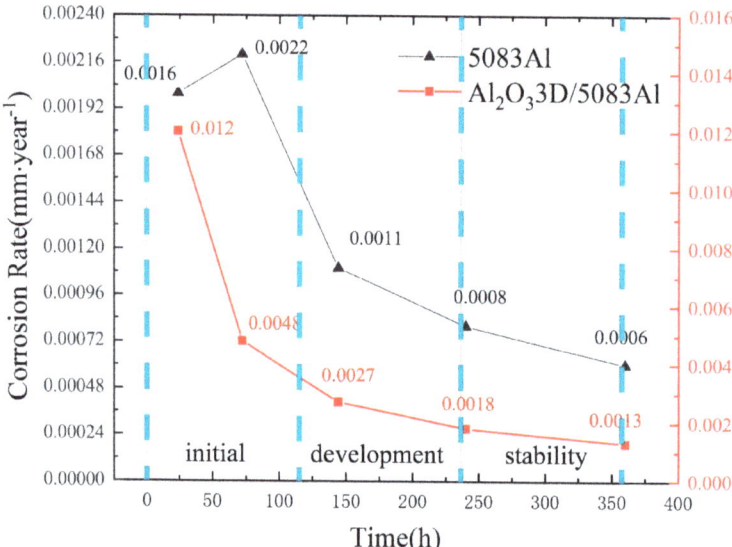

Figure 12. Corrosion rate (CR) of the two materials vs. NSS time.

4. Discussion

The corrosion resistance of $Al_2O_3 3D/5083$ Al was lower than that of 5083 Al alloy during the initial period, higher than that of 5083 Al alloy during the development period, and there is no obvious difference in corrosion resistance during the stability period.

The Al interface prepared by the $Al_2O_3 3D/5083$ Al composite exhibits poor corrosion resistance, uneven elemental division in the 5083 Al matrix, more residual stress in the $Al_2O_3 3D/5083$ Al, and the existence of pores in the Al interface, leading to the decreased corrosion resistance of $Al_2O_3 3D/5083$ Al [28].

Corrosion tendency is expressed by the corrosion potential. The higher the corrosion potential, the lower the corrosion tendency. Electrochemical test results show that the $Al_2O_3 3D/5083$ Al composite exhibits higher corrosion tendency during the initial period, and the corrosion resistance of the material is better than that of the 5083 Al alloy. When the corrosion reaction is implemented, the $Al_2O_3 3D/5083$ Al composite has poorer corrosion resistance than 5083 Al due to the Al matrix. During the corrosion development period, corrosion occurs within the $Al_2O_3 3D/5083$ Al composite products, together with the newly generated $Al(OH)_3$. Corrosion resistance is better than that of the 5083 Al alloy. During the corrosion stability period, the $Al_2O_3 3D$ skeleton combines with the newly generated $Al(OH)_3$, and the material promotes excellent corrosion resistance. The corrosion resistance is close to that of the 5083 Al alloy. The research on the corrosion mechanism of the $Al_2O_3 3D/5083$ Al alloy IPC under Cl^- conditions can explore the methods to protect the composite, and improve the service life of the composite under coastal and salt spray conditions. Therefore, the $Al_2O_3 3D/5083$ Al alloy IPC can not only meet the mechanical and friction properties of the brake disc under different braking conditions, but also ensure its corrosion resistance, which is of great significance for the safe application of composites of brake discs [29].

5. Conclusions

1. The corrosion development process can be divided into the initial period, the development period, and the stability period.
2. The OCP, PDP curves, and EIS tests on the sample of polished surface show that the corrosion resistance of the $Al_2O_3 3D/5083$ Al is better than that of 5083 Al.
3. The NSS test shows that the corrosion resistance of $Al_2O_3 3D/5083$ Al was lower than that of the 5083 Al alloy during the initial period of corrosion and higher than that of the 5083 Al alloy during the corrosion development period.
4. $Al_2O_3 3D$ used as a reinforcement in the $Al_2O_3 3D/5083$ Al composite improves the corrosion resistance of the $Al_2O_3 3D/5083$ Al composite. The interpenetrating structures of $Al_2O_3 3D$ and the 5083 Al matrix, combined with the strong interface, are not easy to corrode. $Al_2O_3 3D$ and the 5083 Al matrix are combined tightly to promote excellent corrosion resistance.

Author Contributions: Conceptualization, C.Z., Y.J., X.C. and L.Y.; writing original draft preparation, C.Z., Y.J. and L.Y.; writing—review and editing, C.Z., L.Y., X.C., P.X., Y.Y., Y.L. and Y.J.; supervision, Y.Y., P.X. and Y.L.; project administration, L.Y., X.C. and Y.J.; funding acquisition, L.Y., X.C. and Y.J. All authors have read and agreed to the published version of the manuscript.

Funding: The work was supported by the National Natural Science Foundation of China (51465014), the Guangxi Innovation Driven Development Project (Grant No. AA17204021), the Foundation of Guangxi Key Laboratory of Optical and Electronic Materials and Devices (No. 20KF-4), and the Foundation of the Introduction of Senior Talents in Hebei Province (H192003015).

Institutional Review Board Statement: Not applicable.

Informed Consent Statement: Not applicable.

Data Availability Statement: Data sharing is not applicable for this article.

Conflicts of Interest: The authors declare no conflict of interest.

References

1. Kota, N.; Charan, M.S.; Laha, T.; Roy, S. Review on development of metal/ceramic interpenetrating phase composites and critical analysis of their properties. *Ceram. Int.* **2022**, *48*, 1451–1483. [CrossRef]
2. San Marchi, C.; Kouzeli, M.; Rao, R.; Lewis, J.A.; Dunand, D.C. Alumina–aluminum interpenetrating-phase composites with three-dimensional periodic architecture. *Scr. Mater.* **2003**, *49*, 861–866. [CrossRef]
3. Xie, F.; Lu, Z.; Yuan, Z. Numerical analysis of elastic and elastoplastic behavior of interpenetrating phase composites. *Comput. Mater. Sci.* **2015**, *97*, 94–101. [CrossRef]
4. Roy, S.; Gibmeier, J.; Kostov, V.; Weidenmann, K.A.; Nagel, A.; Wanner, A. Internal load transfer in a metal matrix composite with a three-dimensional interpenetrating structure. *Acta Mater.* **2011**, *59*, 1424–1435. [CrossRef]
5. Dolata, A.J. Tribological Properties of AlSi$_{12}$-Al$_2$O$_3$ Interpenetrating Composite Layers in Comparison with Unreinforced Matrix Alloy. *Materials* **2017**, *10*, 1045. [CrossRef]
6. Feng, G.H.D. Synthesis of SiC_Al Co-Continuous Composite by Spontaneous Melt Infiitration. *J. Mater. Sci. Technol.* **2000**, *16*, 466–470. [CrossRef]
7. Sahin, I.; Akdogan Eker, A. Analysis of Microstructures and Mechanical Properties of Particle Reinforced AlSi7Mg2 Matrix Composite Materials. *J. Mater. Eng. Perform.* **2010**, *20*, 1090–1096. [CrossRef]
8. Jhaver, R.; Tippur, H. Processing, compression response and finite element modeling of syntactic foam based interpenetrating phase composite (IPC). *Mater. Sci. Eng. A* **2009**, *499*, 507–517. [CrossRef]
9. Jiang, L.; Jiang, Y.-l.; Yu, L.; Su, N.; Ding, Y.-d. Thermal analysis for brake disks of SiC/6061 Al alloy co-continuous composite for CRH3 during emergency braking considering airflow cooling. *Trans. Nonferrous Met. Soc. China* **2012**, *22*, 2783–2791. [CrossRef]
10. Potoczek, M.Ś.R. Microstructure and Physical Properties of AlMg/Al2O3 Interpenetrating Composites Fabricated by Metal Infiltration into Ceramic Foams. *Arch. Metall. Mater.* **2011**, *56*, 1265–1269. [CrossRef]
11. Ji, Y.-y.; Xu, Y.-z.; Zhang, B.-b.; Behnamian, Y.; Xia, D.-h.; Hu, W.-b. Review of micro-scale and atomic-scale corrosion mechanisms of second phases in aluminum alloys. *Trans. Nonferrous Met. Soc. China* **2021**, *31*, 3205–3227. [CrossRef]
12. Tao, J.; Xiang, L.; Zhang, Y.; Zhao, Z.; Su, Y.; Chen, Q.; Sun, J.; Huang, B.; Peng, F. Corrosion Behavior and Mechanical Performance of 7085 Aluminum Alloy in a Humid and Hot Marine Atmosphere. *Materials* **2022**, *15*, 7503. [CrossRef]
13. Liew, Y.; Örnek, C.; Pan, J.; Thierry, D.; Wijesinghe, S.; Blackwood, D.J. In-Situ Time-Lapse SKPFM Investigation of Sensitized AA5083 Aluminum Alloy to Understand Localized Corrosion. *J. Electrochem. Soc.* **2020**, *167*, 141502. [CrossRef]
14. Huang, Q.; He, R.; Wang, C.; Tang, X. Microstructure, Corrosion and Mechanical Properties of TiC Particles/Al-5Mg Composite Fillers for Tungsten Arc Welding of 5083 Aluminum Alloy. *Materials* **2019**, *12*, 3029. [CrossRef] [PubMed]
15. Fatimah, S.; Nashrah, N.; Tekin, K.; Ko, Y.G. Improving Corrosion and Photocatalytic Properties of Composite Oxide Layer Fabricated by Plasma Electrolytic Oxidation with NaAlO$_2$. *Materials* **2022**, *15*, 7055. [CrossRef]
16. Liu, S.; Wang, X.; Tao, Y.; Han, X.; Cui, C. Enhanced corrosion resistance of 5083 aluminum alloy by refining with nano-CeB6/Al inoculant. *Appl. Surf. Sci.* **2019**, *484*, 403–408. [CrossRef]
17. Roseline, S.; Paramasivam, V. Corrosion behaviour of heat treated Aluminium Metal Matrix composites reinforced with Fused Zirconia Alumina 40. *J. Alloys Compd.* **2019**, *799*, 205–215. [CrossRef]
18. Jiang, L.; Jiang, Y.-l.; Yu, L.; Yang, H.-l.; Li, Z.-s.; Ding, Y.-d.; Fu, G.-f. Fabrication, microstructure, friction and wear properties of SiC3D/Al brake disc−graphite/SiC pad tribo-couple for high-speed train. *Trans. Nonferrous Met. Soc. China* **2019**, *29*, 1889–1902. [CrossRef]
19. Yu, L.; Hao, S.; Nong, X.; Cao, X.; Zhang, C.; Liu, Y.; Yan, Y.; Jiang, Y. Comparative Study on the Corrosion Resistance of 6061Al and SiC3D/6061Al Composite in a Chloride Environment. *Materials* **2021**, *14*, 7730. [CrossRef]
20. Liu, X.; Kong, D. Salt spray corrosion and electrochemical corrosion performances of Dacromet fabricated Zn–Al coating. *Anti-Corros. Methods Mater.* **2019**, *66*, 565–572. [CrossRef]
21. Mosleh-Shirazi, S.; Hua, G.; Akhlaghi, F.; Yan, X.; Li, D. Interfacial valence electron localization and the corrosion resistance of Al-SiC nanocomposite. *Sci. Rep.* **2015**, *5*, 18154. [CrossRef] [PubMed]
22. Li, J.; Dang, J. A Summary of Corrosion Properties of Al-Rich Solid Solution and Secondary Phase Particles in Al Alloys. *Metals* **2017**, *7*, 84. [CrossRef]
23. Sun, Y.; Li, C.; Yu, L.; Gao, Z.; Xia, X.; Liu, Y. Corrosion behavior of Al-15%Mg2Si alloy with 1% Ni addition. *Results Phys.* **2020**, *17*, 103129. [CrossRef]
24. Li, Z.; Li, C.; Gao, Z.; Liu, Y.; Liu, X.; Guo, Q.; Yu, L.; Li, H. Corrosion behavior of Al–Mg2Si alloys with/without addition of Al–P master alloy. *Mater. Charact.* **2015**, *110*, 170–174. [CrossRef]
25. Liu, Y.; Ning, X.-S. Influence of α-Al2O3 (0001) surface reconstruction on wettability of Al/Al2O3 interface: A first-principle study. *Comput. Mater. Sci.* **2014**, *85*, 193–199. [CrossRef]
26. Cabrini, M.; Calignano, F.; Fino, P.; Lorenzi, S.; Lorusso, M.; Manfredi, D.; Testa, C.; Pastore, T. Corrosion Behavior of Heat-Treated AlSi10Mg Manufactured by Laser Powder Bed Fusion. *Materials* **2018**, *11*, 1051. [CrossRef]
27. Kim, S.-J.; Kim, S.-K.; Park, J.-C. The corrosion and mechanical properties of Al alloy 5083-H116 in metal inert gas welding based on slow strain rate test. *Surf. Coat. Technol.* **2010**, *205*, S73–S78. [CrossRef]

28. Gopinath K., B.R.; Murthy, V.S.R. Corrosion behavior of cast Al-Al$_2$O$_3$ particulate composites. *J. Mater. Sci. Lett.* **2001**, *20*, 793–794. [CrossRef]
29. Jiang, L.; Jiang, Y.; Yu, L.; Yang, H.; Li, Z.; Ding, Y. Thermo-Mechanical Coupling Analyses for Al Alloy Brake Discs with Al(2)O(3)-SiC((3D))/Al Alloy Composite Wear-Resisting Surface Layer for High-Speed Trains. *Materials* **2019**, *12*, 3155. [CrossRef]

Disclaimer/Publisher's Note: The statements, opinions and data contained in all publications are solely those of the individual author(s) and contributor(s) and not of MDPI and/or the editor(s). MDPI and/or the editor(s) disclaim responsibility for any injury to people or property resulting from any ideas, methods, instructions or products referred to in the content.

Article

Microstructural, Corrosion Resistance, and Tribological Properties of Al₂O₃ Coatings Prepared by Atmospheric Plasma Spraying

Costică Bejinariu [1], Viorel Paleu [2], Ciprian Vasile Stamate [2], Ramona Cimpoeșu [1], Margareta Coteață [3], Gheorghe Bădărău [1], Mihai Axinte [1], Bogdan Istrate [2], Gabriel Dragos Vasilescu [4] and Nicanor Cimpoeșu [1,*]

[1] Materials Science and Engineering Faculty, Gheorghe Asachi Technical University of Iasi, 700050 Iasi, Romania
[2] Mechanical Engineering, Mechatronics and Robotics Department, Mechanical Engineering Faculty, Gheorghe Asachi Technical University of Iasi, 700050 Iași, Romania
[3] Faculty of Machine Manufacturing and Industrial Management, Gheorghe Asachi Technical University of Iasi, 700050 Iasi, Romania
[4] "INSEMEX" National Institute for Research and Development in Mine Safety and Protection to Explosion, 332029 Petrosani, Romania
* Correspondence: nicanor.cimpoesu@tuiasi.ro

Abstract: An usual material, EN-GJL-250 cast iron, used for automotive braking systems, was covered with a ceramic material (105NS-1 aluminium oxide) using an industrial deposition system (Sulzer Metco). The main reason was to improve the corrosion and wear (friction) resistance properties of the cast-iron. Samples were prepared by mechanical grinding and sandblasting before the deposition. We applied two and four passes (around 12–15 μm by layer) each at 90° obtaining ceramic coatings of 30 respectively 60 μm. The surface of the samples (with ceramic coatings) was investigated using scanning electron microscopy (SEM), dispersive energy spectroscopy (EDS) and X-ray diffraction (XRD). Scratch and micro-hardness tests were performed using CETR-UMT-2 micro-tribometer equipment. The better corrosion resistance of the base material was obtained by applying the ceramic coating. The results present a better corrosion resistance and a higher coefficient of friction of the coated samples.

Keywords: coatings; Al₂O₃; plasma spraying; SEM; EDS; AFM

1. Introduction

Over the years, the field of materials destined for friction systems has evolved very much due to the increasingly needed answers to more precious operation requests [1]. The operating conditions have become more and more demanding both because of the increasing weight of the bodies involved in the braking process and of their increasing speed. The materials used for the manufacturing of a braking system make a friction couple that must ensure the transformation of the kinetic energy of the system into thermal energy by using two friction surfaces. The braking discs are parts used for slowing down or stopping a wheel from its rotational movement. These discs are made from Fe-C cast alloys, but in some cases, those that cost more, can be made using composite materials, for example, C—reinforced C or ceramic compounds [2].

A special interest is given to metallic braking discs having applications in the auto, railroad, and aeronautical fields because of the promoted prices and of the recognised already approved technologies. The geometric morphology of the discs, the width and first of all the material used for building them, bring great advantages to such equipment from the commercial perspective. Based on the thermal conductivity, special wear resistance and machinability the braking discs made from Fe-C alloys will continue to represent an important issue in what concerns the developments in the field. The analysis of the braking discs does not represent just immediate gain having applications in the automobile industry, but also an opportunity in various fields such as aeronautics and industrial at every scale [3].

Ceramic materials such as alumina (Al_2O_3) represent a very good solution as coatings in different industrial fields, such as aero spatial elements, building constructions, and electrical and electronic domains, mainly based on their very good properties at high temperatures such as mechanical strength, wear resistance, and in many cases corrosion-resistant properties [4–7].

For the laboratory level, there are different thermal spraying methods used (Vacuum Plasma Spraying—VPS; Combustion Flame Spraying—CFS; Two-Wire Electric Arc Spraying—TWEAS, Plasma Spraying—PS or High-Velocity oxy-fuel spraying—HVOF). Among them, Atmospheric Plasma Spraying (APS) is extensively used for the growth of ceramic coatings, oxides, or non-oxide based on its higher deposition rate and lower cost [8]. For industrial applications, using APS with a robotic arm increases the possibilities of improving the properties of already used metallic parts using ceramic coatings. Along with the high deposition rate of the process, we deal with various quality problems characteristic of APS layers, such as high porosity, low adhesion, and reduced gas jet velocity, resulting in ceramic layers with poor mechanical properties, wear resistance, or decrease in high-temperature corrosion resistance. [9].

Plasma deposition shows the advantages of a higher deposition rate, larger surfaces covered and the possibility of deposition for a wide range of materials (metallic, ceramic or polymers). For industrial applications, it is important to determine the optimum deposition parameters, which directly depend on the nature of the deposited layers and of the substrate, such as the number of deposited layers, the spraying distance, or substrate roughness.

In this paper, there are shown the experimental results obtained on samples of EN-GJL-250 cast iron covered with a ceramic coating of alumina (two and four layers deposited successively) using a plasma jet spraying method after the structural, mechanical, and chemical analysis of new materials. The subject of the paper was proposed by a private industrial company aiming for the optimisation of the APS deposition process. The results encourage the usage of APS of ceramic layer and for a satisfactory covering of metallic surface deposition of more than two layers.

2. Experimental Details

The plasma jet was covered by Ar (pressure 5.2 bar and gas flow 39 NLPM) and H (pressure 3.4 bar and gas flow 6.6 NLPM). We applied two and four passes (around 12-15 μm by layer) [10]. Samples were prepared by mechanical grinding and sandblasting before the deposition. The experimental setup is presented in ref. [10] and is formed by rotational support, an automatic arm for pulverization and a holder for samples. The surface of the experimental samples (with two and four ceramic layers) was investigated using scanning electron microscopy (S.E.M.—Vega-Tescan LMHII, 25kV, high vacuum, 25.5 mm WD, W cathode), dispersive energy spectroscopy (EDX—Bruker, X-Flash detector, automatic mode), atomic force microscopy (A.F.M. EasyscanII-Nanosurf, in contact mode) and X-ray diffraction (X.R.D.—Panalytical equipment). The general surface profile of the ceramic coating was determined using a Taylor-Hobson precision profile meter and the scratch and micro-hardness tests were performed using CETR-UMT-2 micro-tribometer equipment [11]. The scratch test was also evaluated using an acoustic emission sensor to confirm the penetration of the ceramic coating. Four different mechanical stress rates were used on the same material surface and the same load on AMSLER tribometer equipment (the rotating lower disc of the AMSLER machine is made of AISI 52100 bearing steel with a hardness of 62–65 HRc and a diameter of 59 mm both radially and axially). A tensometric data acquisition system was used to monitor the friction torque within the tribosystem. A Vishay P3 strain gauge bridge with four channels was employed with the specific soft program. The acquired data was processed by LabVIEW virtual instrument signal processing application. The mathematical relationships for friction torque and friction coefficient estimation and the LabVIEW program interface are presented in [11].

The electrochemical tests were achieved by linear potentiometry (potentiostat PGP 201) using a cell with three electrodes (1—the sample, 2—Pt electrode, 3—calomel saturated electrode). Before the experiment, the samples were cleaned in an ultrasonic bath using technical alcohol for 60 min [12]. The electrolytic solution used for testing was acid rainwater (mixing solution 1:1 of H_2SO_4 and HNO_3 with pH = 3). These chemical substances are naturally present in the atmosphere; however, before industrialization, the advent of factories and the dependence on hydrocarbon (coal, gasoline etc.) acid rain was a rare event. In recent decades, acid rain has become more and more frequent, especially in the area of highly congested cities and heavily industrialized areas.

3. Experimental Results

3.1. Structural and Chemical Analysis of Ceramic Coatings

Deposition of the ceramic Al oxides (alumina) (two or four passes) using APS produce compact coatings, especially on the sandblasted samples, having an estimated thickness of 30 μm, respectively 60 μm, function almost linearly dependent on the number of passes [13].

Further on, in Figure 1 there are shown, using the scanning—electron microscope, the surfaces of the coatings obtained by plasma spraying. From the viewpoint of the structure, the achieved coating on the grinded sample, but not sandblasted appears discontinuous, Figure 1d, with areas where the cast iron substrate can be seen. The discontinuities of the deposited coating, of any nature: pores, cracks, exfoliations etc. decrease the mechanical and chemical properties. For this reason, it is not recommended to achieve the plasma jet deposition without careful preparation of the surface at the macroscopic level to enhance the adhesion of the layer.

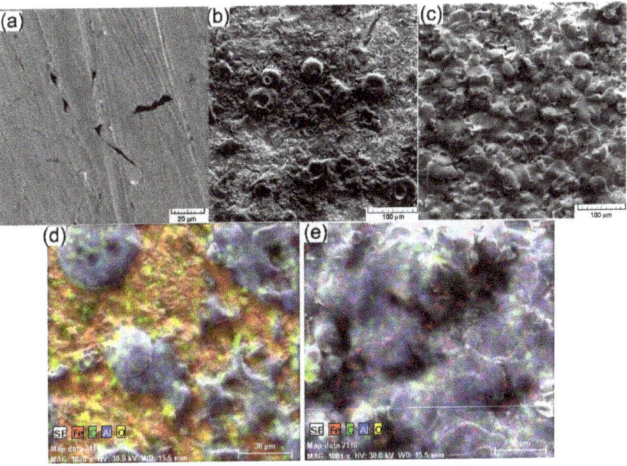

Figure 1. SEM images: (**a**) cast iron EN-GJL-250; (**b**) EN-GJL-250 + 2 layers of ceramic material; (**c**) EN-GJL-250 + 4 layers of ceramic material, in (**d**,**e**) there are shown distributions of elements inside the deposited ceramic coatings using two and four passes, respectively.

Figure 1c presents the morphology of the deposited coating on the sample blasted with glass (showing a higher roughness). On the surface, a compact layer can be seen, without discontinuities or exfoliations, showing local micro-cracks of micrometric sizes. The cracks do not penetrate the whole layer and occur mainly because of the temperature differences between the deposited layer and the colder substrate (layer achieved on the previous passing). The layer is obtained generally by zones with melt material that form a compact mass under the influence of the very high temperature, around 12,000 °C. The microstructure of the ceramic deposited coating of 60 µm, corresponding to four passes can be appreciated as having fewer cracks, fewer pores, and a more homogenous covering on the sample substrate with higher roughness. A higher roughness is an advantage for the adhesion properties of the ceramic coating because it provides greater anchor support.

For the chemical characterisation of thin layers achieved by thermal spraying on cast iron substrate, the distribution of chemical elements inside the deposited ceramic coating was analysed and the results are given in Figure 1d,e. The distribution was analysed for the following chemical elements: Al and O the components of the coating and Fe and C the elements of the substrate EN-GJL-250 cast iron. From the analysis of the morphology of the coating and the distribution of elements Al and Fe, it can be seen, in the case of the sample with two passes, a larger non-uniformity of the ceramic deposited coating. The area covered is a little more than 60% on the analysed surface, in this case, 0.18 mm². The area chosen for analysis, Figure 1b is characteristic of the deposited layer and it shows the same aspect on the entire surface covered with alumina. The non-uniformities of the ceramic coating lead to total exfoliation of it during operation and the promotion of corrosion on the exposed zones to a solution of an electrolyte by comparison with the coated zone.

The surface of the deposited ceramic coating was also analysed using an atomic force microscope (3D). On a surface of 64 µm², Figure 2a, it can also be seen that for the case of four passes a more homogenous surface was obtained. After establishing a suitable sandblasting regime for the cast iron substrate there were achieved experimental samples with two and four passes of ceramic sprayed material (two and four ceramic layers deposited on the same sample).

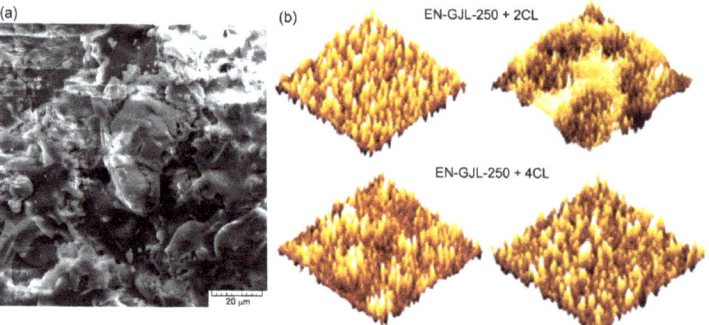

Figure 2. SEM images, in (**a**) and AFM, in (**b**) for the ceramic deposited coating.

Figure 3 shows the spectrum of energies characteristic of the chemical elements identified on the ceramic deposited coating on a cast iron substrate (the spectrum shown was identified for both experimental samples).

From the analysis of the spectrum in Figure 3a it can be seen the qualitative identification of elements of the ceramic coating, respectively, Al and O but also the elements Fe and C specific for the cast iron substrate. The elements characteristic of the substrate (Fe and C) were identified through the pores and cracks that occurred in the ceramic coating during the deposition process. The XRD spectrum, Figure 3b, identifies more characteristic picks for the phase α specific for alumina, the plot is from the sample with four layers. The qualitative phase analysis was performed using the PDXL2 (Rigaku) software and the database ICDD PDF4 + 2022. The qualitative phase analysis indicated the following polycrystalline phases: α-alumina and gamma-alumina (or eta-alumina) in very low percentages.

Table 1 shows in mass percentages (wt%) and atomic percentages (at%) the chemical composition of deposited ceramic coatings on cast iron substrate. In the case of the sample with two layers, it can be seen a large amount of Fe that signifies a lower coating percentage of the metallic material.

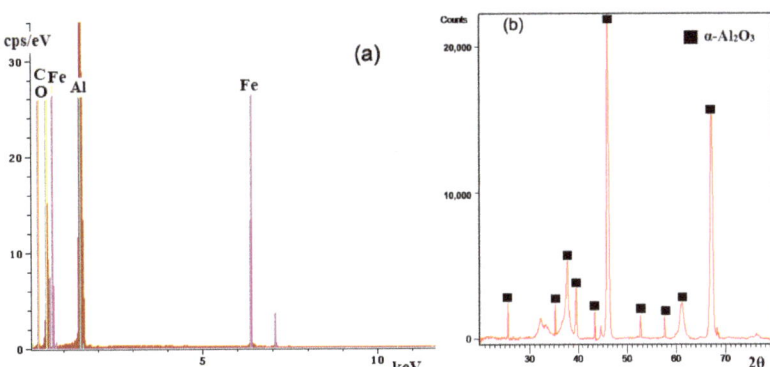

Figure 3. Chemical analysis of the ceramic coatings deposited on the metallic substrate. (**a**) EDS spectrum; (**b**) XRD spectrum.

Table 1. Chemical composition of ceramic coatings deposited on cast iron substrate.

Element/Sample	Al		O		Fe		C	
	wt%	at%	wt%	at%	wt%	at%	wt%	at%
Powders Al_2O_3	54.71	41.73	45.29	58.27	-	-	-	-
EN-GJL-250 + 2 ceramic layers	40.83	32.72	38.01	49.37	16.5	9.7	4.67	8.21
EN-GJL-250 + 4 ceramic layers	50.93	38.11	43.78	55.25	1.7	0.6	3.58	6.02
Error EDS %	1.7		1.1		0.2		0.1	

Standard deviation (SD, from 20 determinations): Al: ± 1.2; O: ± 0.9; Fe: ± 0.1; C: ± 0.2.

For the sample with four passes, the Fe content identified occurs because of the micro-cracks or of the pores existent in the ceramic coating or, of the low thickness of the layer in some areas, a fact depending on the quality of the deposition that can vary as a function of sample geometry. The high percentage of C for both cases is also due to the error of the EDS detector.

The structural and chemical analysis of the ceramic surface obtained after four passes during the deposition process shows a uniform surface of the coating with all areas of the substrate covered, Figure 1d. The distribution of Al and the lack of Fe signal confirm the fact that the metallic surface is completely covered after four passes of the plasma jet.

As a function of the application for which the coating is being deposited, it can be subject to some supplementary technical procedures of rolling or, heat treating for chemical and structural homogenisation, using a furnace or an acetylene flame. In this case, it is intended to maintain the roughness of the surface for increasing the coefficient of friction. For the braking discs, a self-rolling process occurs during the part operation.

3.2. The Analysis of the Behaviour of Tested Materials to Micro-Indentation

For analysing the adhesion of the ceramic coatings to the metallic substrate and for determining the tribological properties, there were performed scratch tests on the experimental samples EN-GJL-250, EN-GJL-250 + 2 ceramic layers, and EN-GJL-250 + 4 ceramic layers. Figure 4 shows the general aspect of the scratches, from the starting point of the test, from the left side to the right side, and at the end of the test. The figures were made using 3 images obtained at the optical microscope, focusing on the areas that characterise the scratch from its beginning to its end. The scratch length achieved was 25 mm using a progressive load. There were several scratching tests aiming at the characterisation of the homogeneity of the tribologic properties of the superficial deposited coatings.

Figure 4. Identification and images of scratches using optical microscopy for the coatings: (**a**) two layers; (**b**) four layers.

At the microscopic level, there were not seen exfoliations of the deposited ceramic coating and the uniform aspect of them shows good structural homogeneity of the layer. Figure 5 shows the characteristics of the scratching behaviour of the sample EN-GJL-250+4 ceramic layers. The scratching equipment was simultaneously operated with an acoustic sensor to record the initial behaviour of the ceramic coating as well as the penetrated layer together with the substrate after a period of time [14].

The experiment started with an initial load of zero Newton (0 N) (F_z) rising up to 8 N on a length of 25 mm. It can be seen in the evolution of the friction force and the acoustic emission a variation of the signals at 10.5 ÷ 11.5 mm, from the beginning of the scratch initiated on the ceramic surface, the area that probably represents the point when the metallic penetrator penetrated the ceramic coating. After this, the friction force increased due to the double effect of the stresses, one of the ceramic coating and the other one, of the EN-GJL-250 metallic substrate.

For the analysis of the behaviour of the coefficient of friction, Figure 4, extracted from the signal of the scratching test, the one obtained on the ceramic coating and after its penetration on the system ceramic coating—cast iron EN-GJI-250 is chemically analysed in certain zones.

The coefficient of friction shows the same behaviour as the friction force, having a variation after 10.5 s from the start of the test, Figure 5a. The scratch obtained after the mechanical testing was analysed using electron microscopy SEM (on the areas from a to g) after characterising the mark at every 2 mm, Figure 5. Figure 5b shows the distribution of the elements Fe and C characterising the cast iron substrate and Al and O characterising the ceramic coating of Al_2O_3 on the areas (a), (b), (d), (f) and (g) exactly on the scratch mark. In the first two distributions from Figure 5b, there is no sign of penetration of the ceramic coating, this being evident in the (d) area by the significant increase in the element Fe signal on the scratch mark. The Fe signal is accompanied by the signal of the element carbon but not so obviously, due to its lower percentage. If in the (d) area, the ceramic coating was only partially penetrated in the following 10–14 mm it was gradually removed. The ceramic coating was removed completely in some areas, especially on the last stress portion. It can be seen in areas (f) and (g) from Figure 5b portions having the ceramic coating present on the scratch marks. Their presence can be explained by a superior adherence to the substrate in these areas or by settling the ceramic material under the force of scratching/pressing and its penetration into the metallic cast iron EN-GJL-250 substrate.

The microstructural analysis was performed starting with the final end of the scratch at each 2 mm until no variation of the ceramic coating microstructure was present, the area considered the starting point of the scratching test and which corresponds to the one obtained from calculus considering the total length of the scratch, respectively 25 mm. From a microstructural viewpoint, a chamfering of the ceramic coating can be seen, in Figure 5b, at 2 mm from the beginning of the scratching test, namely at a loading force of $1 \div 2$ N, confirming the fact that the layers of Al_2O_3 are relatively soft among the ceramics but less brittle compared to very hard ones.

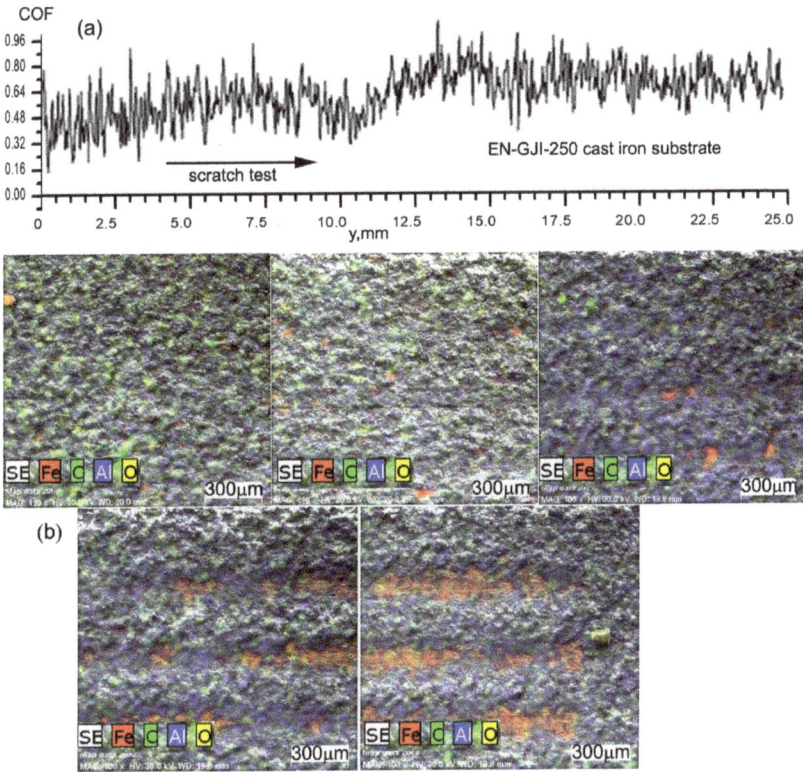

Figure 5. Analysis of the scratching test using the coefficient of friction behaviour: (**a**) variation of coefficient of friction on 25 mm distance; (**b**) distribution of elements Fe, C, Al and O on the surface of the scratched ceramic coating.

The depth of the scratching marks without the penetration of the coating continues up to $10 \div 11$ s of stress i.e., at a force of $4 \div 5$ N. There cannot be seen areas with cracks on the edges of the scratch mark and nor in the areas of ceramic coating between scratches. The analysis at a higher magnification power, Figure 6a–d did not reveal the presence of the cracks or pores on the pressed ceramic surface, nor their appearance on the metallic substrate.

The integrity of the ceramic coating is very little affected on the edges of the scratch mark, showing the high stability of the ceramic deposited layer. Inside the analysed scratch marks, they appear exfoliation areas of the ceramic coating, Figure 5b but also the presence of areas with a compressed ceramic layer. In practical applications where the increase in the wear coefficient is not especially aimed, additional processing of the coating, carried out mechanically or by heat treatment, is recommended, for the uniformity of the surface, the reduction of roughness and the homogenization of the coatings.

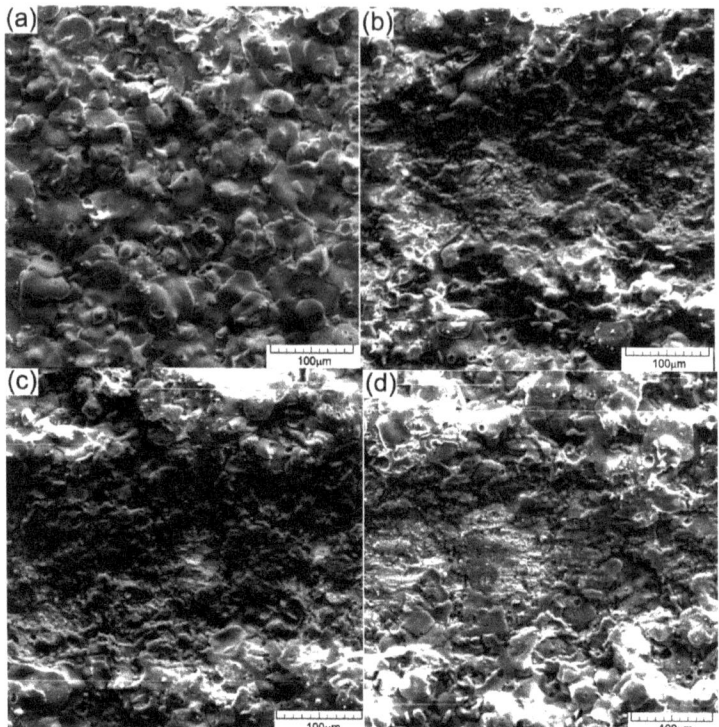

Figure 6. SEM images—details of the scratched areas on the ceramic coating (**a**) initiation of scratch, (**b**) first area of ceramic layer deformation, (**c**) local compaction of the ceramic layer and (**d**) end of the scratch mark.

Figure 7 shows the behaviour of experimental materials EN-GJL-250, EN-GJL-250 + 2 ceramic layers, and EN-GJL-250 + 4 ceramic layers for the variation: (a) friction force, (b) acoustic emission and (c) coefficient of friction during the scratching test.

From Figure 7a one can see that the friction forces are greater in the case of the samples with ceramic deposited layers by comparing with the friction force occurring on the EN-GJL-250 cast iron and which shows only slight variations in behaviour due to the differences in hardness between the metallic substrate characteristic of cast iron and graphite formations. In both cases of coatings (with two, respectively four layers) one can see an increase in the friction force after the penetration of the ceramic coating and the complex friction among the indenter on one hand and the ceramic coating and the substrate on the other hand. Additionally, it can be seen that increasing 2 ÷ 3 times the friction force in the case of the sample with the coating made by 4 ceramic layers compared with that having only two deposited ceramic layers as a coating.

In the case of acoustic emission (acoustic emission—AE), Figure 7b, the signal of the substrate is also an almost straight line compared with the emissions of the samples with ceramic coating.

Producing and developing cracks can be events having very short periods of time to occur and grow. The acoustic emissions were conceived for detecting the behaviour at fracturing and cracking of materials. Wakayama and Ishiwata [15] used AE detection to analyse and evaluate the ceramic micro-cracks, [16] for detecting the deterioration in composite ceramics and composites reinforced with fibres. The same authors used the AE detection method [17] to detect the part fracturing during the processing of the surface of engineered ceramics.

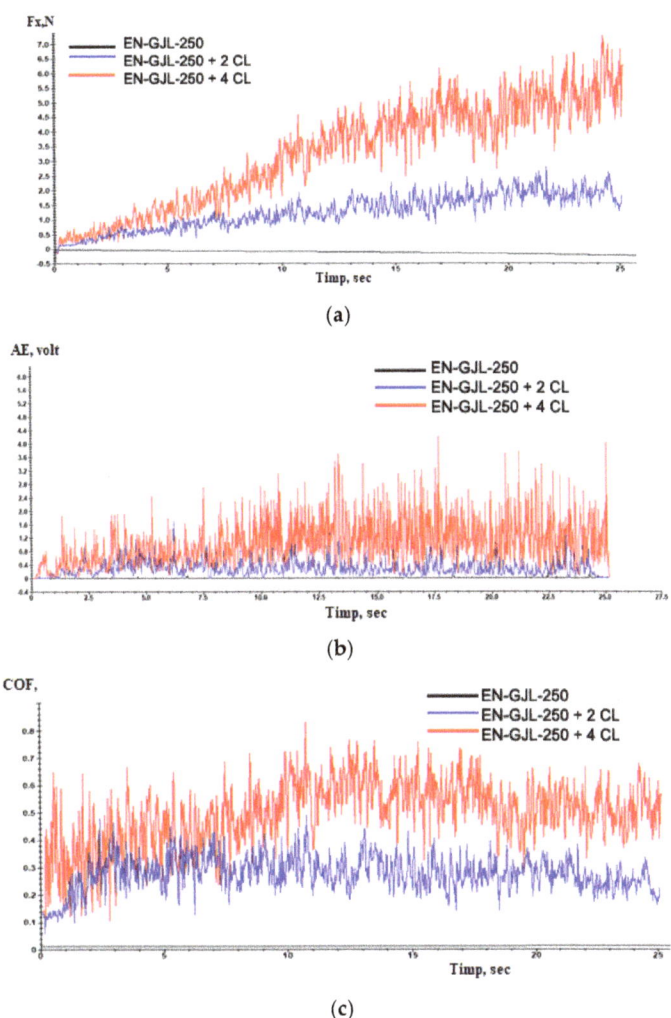

Figure 7. The behaviour of experimental materials EN-GJL-250, EN-GJL-250 + 2 ceramic layers and EN-GJL-250 + 4 ceramic layers at scratching: (**a**) friction force, (**b**) acoustic emission and (**c**) coefficient of friction.

During the scratching tests, the AE technique was also used for monitoring fragile breaking [14]. The level of acoustic emissions is higher in the case of the sample with 4 ceramic layers, having a visible increase in the areas where the deposited coating was penetrated. The amplitude of the acoustic emission signal has increased significantly because of the severe vibrations of the indented that resulted from the initiation and propagation of cracks and/or during the removal of the material through plastic deformation of fragile breaking of the Al_2O_3 coating during scratching. In the case of a large fluctuation in the AE signal, as the amplitude is higher the damages caused to the deposited ceramic coating or substrate, in the case the substrate is reached, is more severe.

The coefficient of friction, Figure 8 shows a substantial increase in the case of the samples having ceramic coatings compared with the cast iron substrate. This increase is due to both the roughness of the ceramic layers and their nature. After the penetration of the ceramic coating, at the value of the coefficient of friction, it is added also the friction with the substrate that contributes in this manner to the identified increase in the coefficient of friction. The values of the coefficient of friction COF vary in a wide range because of the high resistance that the ceramic coating presents, determining a greater capacity for plastic deformation than the substrate. Despite these relatively high coefficients of friction —COF values, plasma-sprayed coatings are still accepted for many applications [18]. For applications that operate under severe wear conditions, these coatings may be supplemented by various additional treatments such as laser reshaping, sealing treatment, or surface grinding to improve the surface modification and therefore the coefficient of friction values [19,20]. If we consider the variation of the coefficient of friction of ceramic materials as coatings until their penetration (in the time range 9 ÷ 11 s) from 0 to 10 s of testing, one can see two zones of variation. Initially, the coefficient of friction grows suddenly up to values of 0.6 with a period of stability, after which it follows a slight decrease in the coefficient of friction down to 0.3 ÷ 0.4 probably because of the large degree of flatness of the deposited ceramic coating and the diminishing of roughness values of the coated surface. According to the model Czihos [21,22] the curves of the coefficient of friction–COF consist of three stages: initial wear, steady state and accelerated wear up to the point of penetration point and contact with the metallic substrate in this case. The third stage in the coefficient of friction—COF curves for the coated and worn samples in this study opposes the Czihos model. It seems that some tribological interaction due especially to the joint influence of the substrate and the edges of the ceramic coating in the penetrated zone affects the stabilization of the coefficient of friction—COF.

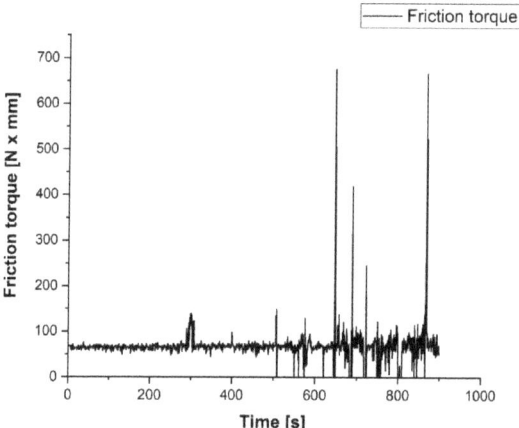

Figure 8. The friction test results performed on the AMSLER tribometer.

3.3. Wear Resistance Analysis

The coated sample was tested on an AMSLER apparatus using a disc made from ASTM 52100 bearing steel. The data collection was performed by a strain-meter that monitored the friction torque in the tribo-system. For data acquisition, it was coupled with a strain indicator and recorder Vishay P3 with 4 channels using its specific software. The acquired data were processed in Lab VIEW for virtual signal processing. The mathematical relations for estimating the friction torque and the coefficient of friction as well as the Lab VIEW interface are given in [23].

A friction test was performed on the AMSLER equipment at a rotation speed of 100 rpm and a constant axial load of about 60 N (6 kg). The evolution of the friction torque T_f in N x mm and of the coefficient of friction μ is shown in Figure 8. As one can see, in the first 5 min, the coefficient of friction between the coating layer and the ASTM 52100 steel disc was about 0.16 ÷ 0.18, the friction process, and wear being smooth and continuous. After 5 min, the coating layer was partially removed and the first metallic contact with a small surface raised the coefficient of friction up to 0.35, but just for seconds. The friction force of the contact became unstable, but within reasonable limits until the worn surface extended and most of the contact became metal on metal.

After 500 s from the start of the test, the friction became dynamic and this can be explained by the intensification of some micro-seizure phenomenon on the metallic contact surface. In the last 5 min of testing, the dynamic phenomenon of sliding on the contact area of the tribological contact manifested itself by strong vibrations and collisions on the tested samples. Consequently, the variation in the coefficient of friction was very large. The test was stopped after 15 min due to the simple observation of the wear zone of the coated sample which revealed the complete removal of the coating layer in the contact area. The statistical analysis of the data acquisition process shows a value of the signal-to-noise ratio SNR = 1.67, which confirms the good quality of the acquired signal. For the last period of the test, the difficulty of the acquisition and the standard deviation have high values, confirming the fluctuation of data acquisition, the causes have already been mentioned. Comparing the results obtained with those previously reported for EN-GJL-250 one can see a coefficient of friction increased up to 0.17 for the entire test. These results recommend the coatings using Al_2O_3 for lighter-duty applications [24].

Future tests should be carried out in this direction, with a lower contact pressure and used as wear material instead of the steel disc, ferodou discs, or special materials used in braking systems.

Figure 9 shows SEM images of the wear trace obtained after the test performed on the AMSLER equipment. The trace is about 4mm long and 2 mm wide. A complete removal of the ceramic coating from the contact area can be seen in Figure 10a.

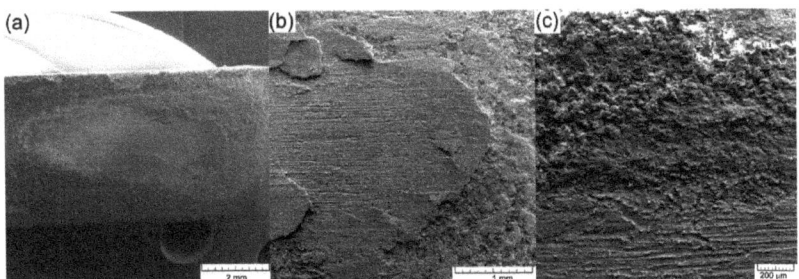

Figure 9. SEM images of the worn zone during the test: (**a**) wear trace; (**b**) detail of the wear end area; (**c**) wear edge.

The contact achieved during the experiment was extremely hard because it also engaged material from the substrate, Figure 9b. The ceramic material was subjected to advanced wear by being placed between two metallic materials, cast iron as a substrate and the steel disc as a wear material. The relatively brittle nature of the coating resulted in its exfoliation at the contact area but without further affecting the integrity of the coating near the wear trace, Figure 9c.

To highlight the wear zone in Figure 11, it is shown the distribution of Al, O, Fe and C elements in the contact zone of the wear test (a) distribution of all elements, (b) distribution of aluminium, (c) distribution of oxygen and (d) distribution of iron.

Complete removal of the ceramic coating from the contact area can be seen. No compacted parts of the ceramic material were identified in the contact area. The ceramic coating shows exfoliations of micro-cracks and pores only at the beginning and end of the contact zone, the sides of the wear trace being unaffected by the test. The removal of the ceramic coating was achieved following strong mechanical shocks that primarily targeted the wear trace and less the surrounding zones that do not show affected surfaces.

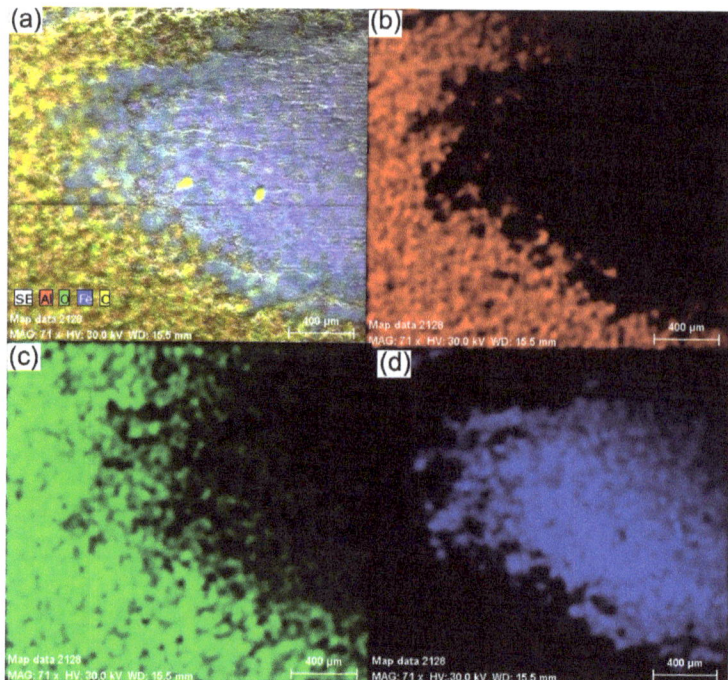

Figure 10. The chemical elements distribution for Al, O, F and C in the contact zone at the wear test: (**a**) distribution of all elements; (**b**) distribution of aluminium; (**c**) distribution of oxygen; (**d**) distribution of iron.

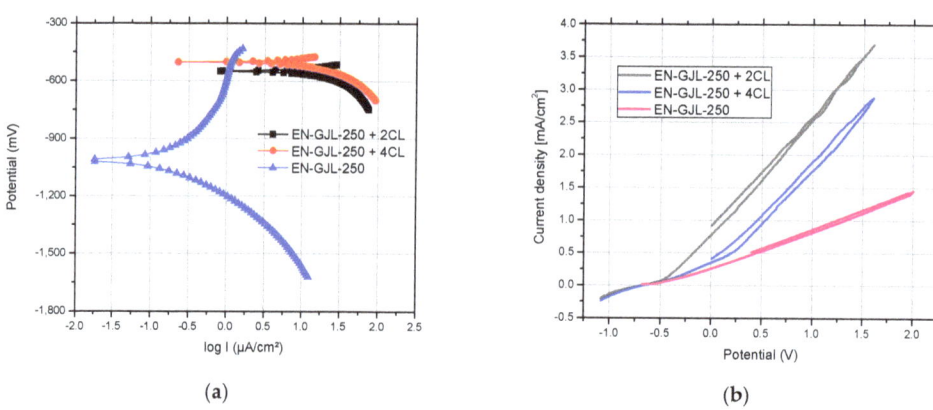

Figure 11. Potentiodynamic graphs for coated samples with Al_2O_3 of different thicknesses on EN-GJL-250 cast iron substrate compared with the free substrate EN-GJL-250 cast iron: (**a**) linear polarisation curves; (**b**) cyclic polarisation curves.

3.4. Electro-Corrosion of Metal-Ceramic Systems

The experimental results show the electro-corrosion resistance of three samples (cast iron substrate EN-GJL-250, EN-GJL-250 + 2 layers of ceramic material Al_2O_3~30 μm and EN-GJL-250 + 4 layers of ceramic material Al_2O_3~60 μm) in electrolytic acid rain solution. Figure 11 shows the linear potentiodynamic graphs for Al_2O_3 coatings of different thicknesses on EN-GJL-250 cast iron compared with the substrate EN-GJL-250 cast iron and in Figure 12b the cyclic polarisation curves. The liner potentiodynamic graphs were represented in the range of potential: $-0{,}8 \div 1$ V using a scanning speed of 1 mV/s [25]. The corrosion speed can be correlated with the corrosion current intensity or the current density based on the law of Faraday [26]. For the experimental cases corrosion speeds of the order of millimetres per year for the cast iron EN-GJL-250 and micrometres per year for the coated metallic materials. From Figure 11a one can see a big difference in behaviour between the cast iron material and the cast iron coated with ceramic layer material. No significant difference was noticed for the curves of cyclic polarisation, Figure 12b. The samples coated with ceramic layers show similar behaviour with an almost non-existent anodic reaction.

Figure 12. SEM images on the surface after the electrochemical test: (**a,b**) cast iron EN-GJL-250; (**c,d**) EN-GJL-250 + 2 ceramic layers; (**e,f**) EN-GJL-250 + 4 ceramic layers.

The cathodic curve of cyclic graphs, Figure 11b, shows a similar trajectory to the anodic curve—having a reduced hysteresis loop, and the current density in the passive region is similar to that registered during direct scanning (anodic) at the same potential [12]. The slight difference between the anodic line and the cathodic line (the lack of a loop) is related to the stability of the surface and the competition between diffusion and dissolution in the case of pitting corrosion. The pitting corrosion occurs based on a very fast diffusion process having a semi-circle dimension appearance. In the first part of the cathodic process (the reversal line), the effects of the dissolution process are reduced and the time for the continuation of the diffusion is limited and usually insufficient.

The main parameters of the corrosion process (E_0 and j_{cor}) obtained by processing the linear polarization graphs are centralized in Table 2. The corrosion current thus determined is, in fact, the corrosion current that occurs at the metal/corrosive environment interface when the metal is introduced into the solution and cannot be directly measured by electrochemical methods. The open circuit potential—(OCP) shows large differences between the EN-GJL-250 cast iron and the metallic material coated with ceramic layers due to the influence of the inert material layer on the corrosion resistance of the entire assembly. The bias resistance validated the OCP values and the corrosion current values.

Table 2. Electrochemical parameters after the corrosion tests in acid rain solution.

Sample	OCP mV	E_0 mV	ba mV	bc mV	Rp ohm.cm^2	Jcor μA/cm^2	Vcor mm/Year
EN-GJL-250 + 2 ceramic layers	−488	551.1	-	−468.8	1448	30.01	0.11
EN-GJL-250 + 4 ceramic layers	−430	505.2	-	−337.6	1967	26.15	0.09
EN-GJL-250	−716	−1016.8	659.8	−347.9	324.3	138.3	3.59

The corrosion current of the starting material (EN-GJL-250) is four to five times greater compared with the value recorded for the ceramic coating samples. The corrosion rate was 30 to 40 times higher for EN-GJL-250 compared to coated samples.

Scanning electron microscopy (SEM) (VegaTescan LMH II) was applied for the morphologic analyse of coatings and structure of the substrate material before the electrochemical tests. The results obtained are given in Figure 12a. In Figure 12b,c, the micrographs of the coatings show a dense microstructure with high cohesion and small surface cracks. Furthermore, some areas with pores can be seen in both coated samples. During the deposition process, micro-cracks and pores gather forming larger cracks. The main reason for the occurrence of these defects is the very short solidification time of the material in the atmosphere and the temperature difference between the deposited layers, the phenomena of dilation/constriction involving very important thermal stresses.

The coated surfaces in both cases show a complete melted zone. Further on, some formations of solidified material can be seen, but for these, solidification occurred later than in the underlying layers. A relative degree of homogeneity of the coating layer is crucial for the increase of the corrosion resistance of the substrate.

Figure 12 shows SEM images of the experimental materials after the electrochemical tests (a) and (b) cast iron EN-GJL-250, (c) and (d) EN-GJL-250 + 2 ceramic layers and (e) and (f) EN-GJL-250 + 4 ceramic layers at two different magnifications 200× and 1000× respectively.

In all cases, shown in Figure 12, the general corrosion detected on the cyclic polarization curves is confirmed, Figure 12b, without specific areas of corrosion (pitting). Some coatings from Figure 12d,f show, especially on the areas represented by the ceramic material, pitting corrosion at the outer part of the particles. This behaviour does not represent the entire surface, being localised only zonally and not being recorded by the potentiostat in the cyclic curves [27]. It can thus be stated that micro-zonally the larger agglomerations on the surface of the coating morphologically manifest a behaviour specific for pitting corrosion. If the environment continues to be aggressive (the dissolution rate is high enough to overcome diffusion) the pitting pits appearing at the surface of the ceramic coating can penetrate through the ceramic layer and the electrolyte will come into contact with the metallic substrate, which is much more susceptible to corrosion.

In the case of samples coated with ceramic materials, an aggressive surface attack can be seen even though the resistance of the outer oxide layer contributes a very high corrosion resistance, oxide being a material with good chemical inertia. Normally, the inert behaviour of ceramic materials that protect the substrate (such as alumina) should keep the surface intact. The pores and the micro-cracks present in the coatings became larger after the electro-corrosion tests because the original micro-pores and micro-cracks were damaged and chemically attacked. The main reason for the corrosion is given by the initial existence of pores and cracks in the coated surfaces. The SEM images shown in Figure 12d,f suggest that the corrosion damage was mainly confined to the coating defects (i.e., pores and cracks) [27,28]. It can be seen that some spherical corrosion products have formed around the coating defects. The results of the EDS analysis, Table 3 indicated that the corrosion products were mainly composed of Fe and O. It was shown that electrochemical corrosion produced compounds on the cast iron substrate during the electrochemical experiments.

Table 3. Chemical composition of the experimental materials after the electro-corrosion [27].

Element/Sample	Fe		O		Al		C		Si	
	wt%	at%	wt%	at%	wt%	at%	wt%	at%	wt%	at%
EN-GJL-250	49.01	22.39	42.13	65.13	-	-	3.4	6.9	3.9	3.7
EN-GJL-250 + 2 layers	31.98	14.7	37.16	49.52	23.93	18.75	4.53	10.85	2.33	1.78
EN-GJL-250 + 4 layers	34.09	16.77	30.67	45.73	26.93	27.58	4.86	11.12	3.3	3.2
EDS Error %	0.65		1.0		0.5		0.8		0.1	

SD: Al: ± 1.2; O: ± 0.9; Fe: ± 0.1; C: ± 0.2, Si ± 0.1.

The corrosion process occurs mainly through the cracks and pores in the ceramic layer that allow the contact of the electrolytic solution with the metallic substrate. In all three cases, the materials show a pronounced oxidation on the surface, especially on the cast iron EN-GJL-250. This is because a part of the oxygen, in the other two experimental cases, is part of the coating and just a percentage of it participates in the formation of oxides. Generally, the ceramic layer was penetrated by the electrolyte to the substrate because iron oxides appear at the surface. Since the ceramic top layer and the metallic bonding layer are very passive, there is not much difference in their electrical potential and no electrical micro-piles formed between the two materials.

4. Conclusions

After analysing the experimental results, the following conclusions can be drawn:
- Microstructurally a chamfering of the ceramic coating can be seen starting at 2 mm from the start of the scratch test i.e., at a force of about 1 ÷ 2 N, which confirmed the fact that the Al_2O_3 coatings are relatively soft compared with other ceramics but less brittle than the very hard ones;
- No area with macro-cracks was noticed on the edges of the scratch marks and nor in the ceramic material between scratches. Analysis at a higher magnification of the surface image did not reveal cracks or pores on the pressed ceramic surface, nor did they appear in the metallic substrate;
- We observed that the friction forces are higher in the case of the coated samples compared with the friction force specific for the cast iron EN-GJL-250 and which shows only slight variations in behaviour due to hardness differences between the metallic matrix and the graphite formations characteristic for cast iron. In both cases of the coatings (with two deposited ceramic layers, respectively four ceramic layers) one

- can see an increase in the friction force after the penetration of the ceramic coating and the complex friction between the indenter on one hand and the ceramic penetrated layer and substrate on the other hand. Furthermore, a 2 ÷ 3 times increase in the friction force can be seen in the case of the sample coated with four ceramic layers compared to the one coated with two ceramic layers;
- In the case of experimental samples with deposited ceramic coatings, the general corrosion detected on the cyclic polarization curves was confirmed, without specific areas of corrosion (pitting). Some coatings show, especially on the areas represented by the ceramic material, pitting corrosion at the outer part of the particles. This behaviour does not represent the entire surface, being localised only zonally and not being recorded by the potentiostat in the cyclic curves. It can be stated that zonally the larger agglomerations on the surface of the coating morphologically manifest a behaviour specific for pitting corrosion. If the environment continues to be aggressive (the dissolution rate is high enough to overcome diffusion) the pitting pits appearing at the surface of the ceramic coating can penetrate through the ceramic layer and the electrolyte will come into contact with the metallic substrate which is much more susceptible to corrosion. The corrosion current of the starting material (cast iron EN-GJL-250) is four to five times greater compared with the value recorded for the ceramic coating samples. The corrosion rate is 30 to 40 times higher for cast iron EN-GJL-250 compared to coated samples.

Author Contributions: Conceptualization, C.B. and N.C.; methodology, G.B.; software, R.C.; validation, V.P., C.V.S. and M.C.; formal analysis, B.I.; investigation, N.C.; resources, M.A.; data curation, R.C.; writing—original draft preparation, C.B.; writing—review and editing, G.D.V.; visualization, N.C.; supervision, G.D.V. All authors have read and agreed to the published version of the manuscript.

Funding: Part of this work was supported by a grant of the Ministry of Research, Innovation and Digitization, CNCS—UEFISCDI, project number PN-III-P1-1.1-PD-2021-0208, within PNCDI III; Ministry of Research, Innovation and Digitization, project FAIR_09/24.11.2020, the Executive Agency for Higher Education, Research, Development and Innovation, UEFISCDI, ROBIM, project number PN-III-P4-ID-PCE2020-0332.

Data Availability Statement: Not applicable.

Conflicts of Interest: The authors declare no conflict of interest.

References

1. Puhn, F. *Brake Handbook*; HP Books: Tucson, AZ, USA, 1987; ISBN 0-89586-232-8.
2. Florea, C.D.; Bejinariu, C.; Cimpoesu, N.; Cimpoesu, R. *Automotive Brake Disc Materials*; MRF (Materials Research Foundations): Millersville, PA, USA, 2021; ISBN 9781644901441.
3. Cueva, G.; Sinatora, A.; Guesser, W.; Tschiptschin, A. Wear resistance of cast irons used in brake disc rotors. *Wear* **2003**, *255*, 1256–1260. [CrossRef]
4. Arslan, E.; Totik, Y.; Demirci, E.; Vangolu, Y.; Alsaran, A.; Efeoglu, I. High temperature wear behavior of aluminum oxide layers produced by AC micro arc oxidation. *Surf. Coat. Technol.* **2009**, *204*, 829–833. [CrossRef]
5. Jiang, X.-Y.; Hu, J.; Jiang, S.-L.; Wang, X.; Zhang, L.-B.; Li, Q.; Lu, H.-P.; Yin, L.-J.; Xie, J.-L.; Deng, L.-J. Effect of high-enthalpy atmospheric plasma spraying parameters on the mechanical and wear resistant properties of alumina ceramic coatings. *Surf. Coat. Technol.* **2021**, *418*, 127193. [CrossRef]
6. Song, R.G.; Wang, C.; Jiang, Y.; Lu, H.; Li, G.; Wang, Z.X. Microstructure and properties of Al_2O_3/TiO_2 nanostructured ceramic composite coatings prepared by plasma spraying. *J. Alloys Compd.* **2012**, *544*, 13–18. [CrossRef]
7. Lampke, T.; Meyer, D.; Alisch, G.; Wielage, B.; Pokhmurska, H.; Klapkiv, M.; Student, M. Corrosion and wear behavior of alumina coatings obtained by various methods. *Mater. Sci.* **2011**, *46*, 591–598. [CrossRef]
8. Ruys, A. *Metal-Reinforced Ceramics*; Elsevier: Amsterdam, The Netherlands, 2020; ISBN 9780081028704.
9. Szkodo, M.; Bień, A.; Antoszkiewicz, M. Effect of plasma sprayed and laser re-melted Al2O3 coatings on hardness and wear properties of stainless steel. *Ceram. Int.* **2016**, *42*, 11275–11284. [CrossRef]
10. Munteanu, C.; Chicet, D.-L.; Bistricianu, I.-L.; Pintilei, G.-L. Inovative Character of Plasma Spray Deposited Method. *Acta Tech. Napoc. Ser. Appl. Math. Mech.* **2011**, *54*, 2.
11. Cretu, S.; Benchea, M.; Iovan-Dragomir, A. On basic reference rating life of cylindrical roller bearings. part 1. elastic analysis. *J. Balk. Tribol. Assoc.* **2016**, *21*, 820–830.

12. Cimpoesu, R.; Vizureanu, P.; Stirbu, I.; Sodor; Zegan, G.; Prelipceanu, M.; Cimpoesu, N.; Ioanid, N. Corrosion-Resistance Analysis of HA Layer Deposited through Electrophoresis on Ti$_4$Al$_4$Zr Metallic Substrate. *Appl. Sci.* **2021**, *11*, 4198. [CrossRef]
13. Florea, C.; Bejinariu, C.; Munteanu, C.; Cimpoesu, N. Preliminary Results on Complex Ceramic Layers Deposition by Atmospheric Plasma Spraying. *Adv. Mater. Eng. Tech. V* **2017**, *1835*, 020053. [CrossRef]
14. Tan, Y.; Jiang, S.; Yangn, D.; Sheng, Y. Scratching of Al$_2$O$_3$ under pre-stressing. *J. Mater. Process. Tech.* **2011**, *211*, 1217–1223. [CrossRef]
15. Wakayama, S.; Ishiwata, K. Fracture analysis based on quantitative evaluation of microcracking in ceramics using AE source characterization. *J. Solid Mech. Mater. Eng.* **2009**, *3*, 96–105. [CrossRef]
16. Kaya, F. Damage detection in fibre reinforced ceramic and metal matrix composites by Acoustic Emission. *Key Eng. Mater.* **2010**, *434–435*, 57–60. [CrossRef]
17. Akbari, J.; Saito, Y.; Hanaoka, T.; Higuchi, S.; Enomoto, S. Effect of grinding parameters on acoustic emission signa ls while grinding ceramics. *J. Mater. Process. Tech.* **1996**, *62*, 403–407. [CrossRef]
18. Singh, V.P.; Sil, A.; Jayaganthan, R. Wear of plasma sprayed conventional and nanostructured Al$_2$O$_3$ and Cr$_2$O$_3$, based coatings. *Chem. Mater. Sci.* **2012**, *65*, 1–12. [CrossRef]
19. Federici, M.; Menapace, C.; Moscatelli, A.; Gialanella, S. Effect of roughness on the wear behaviour of HVOF coatings dry sliding against a friction material. *Wear* **2016**, *368*, 326–334. [CrossRef]
20. Pantelis, D.I.; Psyllaki, P.; Alexopoulos, N. Tribological behaviour of plasma-sprayed Al$_2$O$_3$ coatings under severe wear conditions. *Wear* **2000**, *237*, 197–204. [CrossRef]
21. Czichos, H. *Tribology—A Systems Approach to the Science and Technology of Friction Lubrication and Wear*; Elsevier Sci. Pub. Co.: Amsterdam, The Netherlands, 1978; ISBN 0-444-41676-5.
22. Stebut, J.V.; Lapostolle, F.; Bucsa, M.; Vallen, H. Acoustic emission monitoring of single cracking events and associated damage mechanism analysis in indentation and scratch testing. *Surf. Coat. Technol.* **1999**, *116–119*, 160–171. [CrossRef]
23. Florea, C.D.; Bejinariu, C.; Carcea, I.; Paleu, V.; Chicet, D.; Cimpoeşu, N. Preliminary results on microstructural, chemical and wear analyze of new cast iron with chromium addition. *Key Eng. Mater.* **2015**, *660*, 97–102. [CrossRef]
24. Florea, C.D.; Bejinariu, C.; Paleu, V.; Chicet, D.; Carcea, I.; Alexandru, A.; Cimpoesu, N. Chromium Addition Effect on Wear Properties of Cast-Iron Material. *Appl. Mech. Mater.* **2015**, *809–810*, 572–577. [CrossRef]
25. Aelenei, N.; Lungu, M.; Mareci, D.; Cimpoeşu, N. HSLA steel and cast iron corrosion in natural seawater. *Environ. Eng. Manag. J.* **2011**, *10*, 1951–1958. [CrossRef]
26. Nejneru, C.; Cimpoeşu, N.; Stanciu, S.; Vizureanu, P.; Sandu, A.V. Sea water corrosion of a shape memory alloy type cuznal. *Metal. Int.* **2009**, *14*, 95–105.
27. Bejinariu, C.; Munteanu, C.; Florea, C.D.; Istrate, B.; Cimpoesu, N.; Alexandru, A.; Sandu, A.V. Electro-chemical Corrosion of a Cast Iron Protected with a Al2O3 Ceramic Layer. *Rev. Chim.* **2018**, *69*, 12.
28. Florea, C.D. Contributions Regarding the Characterization of Some Materials Used for the Construction of Car Brake Discs. Ph.D. Thesis, Technical University Gheorghe Asachi Iasi, Iasi, Romanian, 2019.

Article

Influence of Friction Stir Surface Processing on the Corrosion Resistance of Al 6061

Ibrahim H. Zainelabdeen [1], Fadi A. Al-Badour [1,2,*], Rami K. Suleiman [2], Akeem Yusuf Adesina [2], Necar Merah [1,2] and Fadi A. Ghaith [3]

1 Mechanical Engineering Department, King Fahd University of Petroleum and Minerals, Dhahran 31261, Saudi Arabia
2 Interdisciplinary Research Center for Advanced Materials, King Fahd University of Petroleum and Minerals, Dhahran 31261, Saudi Arabia
3 School of Engineering and Physical Sciences, Heriot-Watt University, Dubai 38103, United Arab Emirates
* Correspondence: fbadour@kfupm.edu.sa

Abstract: In this work, friction stir processing using a pinless tool with a featured shoulder was performed to alter the surface properties of Al 6061-O, focusing on the effect of tool traverse speed on surface properties, i.e., microstructure, hardness, and corrosion resistance. All processed samples showed refinement in grain size, microhardness, and corrosion resistance compared to the base material. Increasing tool-traverse speed marginally refined the microstructure, but produced a significant reduction in microhardness. Electrochemical impedance spectroscopy, linear polarization resistance, and potentiodynamic polarization were used to evaluate the effect of the processing conditions on corrosion behavior in a saline environment. All corrosion test results are found to agree and were supported with pictures of corroded samples captured using a field emission scanning electron microscope. A remarkable reduction in the corrosion rate was obtained with increasing traverse speed. At the highest traverse speed, the corrosion current density dropped by approximately 600 times when compared with that of the base alloy according to potentiodynamic polarization results. This is mainly due to the grain refinement produced by the friction stir process.

Keywords: friction stir processing; pinless tool; surface processing; corrosion resistance; saline environment; Al 6061

1. Introduction

Aluminum-based alloys are utilized in a variety of advanced commercial applications such as automotive, defense, marine, and aerospace. This is ascribed to their outstanding characteristics, such as a high strength-to-weight ratio, high ductility, and good corrosion resistance. In addition, the resistivity of aluminum alloys against corrosion is mainly attributed to the oxide layer formed when the alloy is exposed to the environment. It is, however, believed that aluminum-based alloys exhibit low corrosion resistance when exposed to saline media [1]. Along with bad corrosion resistance in saline media, aluminum-based alloys suffer from bad surface characteristics such as wear resistance and low hardness. Therefore, the adoption of aluminum alloys in saline environments is restricted [2]. The impact of grain size on the mechanical and physical characteristics of different alloys is well documented and it is known to be a vital determinant of many microstructure-dependent properties [3]. Therefore, grain size refinement is among the most adopted approaches for improved material properties. Thus, several alternatives have been developed over the last few decades to produce refined grains in a feasible way such as severe plastic deformation techniques [4]. One such technique is friction stir processing (FSP), which has drawn more attention because of its many advantages [5]. Most severe plastic deformation techniques result in bulk material processing or the altering of the initial part shape, Moreover, a special surface pre-processing/treatment is required [6].

FSP was developed on the same working principles as friction stir welding (FSW), where a rotating tool, consisting of two parts—a shoulder with a larger diameter and a pin—is pressed against neighboring ends of sheets or plates that need to be linked and then moved along the line of the joint [7]. FSP uses the same principle in processing the surface and subsurface of a single bulk workpiece, thereby modifying the surface properties by the dynamic recrystallization process encountered in FSW. The microstructural evolution and mechanical properties after FSP of various alloys have been extensively studied over the last two decades. Nevertheless, the effect on corrosion resistance of FSP has not received the same attention. Several studies have indicated that the corrosion susceptibility may either diminish or increase for different aluminum alloys after severe plastic deformation, as reported by Mehdi et al. [8]. It was suggested that the corrosion behavior primarily depends on the severe plastic deformation processes, working material alloying elements, and process conditions. For example, Surekha et al. [9,10], in two separate studies, have investigated the influence of traverse speed, rotation speed, and the number of passes on the corrosion resistance of FS-processed 2219 AA. Their results showed that the corrosion susceptibility was influenced by changing the number of passes and rotating speeds, rather than the traverse speed. They reported an enhancement in corrosion resistance with the increase in tool rotational speed and number of passes which was ascribed to the reduction in the intermetallic phases. Conversely, Eldeeb et al. [11] reported that the traverse speed has more influence on the corrosion resistance of the stirred zone in comparison to the tool rotational speed of friction-stir-processed Al6061-O. Esmaily et al. [12] have reported enhancement in the corrosion behavior of 6005 AA after deploying multi-pass FSP. Reddy et al. [13] also compared the corrosion behavior and wear resistance of FS-processed A356 with that of cast material. It was revealed that FSP resulted in outstanding corrosion and wear resistance. They explained the improvement in the wear resistance by the homogeneous dispersion of silicon particles through the aluminum matrix, whereas the reduction in corrosion rate was attributed to the formation of an intact passive film.

It should be noted that most of the literature on FSP focuses on studying the impact of changing parameters on the microstructural features, mechanical, and corrosion characteristics of full penetration FSP. However, to the authors' knowledge, few investigations have been carried out on the effect of friction stir surface processing parameters using a pinless tool on the microstructure, and consequently, the corrosion resistance, associated with the newly developed surface, with a focus on aluminum and its alloys. To rectify this, the impact of varying the tool traverse speed during FSP on the grain structure, microhardness, and corrosion behavior of 6061-O AA base metal is studied herein in detail.

2. Materials and Methods

The workpieces utilized in the current study are an aluminum alloy 6061-O grade with dimensions of 190 × 90 × 6 mm^3 (Figure 1). The chemical composition of the workpieces was obtained using spark spectrometry and is shown in Table 1. A novel pinless FSP tool was designed and fabricated for this study from 4140 alloyed steel. The tool was heat-treated to a hardness ranging between 52 and 54 HRC. The FSP tool has a shoulder diameter of 23 mm with two circular groves, which were machined such that they will enhance the material flow under the tool shoulder and increase the contact area. Figure 1 is a schematic demonstration of the FSP process, showing the pinless rotating tool traversing the workpiece to be processed. A literature search showed that a rotation speed of 1000 rpm yields better properties for different grades of aluminum alloys [14–18]. Therefore, the rotation speed was fixed at 1000 rpm, while the traverse speed was varied from 100 to 250 mm/min. To improve the forging action, the tool was tilted by 3° towards the trailing side [19], with a plunge depth of 0.2 mm to assure sufficient forging force. FSP was carried out using a research-based three-axis RM-1 friction stir welder (MTI, West Washington, South Bend, IN, USA) under position control mode. Accordingly, process conditions and sample designations are shown in Table 2. After FSP, the base metal and processed samples were sectioned from the processed zone into a suitable size for subsequent characterization and corrosion testing; thereafter prepared by grinding using silicon carbide papers with grit sizes varying from 240 to 1200. Consequently, a polishing process using alumina and diamond paste was performed to obtain a mirror-like finish followed by etching using 2 g of sodium hydroxide in 100 mL of distilled water to reveal the grain structure.

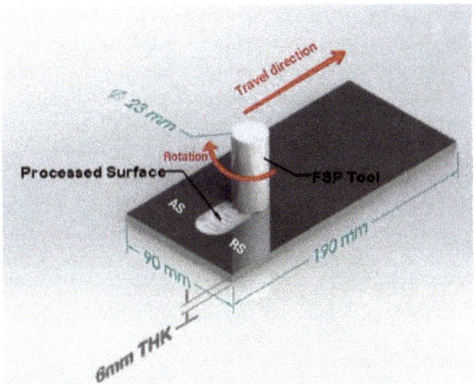

Figure 1. The schematic diagram for friction stir processing.

Table 1. Chemical composition of 6061 AA.

Element	Mg	Si	Cu	Fe	Mn	Al
wt.%	0.82	0.71	0.23	0.63	0.14	Bal.

Table 2. Process conditions and sample designations.

Samples	Rotation Speed (rpm)	Traverse Speed (mm/min)
FP10		100
FP15	1000	150
FP25		250

2.1. Characterization and Mechanical Properties

An optical digital microscope (Olympus, DSX 510, Tokyo, Japan) was utilized to capture images of the etched surfaces, while Olympus stream software was used to determine the grain size of the friction-stir-processed surface as well as the base material, following ASTM 112-13 [20]. A line intercept procedure (average of horizontal and vertical intercept lengths) was adopted in measuring the average grain size as per the recommendation of the ASTM-112-13 standard. More insights into grain morphology parameters, as well as the method used to measure grain size, are shown in Figures and Tables S1–S4 (Supplementary Material). In addition, images of corroded samples were acquired using a field emission Scanning Electron Microscope (FESEM) (Quanta 250, Bruno, Czech Republic). Analysis of the present phases for the base and all processed samples was performed using X-ray Diffraction (XRD) in a Bruker diffractometer (D2 PHASER) that operated at 30 kV with CuK$_\alpha$ (λ_α = 1.54 Å). A micro indenter (Micro Combi, CSM Instruments, Peseux, Switzerland) was utilized to measure the hardness and elastic modulus of the base and processed samples. A pyramidal indenter was used at a normal force of 3 N for 10 s dwell time. The test was repeated six times and the average values of microhardness and elastic modulus were recorded.

2.2. Electrochemical Analysis

All electrochemical measurements, consisting of electrochemical impedance spectroscopy (EIS), linear polarization (LPR), and potentiodynamic polarization (PDP) measurements, on the base sample as well as the processed samples were carried out on the traditional three-electrode configuration which includes a graphite counter electrode and a saturated calomel electrode (SCE) reference electrode. A Gamry Instruments Reference 3000 potentiostat/galvanostat was employed for data collection, while Echem Analyst 6.0 software was used for electrochemical data analysis and fitting. All working electrode samples i.e., the processed samples, were machined from the middle of the processed zone into a suitable size and ground using silicon carbide papers up to 800 grit and thereafter masked such that only 1 cm^2 of the processed zone was exposed to the 3.5 wt.% NaCl at 23 °C for 15 days. Before conducting the corrosion measurements, all samples were cleaned using distilled water and acetone. EIS tests were conducted using a frequency range of 10^{-2} to 10^5 Hz and an AC voltage of 10 mV. PDP tests were carried out at a scanning rate of 0.125 mV/s using an applied potential varying between –0.20 V, cathodically, and 0.50 V, anodically, under open circuit potential (OCP). Thereafter, Tafel parameters were obtained from PDP curves. LPR measurements were acquired at a scanning rate of 100 mV/s between +25 and −25 mV under OCP.

3. Results and Discussion

3.1. Effect of Traverse Speed on Forging Force and Torque

The effect of increasing traverse speed from 100 to 250 mm/min on forging force is presented in Figure 2a. It is obvious from these results that the forging force increases with increasing traverse speed, reaching 22 kN at 250 mm/min. The increase in forging force with traverse speed is attributed to the heat input reduction, which makes the area around the tool harder and thereby raises the material flow stress and consequently the forging forces [21–23] for maintaining the tool plunging depth. Figure 2b shows that the traverse speed has an even higher impact on the generated torque. The considerable increase in the measured torque with the rise in traverse speed is also attributable to higher FSP-induced flow stresses [24]. It is evident that increasing the tool traverse speed under constant rotation speed results in a heat-input reduction, whereas increasing the tool rotation speed and fixing the tool traverse speed results in increasing the heat input during the friction stir processing, and vice versa. Accordingly, the heat input for each process parameter is obtained from the measured spindle torque, tool rotation, and traverse speeds (Equation (1)) [25].

$$H = \frac{P}{v} = \frac{T\omega}{v} \qquad (1)$$

where H is the heat input in J/mm, T is the average measured spindle torque in N·m, ω is the tool rotation speed in rad/s, and v is the tool traverse speed in mm/s.

It is worth noting that the heat input equation (Equation (1)) neglects the efficiency of the heat transfer process. Moreover, it assumes that the frictional work of the tool is completely converted into heat, resulting in material softening.

3.2. Microstructural and Mechanical Properties Analysis

The optical microscopic images of as-received (base), and FS-processed for the samples at various traverse speeds, 100, 150, and 250 mm/min, are presented in Figure 3a, Figure 3b, Figure 3c, and Figure 3d, respectively. It is worth noting that the only source of heat during this investigation was that of the tool shoulder interaction with the workpiece material. Compared to the base metal, samples processed at different traverse speeds exhibited substantial grain refinement. Adopting FSP at a traverse speed of 100 mm/min significantly reduces the average grain size of the base specimen from 93.93 µm to 17.74 µm. Additionally, the grain size exhibits a slight reduction with increasing traverse speed, from 16.38 µm for the sample processed at 150 mm/min to 15.58 µm for the sample processed at the highest speed of 250 mm/min. Grain refinement with increasing traverse speed during friction stir processing has been extensively reported in the literature and has been attributed to a heat-input reduction [26,27]. It is worth noting that microstructure evolution is sensitive to heat input, where higher heat input is not only accompanied by grain growth but also a significant change in grain morphology [28]. In addition, the relationship between the grain growth upon cooling and heat input induced by the severe plastic deformation during friction stir processing can be explained through the following equation (Equation (2)) [29].

$$D^2 - D^2_0 = At \exp(-Q/RT) \tag{2}$$

where D and D_0 are the initial and deformed grain size, respectively, A is a constant, T is the peak temperature, Q is the activation energy, and t is the time to cool down to 448 K [29]. Furthermore, it was reported that dynamic recrystallization is the main mechanism of grain refinement during FSP where the processed zone experienced high strain rates and high peak temperatures [30].

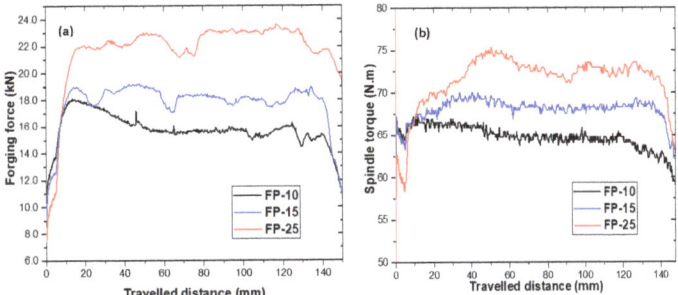

Figure 2. Effect of traverse speed on (**a**) forging forces and (**b**) spindle torque.

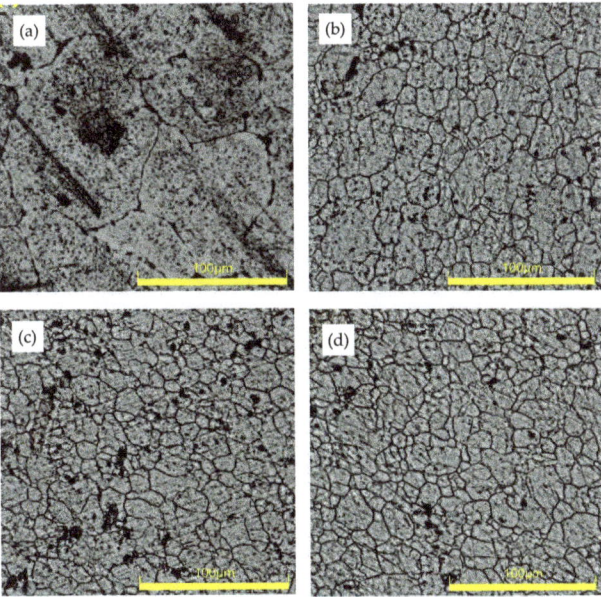

Figure 3. Optical micrograph of the samples showing grain morphology of (**a**) unprocessed base, (**b**) FP-10 (**c**) FP-15, and (**d**) FP-25.

Figure 4 presents the effect of tool traverse speed on calculated heat input using (Equation (1)) and the measured grain size.

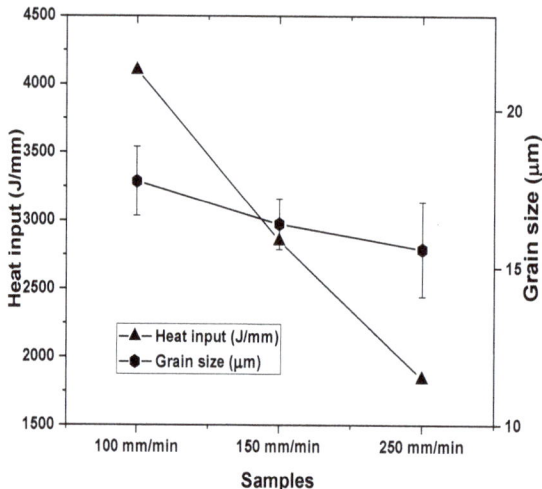

Figure 4. Effect of tool traverse speed on heat input and processed surface grain size.

Figure 5 shows the influence of varying FSP traverse speed on the XRD patterns. As Figure 5a shows, only the α-Al phase was detected for the base material and processed samples. Furthermore, little change in diffraction peaks was observed, which confirmed that no secondary intermetallic phases precipitated because of plastic deformation induced by FSP. Interestingly, the samples after processing showed an appreciable shift toward high diffraction angles, and the sample processed at the highest processing speed showed the most pronounced shift, as can clearly be seen in Figure 5b. To the best of the authors' knowledge, a shift in aluminum peaks after FSP was not reported before; it has, however, been reported for magnesium and copper after FSP. This phenomenon of peak shifting was ascribed to the reduction in lattice parameters that causes compressive stresses [31,32].

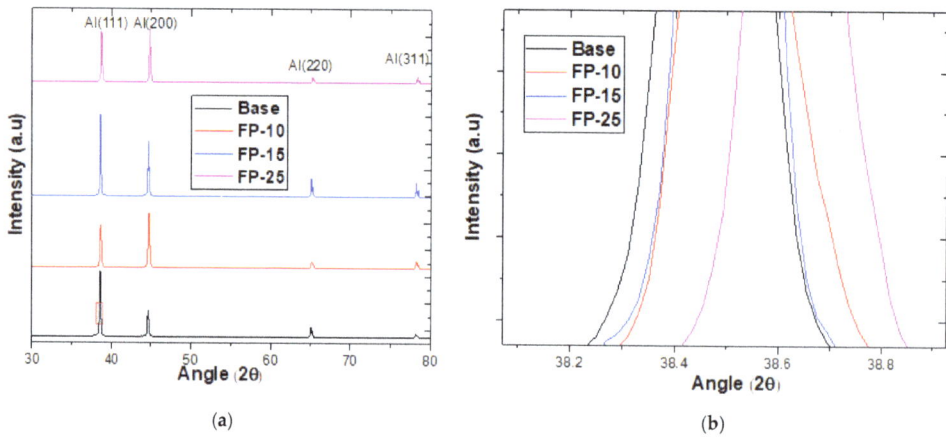

Figure 5. (a) XRD spectra for all samples before and after FSP at different traverse speeds. (b) zoom in at (111) peak.

The microhardness and elastic modulus of the base sample was clearly much lower compared to that of the processed samples under all processing conditions, as shown in Figure 6. The microhardness of samples FP-10, FP-15, and FP-25 were improved by 96, 48, and 40%, respectively, compared to that of the base metal. The enhanced hardness is ascribed to the high heat input that may have resulted in the reduction of coarse secondary strengthening precipitate [33]. It should be mentioned that, though the grain sizes of the processed materials at different speeds are comparable (Figure 3), the measured hardnesses are different. Researchers have shown that, in precipitation-hardened aluminum alloys, grain refinement is not the dominant strengthening mechanism [34]. Other researchers have found that the effect of grain size of precipitated aluminum alloys on hardness is insignificant [35]. In addition, the elastic modulus of the base sample was around 64 GPa while, after performing FSP at the lowest speed, the elastic modulus increased to around 68 GPa. Moreover, the elastic modulus for the samples processed at moderate and high speeds showed only a small further improvement to approximately 70 GPa. The improvement in the elastic modulus may be attributed to dynamic recrystallization [36], which leads to reduced dislocation mobility, therefore, minimizing elastic strain.

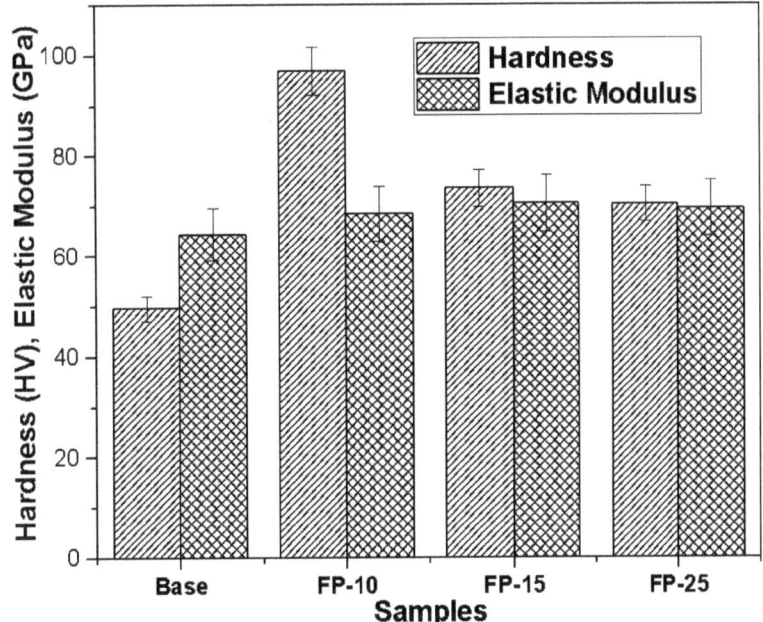

Figure 6. Impact of increasing tool traverse speed on microhardness and elastic modulus.

3.3. Electrochemical Behavior

3.3.1. Electrochemical Impedance Spectroscopy

Electrochemical impedance spectroscopy (EIS) has been extensively used in the investigation of the metal-electrolyte interface, surface responses, oxides films formation, passivation, corrosion kinetics and mechanism, and coating protective effectiveness to metallic substrates [37,38]. Thus, in order to investigate the effects of FSP parameters on the kinetics and characteristics of the electrochemical process, EIS measurements were carried out on the base sample and all processed samples.

Figure 7a–d displays the Nyquist curves, phase angle, and modulus of the base sample and all processed samples. The Nyquist plots depicted in Figure 7a,b show a relatively large incomplete semicircle at the high-frequency region, while the low-frequency region demonstrates a diffusion tail. The larger incomplete semicircle at the high-frequency region in the Nyquist plot of processed samples may be ascribed to a charge transfer reaction at the boundary between the thin film formed on the aluminum surface and the NaCl solution [39]. However, the occurrence of the small tail at the low-frequency region of the spectra is attributed to diffusion-controlled processes [40]. The depression phenomenon under the x-axis can be clearly identified through the incomplete semicircle behavior of the arcs, which is probably caused by surface heterogeneities resulting from possible corrosion products and frequency dispersion [41]. As can be observed in Figure 7a, the arc diameter rises with increasing traverse speed. In addition, the smallest diameter was observed for the base sample, while the sample processed with the highest traverse speed (FP-25) demonstrates the highest semicircle diameter. Moreover, the variation in the semicircle diameter between the base sample and the sample processed at the lowest traverse speed (FP-10) was remarkably low compared to the deviation in semicircle diameter between the sample processed at moderate velocity (FP-15) and that fabricated at the highest velocity (FP-25). Interestingly, the variance in semicircle diameter between the base sample and (FP-25) sample was not even distinguishable without zooming the base sample area in the Nyquist plot, as seen in Figure 7b. A huge increment in the semicircle diameters revealed the significant impact of process conditions on corrosion behavior. Furthermore, the increase in the radius of the semicircle with the processing can be interpreted by an enhancement in the surface protection due to the rising stability and compactness of the passive film formed [42,43]. Therefore, the enhancement in the corrosion resistance was found to follow the rank of FP-10 < FP-15 < FP-25, and demonstrated a markedly improved resistance compared to that of the unprocessed sample.

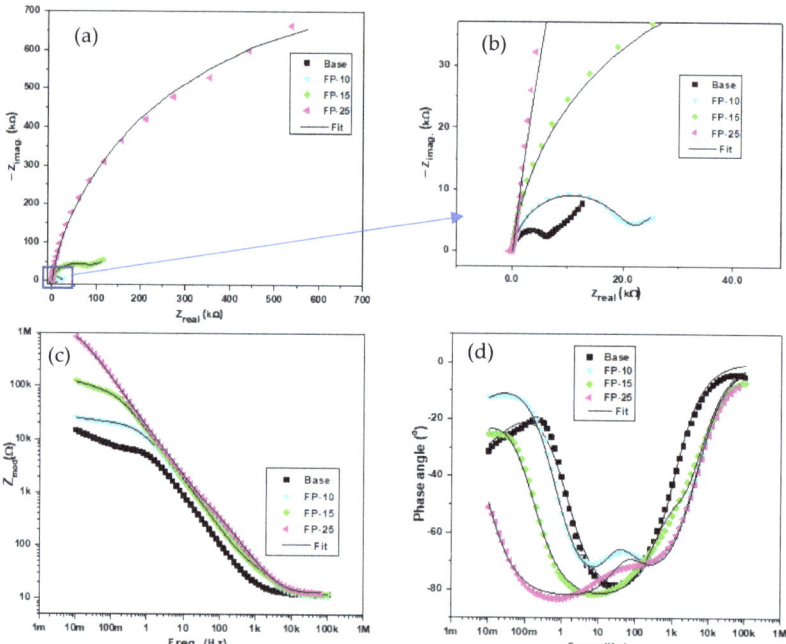

Figure 7. Plots of the EIS experimental and fitted data for all samples showing the (**a,b**) Nyquist, (**c**) bode, and (**d**) phase angle plots.

The bode plots for all samples are presented in Figure 7c. It is evident from the figure that the total impedance for all the processed samples is remarkably higher than that of the base sample, which reflects the increased modulus impedance of the processed samples. In addition, the total impedance revealed a significant increment by increasing the traverse speed. Moreover, the deviation in the total impedance was at its lowest when comparing the base material with the sample processed at the lowest traverse speed; however, with increasing traverse speed, the deviation between samples processed at the intermediate and lowest speed was much higher. Furthermore, a huge variation was observed between the intermediate-speed sample (FP-15) and the sample processed at the highest speed (FP-25), which indicates a superior corrosion resistance for the sample processed at 250 mm/min compared to all other samples.

Figure 7d displays the typical Phase angle-frequency plots for both the processed and base samples. The frequency of the phase angle maxima of the base sample is 25 Hz at $-78°$. In contrast, the (FP-10) sample demonstrated two time constants, the first shifted toward a frequency of 6.4 Hz at approximately $-71°$ and the second shifted toward a frequency of 241 Hz at $-69°$. Additionally, samples (FP-15) and (FP-25) exhibited phase angle maxima that are shifted toward frequencies of 8 and 0.78 Hz at -81 and $-83°$, respectively. It is worth noting that observing the phase angle maxima for the base sample at higher frequency suggests weakness in the protective barrier [44]. Moreover, it can be noticed that the base sample demonstrated a single time constant and upon increasing the traverse speed two time-constant characteristics can be observed. Similar behavior was also observed in [45]. Furthermore, more peak broadening indicates an enhancement in the passive protective barrier over a wide frequency range and a consequent corrosion resistance improvement [44,46].

All extracted impedance curves were fitted to appropriate corresponding circuits for further numerical estimation of the barrier characteristics of processed surfaces and properties of active corrosion protection. Accordingly, two equivalent circuits were proposed; the first one (Circuit a) was used to stimulate the corrosion behavior at the base surface as shown in Figure 8a. This circuit included a charge transfer resistance (R_{ct}) that serially connected to a Warburg element (W) and was in parallel to a constant phase element for double-layer capacitance (CPE_{dl}), where both (W, $R_{ct}//CPE_{dl}$) components were connected in series with the solution resistance (R_S) component between the reference and working electrodes. The second circuit illustrated in Figure 8b (Circuit b) was used to simulate all the electrochemical processes at the surface of processed samples and contained an R_{ct} that linked in series to a Warburg element (W) and both components were in parallel with a constant phase element accounting for the double-layer capacitance of the inner barrier layer (CPE_{dl}), and the aforementioned three components were connected in series with a film resistance (R_f) component. Further, W, R_{ct}, CPE_{dl}, and R_f were connected in parallel with the constant phase element of the outer passive film (CPE_f), and the previous circuit was connected in series with the solution resistance between the reference and working electrodes (R_S).

The charge transfer resistance (R_{ct}) depicted in Table 3 exhibited a rapid increment for all processed samples over the base material. In addition, it is evident from the fitted results that the charge transfer resistance increases with increasing traverse speed, which can be ascribed to the passive film formed at the metal/electrolyte interface [47]. The presence of this film isolated the aluminum metal surface from the corrosive ions attacks and thereby obstructing any further transfer of charge or mass. It should be noted that there is a remarkable difference between the charge transfer resistance (R_{ct}) and film resistance (R_f) particularly if the processing was carried out at high speeds and this may be attributed to the key role of the formed oxide film on the metal surface in mitigating the corrosion process at a particular speed [48]. Moreover, the huge variation between the outer and inner resistances indicates that the resistance provided by the inner barrier is much higher when compared with the outer film [49]. Generally, the Warburg element is added to simulate the diffusion effect as indicated by the straight line of a slope close to $45°$ to the impedance real axis. This can be observed with the unprocessed sample and gradually reduces for samples processed at 100 and 150 mm/min as seen in Figure 7a,b. It is expected that a further decrease will be experienced at the highest traverse speed sample i.e., FP-25 if the test was allowed for a much lower frequency. Considering the Warburg impedance, it is obvious from the table that the lowest Warburg impedance was obtained at the highest traverse speed of 250 mm/min and, by reducing the traverse speed to 150 mm/min, the Warburg impedance demonstrated a slight increment. Moreover, with a further reduction in the traverse speed down to 100 mm/min, a considerable increase in Warburg impedance was observed, which is relatively comparable with the base metal. A lower Warburg value indicates a reduction in the diffusion of chloride ions through an oxide passive film and thereby improving the corrosion resistance [50,51].

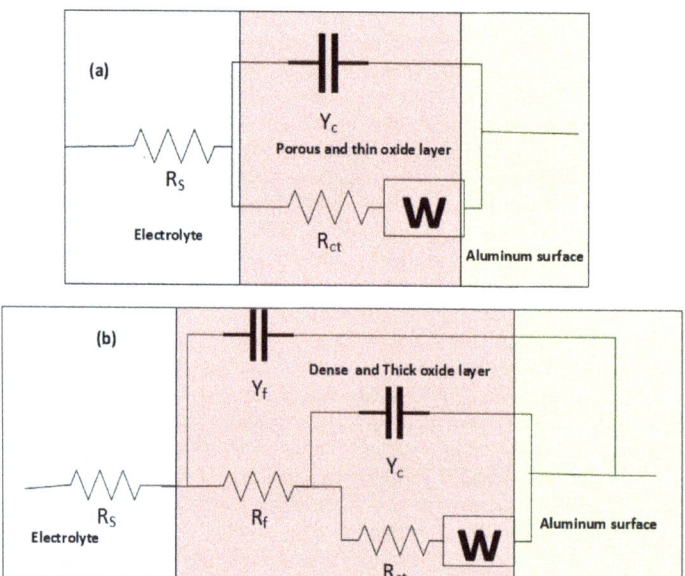

Figure 8. Proposed equivalent circuits of (**a**) base, and (**b**) processed samples.

Table 3. Electrochemical parameters obtained from simulating the experimental EIS data.

Samples	R_s (Ω)	Y_n (10^{-6}) (Ss^a)	n	R_f ($k\Omega$)	Y_c (10^{-6}) (Ss^a)	c	R_{ct} ($k\Omega$)	W (10^{-5}) $Ss^{1/2}$	χ (10^{-4})
Base	12.38	23.5	0.924	-			6.14	41.4	29.10
FP-10	11.26	10.94	0.878	1.59	3.92	0.98	18.7	56.7	4.68
FP-15	12.24	6.94	0.921	0.09	5.25	0.93	83.9	7.26	15.96
FP-25	12.79	4.43	0.892	1.72	2.16	0.97	1500	6.80	15.29

It is worth noting that the constant phase element was utilized as an alternative to pure capacitance because the non-ideal capacitive behavior indicated by the deviation in phase shifts from 90°, which might be attributed to a heterogeneity property on the surface of the metal. Additionally, several models have been developed to correlate the passive film capacitance knowing the constant phase element component (admittance). Hsu and Mansfeld introduced a model (Equation (3)) to show the relationship between capacitance and admittance [52].

$$C_{dl} = Y_0 \, (\omega_{max})^{n-1} \quad (3)$$

where, ω_{max} is the frequency when the imaginary part of the impedance is at its max value, Y_0 is the admittance of the CPE_{dl}, and n is the value corresponding to the surface roughness.

It can be seen from Table 4 that the double-layer capacitance (C_{dl}) exhibits a significant reduction for all processed samples when compared to the unprocessed sample and the values are in the order of (FP-15) > (FP-10) > (FP-25). The electrochemical behavior between the charged aluminum surface and 3.5 wt.% NaCl is reflecting an electrical dual-layer capacitance behavior [53]. Accordingly, the reduction in the double-layer capacitance of the processed surfaces indicates a corrosion resistance enhancement with increasing traverse speed, which can be attributed to a rapid decrease in surface activeness resulting from the formation of a thicker passive film [54,55]. Accordingly, the thickness of the passive film (t) can be calculated using the Helmholtz model [56] (Equation (4)).

$$t = \frac{\varepsilon \varepsilon_0}{C_{dl}} x \qquad (4)$$

where C_{dl} is the double-layer capacitance (μF), ε_0 is the vacuum permittivity (8.85×10^{-14} F cm^{-1} [57]), and ε is the passive layer dielectric constant (for aluminum, $\varepsilon = 10$) [57].

Table 4. Variation in double-layer capacitance and passive film thickness for different samples.

Samples	C_{dl} (μF)	t (nm)
Base	33.3	2.65
FP-10	11.6	7.63
FP-15	9.99	8.86
FP-25	7.28	12.16

3.3.2. Potentiodynamic Polarization (PDP)

The typical PDP plots of all samples after exposure to the NaCl electrolyte are presented in Figure 9. Various electrochemical parameters, such as the anodic Tafel slope (β_a), cathodic Tafel slope (β_c), corrosion current density (I_{corr}), and corrosion potential (E_{corr}) for the base sample as well as the processed samples, were derived from the potentiodynamic polarization curves and are listed in Table 5. The polarization curves exhibited significant differences between all samples. As can be noted from the data in Table 5, there is a remarkable difference in corrosion potential between the base and processed samples. It is clear from Figure 9 and Table 5 that the corrosion potential (E_{corr}) is shifted to a more positive noble potential with increasing tool traverse speed. Moreover, the FP-10 sample exhibited the lowest corrosion potential i.e the most negative, whereas the FP-25 sample showed the noblest behavior among all processed and base samples. This shift to a noble direction is an indication of an improvement in corrosion resistance. Further, as can be inferred from Figure 9 and Table 5, the corrosion current densities (I_{corr}) of all processed samples were markedly lower than that of the unprocessed alloy. Additionally, a further reduction in I_{corr} can be observed as the traverse speed increases. The reduction in I_{corr} indicates that the friction stir processing for 6061 AA has enhanced its corrosion resistance. Moreover, the corrosion current density for the base sample is approximately 600 times higher than that of the FP-25 sample, which indicates the outstanding corrosion protectiveness of the FP-25 sample. Also, as compared to a base sample, the pitting potential of the FP-25 sample raised from −780 mV to −600 mV, which suggested an enhancement in the pitting resistance. Furthermore, the pitting resistance of the other two processed samples is also high despite their lower corrosion potential. Enhancing pitting resistance was also reported after the FSP of AA 7075 [58].

A remarkable variation in the cathodic and anodic Tafel slopes can be detected through the inspection of Table 5. In particular, the anodic behavior revealed that the unprocessed sample had less tendency to be oxidized compared to all processed samples, where β_a value for the unprocessed sample was the highest, and a further reduction in the anodic slope can be noticed with increasing tool traverse speed, which suggests an enhancement in the oxidation tendency [59]. In addition, the cathodic Tafel slope β_c revealed a remarkable reduction upon deploying FSP. Moreover, the cathodic slope is further reduced with increasing traverse speed, which suggests a diffusion of oxygen molecules to form (OH$^-$) in the processed samples [60].

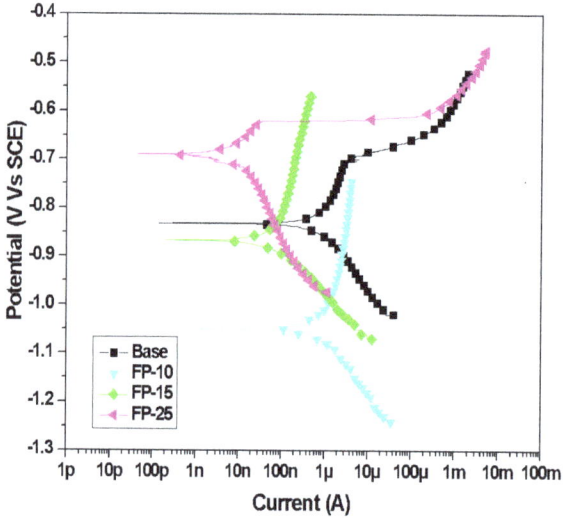

Figure 9. Typical PDP curves for all samples after 15 days 3.5 wt.% NaCl.

Table 5. Electrochemical parameters obtained from PDP plot.

Samples	β_a	β_c	I_{corr} (μA)	E_{corr} (V)	Corrosion Rate (mpy)
Base	1.994	2.78×10^{-1}	3.28	-0.832	1.498
FP-10	7.38×10^{-1}	1.68×10^{-1}	1.7	-1.05	7.79×10^{-1}
FP-15	6.57×10^{-1}	1.10×10^{-1}	1.12×10^{-1}	-0.867	5.11×10^{-2}
FP-25	8.69×10^{-2}	7.60×10^{-2}	5.47×10^{-3}	-0.693	2.50×10^{-3}

3.3.3. Linear Polarization Resistance (LPR)

Linear polarization plots for different samples (processed and base) in NaCl solution are presented in Figure 10. The polarization resistance (R_P) was obtained as a slope of potential versus current. To calculate the corrosion current density based on the linear polarization resistance approach, the Stern–Geary equation was utilized, which correlates the current density (I_{corr}) with polarization resistance (R_P) (Equation (5)) [61].

$$I_{corr} = \frac{\beta_a \beta_c}{2.303\, R_P (\beta_a \beta_c)} \quad (5)$$

where R_P is the polarization resistance obtained from the LPR slope, and β_a and β_c are the cathodic and anodic Tafel slopes derived from PDP curves, respectively.

Subsequently, to calculate the corrosion rate (mpy), Equation (6) was used:

$$CR = \frac{0.131\, I_{corr}\, EW}{\rho} \quad (6)$$

where ρ is the sample density and EW is the sample equivalent weight.

As indicated in Table 6, the corrosion current density of the unprocessed sample is significantly higher than that of all processed samples. Additionally, the highest polarization resistance was obtained for the sample processed at the highest speed and a further reduction can be observed with a reduction in the traverse speed. It should be emphasized here that the polarization resistance is directly related to the corrosion resistance, therefore, a sample with the highest polarization resistance exhibited the highest corrosion resistance.

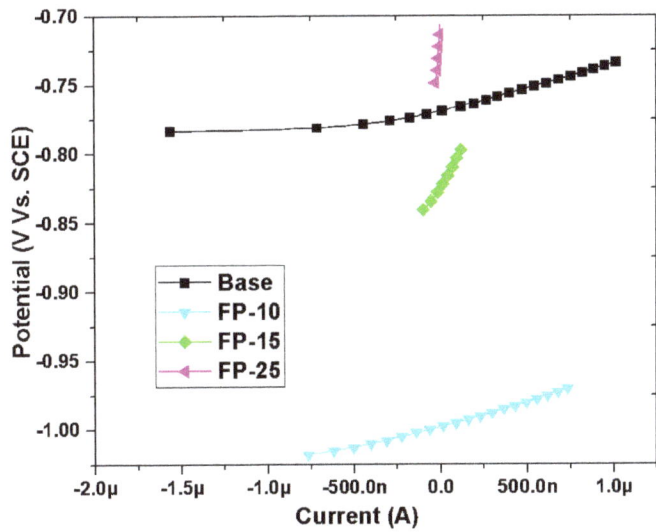

Figure 10. Typical LPR plots for all samples in 3.5 wt.% NaCl electrolyte.

Table 6. LPR data obtained after 15 days of exposure of all samples to 3.5 wt.% NaCl solution.

Sample	R_p (kΩ)	I_{corr} (μA)	CR (mpy)
Base	35.87	2.724	1.168
FP-10	34.63	1.72	0.736
FP-15	220.1	0.185	0.0795
FP-25	2087	0.00844	0.00362

Results obtained from the different electrochemical techniques (LPR, EIS, and PDP) demonstrate a strong agreement. In addition, the base sample shows the highest corrosion susceptibility, while, after surface processing by employing FSP, a further increment in the corrosion resistance was observed on increasing the tool traverse speed. It should be emphasized that many studies have indicated a direct relationship between grain size, the passive film formed, and the corrosion resistance of aluminum-based alloys [62,63]. Due to the unique properties of grain boundaries compared to bulk materials, such as diffusion rates, atomic coordination, and reactivity, it is expected that a reduction in grain size will result in a substantial change in electrochemical behavior [63]. Additionally, breaking the intermetallic phases is another consequence of grain refinement [64]. These intermetallic compounds have a cathodic nature relative to the aluminum matrix and therefore their refinement and dissolution will remarkably improve the corrosion resistance. Ralston et al. [62] reported that a surface with a higher density of grain boundaries and finer grains is more likely to attract an intact passive protective film. Jilani et al. [65] have also studied the impact of grain refinement on the corrosion resistance of 1-XXX aluminum alloy. Their results confirmed that grain size reduction along with precipitate redistribution markedly improved the corrosion resistance by forming a continuous protective passive film. Likewise, the influence of severe plastic deformation of 2099 Al-Li on the formation of passive film was studied by Jinlong et al. [66]. The results illustrated that the corrosion susceptibility reduces due to refinement in both grain and precipitates, where grain refinement leads to electron work function decline and, thus, a thicker passive film.

Generally, various studies in the literature have shown that FSP resulted in an increase in the fraction of high-angle grain boundaries [67–69]. For instance, Li et al. [69] studied the impact of heat input during FSP on Mg-Li alloy on the fraction of high-angle grain boundaries. Their results indicated that the reduction in grain size and the increase in the fraction of high-angle grain boundaries are associated with lower heat input.

The consequence of this shift to a higher fraction of high-angle grain boundaries (HAGBs) on corrosion behavior was intensively investigated by Argade et al. [70], Dan et al. [71], and Rao et al. [60]. For example, Argade et al. [70] have concluded that the corrosion enhancement after friction stir processing of AA 5083 was attributed to grain refinement, which improved the polarization resistance, passivation, and pitting potential. Moreover, high-angle grain boundaries provided by the process raise the corrosion resistance by accelerating the passivation re-passivation phenomenon. Dan et al. [5] have studied the impact of grain refinement on the corrosion behavior of pure aluminum and have reported a significant enhancement in corrosion and pitting resistance for the sample with finer grain, which was attributed to a denser passive film. Further, it was also suggested that the oxide film favored a higher grain boundary, which protects the surface of the processed sample against chloride attack, thereby enhancing the corrosion resistance. Rao et al. [60] have demonstrated that the transformation from low-angle grain boundaries to high-angle counterparts after friction stir processing of Al–30Si alloy is a fundamental reason for the stability of the oxide film formed. Therefore, the results obtained in the current study may follow the same behavior, whereby employing FSP significantly diminishes grain size and raises the fraction of high-angle grain boundaries. Moreover, a further reduction in grain size was detected with increasing tool traveling speed, which resulted in superior corrosion resistance due to the more adhered and compacted passivation film.

3.4. SEM Images of the Corroded Samples

The morphology of the intense corrosion attack of the processed and base samples after 15 days of exposure to the electrolyte is depicted in Figure 11. For the base sample, a severe corrosion attack on the surface can be clearly noticed and indicated by corrosion products adhered to the surface. Additionally, within the corrosion products, large pits are observed. Surprisingly, microcracks can also be detected beneath the corrosion products, which may be ascribed to the initiation of intergranular corrosion. The corrosion marks of the friction-stir-processed samples demonstrated a different appearance, as shown in Figure 11b–d. The size and the density of pits became noticeably lower compared with that of the base sample. Moreover, the adsorption of corrosion products is hardly detected on samples processed at 100 and 250 mm/min, as shown in Figure 11b,d, which is attributed to a grain refinement that enhanced the passivation film formed. Interestingly, the adhesion of corrosion products can be seen on the surface of the FP-15 sample in Figure 11c; however, one with completely distinctive characteristics compared to the base counterpart which can be interpreted by the reduction in the outer film resistance of this sample (90 Ω). In addition, a dense layer of corrosion products may act as a barrier to impede a further corrosion attack.

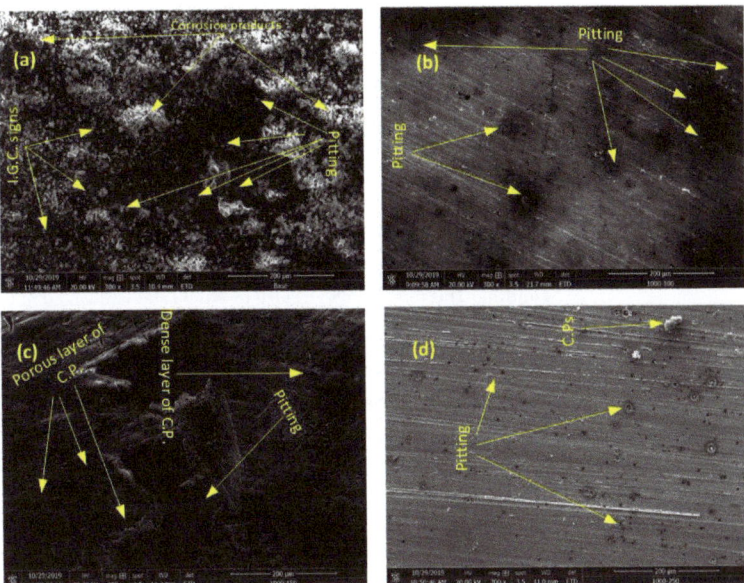

Figure 11. SEM images of the corroded surface of (**a**) base sample, (**b**) FP-10 sample, (**c**) FP-15 sample, and (**d**) FP-25 sample in 3.5 wt.% NaCl.

4. Conclusions

Friction stir processing using a pinless tool was successfully applied under different tool traverse speeds on structural aluminum alloy 6061. The results have shown a remarkable influence of FSP on the modified surface microstructure, mechanical properties, and corrosion resistance. From the current study, the following conclusions can be drawn:

1. It was found that friction stir processing resulted in considerable grain refinement of the base material of up to 81, 82.5, and 83.4 % reduction in the grain size at a traverse speed of 100, 150, and 250 mm/min, respectively. In addition, increasing the traverse speed from 100 to 250 mm/min has only a slight influence on the grain size refinement. Consequently, the microhardness was generally enhanced compared to the base material. However, increasing the traverse speed between 150 and 250 mm/min resulted in a reduction in the microhardness properties. This was attributed to a lower heat input at higher traverse speeds.
2. All the employed electrochemical techniques used in testing the corrosion resistance of all samples proved the achieving of an outstanding corrosion resistance behavior after deploying FSP on the 6061-aluminum base substrate, and an increase of the traverse speed caused a rapid improvement in the corrosion resistance.
3. Estimation of the passivation film thickness from the EIS measurements showed that the film thickness increased from 2.7 nm of the base materials to 7.6, 8.9, and 12.2 nm for traverse speeds of 100, 150, and 250 mm/min, respectively. This led to an improvement in the corrosion behavior of the samples by reducing the corrosion rate by 47.9, 96.6, and 99.8% for traverse speeds of 100, 150, and 250 mm/min, respectively, according to the potentiodynamic measurement.

Supplementary Materials: The following supporting information can be downloaded at: https://www.mdpi.com/article/10.3390/ma15228124/s1, Figure S1: Optical micrograph for FP-10 sample to demonstrate the method adopted to measure the mean grain size; Figure S2: Optical micrograph for FP-15 sample to demonstrate the method adopted to measure the mean grain size; Figure S3: Optical micrograph for FP-25 sample to demonstrate the method adopted to measure the mean grain size;

Figure S4: Optical micrograph for base sample to demonstrate the method adopted to measure the mean grain size; Table S1: Optical micrograph for FP-10 sample to demonstrate the method adopted to measure the mean grain size; Table S2: Typical grain morphology information for FP-15 sample drawn from the Olympus stream analysis software; Table S3: Typical grain morphology information for FP-25 sample drawn from the Olympus stream analysis software; Table S4: Typical grain morphology information for the base sample drawn from the Olympus stream analysis software.

Author Contributions: F.A.A.-B.: Conceptualization, Methodology, Supervision. I.H.Z.: Investigation Data curation, Writing—original draft. R.K.S.: Data curation, Validation, A.Y.A.: Data curation, Visualization, Validation, N.M.: Writing—review & editing. F.A.G.: Visualization, Writing—review & editing. All authors have read and agreed to the published version of the manuscript.

Funding: This research was funded by King Fahd University of Petroleum and Minerals (KFUPM) Project No. DF201013.

Institutional Review Board Statement: Not applicable.

Informed Consent Statement: Not applicable.

Data Availability Statement: The raw/processed data required to reproduce these findings cannot be shared at this time as the data also forms part of an ongoing study.

Acknowledgments: The authors thank King Fahd University of Petroleum and Minerals (KFUPM) for providing all support. The financial support by the deanship of scientific research at KFUPM through project no. DF201013 is gratefully acknowledged.

Conflicts of Interest: The authors declare no conflict of interest.

References

1. Davis, J.R. *Corrosion of Aluminum and Aluminum Alloys*; ASM International: Cleveland, OH, USA, 2000. [CrossRef]
2. Kheirkhah, S.; Imani, M.; Aliramezani, R.; Zamani, M.H.; Kheilnejad, A. Microstructure, mechanical properties and corrosion resistance of Al6061/BN surface composite prepared by friction stir processing. *Surf. Topogr. Metrol. Prop.* **2019**, *7*, 035002. [CrossRef]
3. El-Garaihy, W.H.; Fouad, D.M.; Salem, H.G. Multi-channel Spiral Twist Extrusion (MCSTE): A Novel Severe Plastic Deformation Technique for Grain Refinement. *Metall. Mater. Trans. A* **2018**, *49*, 2854–2864. [CrossRef]
4. Orłowska, M.; Ura-Bińczyk, E.; Olejnik, L.; Lewandowska, M. The effect of grain size and grain boundary misorientation on the corrosion resistance of commercially pure aluminium. *Corros. Sci.* **2018**, *148*, 57–70. [CrossRef]
5. Ma, Z.Y. Friction stir processing technology: A review. *Metall. Mater. Trans. A* **2008**, *39*, 642–658. [CrossRef]
6. Faraji, G.; Kim, H.S.; Kashi, H.T. *Severe Plastic Deformation: Methods, Processing and Properties*; Elsevier: Amsterdam, The Netherlands, 2018.
7. Mishra, R.S.; Ma, Z.Y.; Charit, I. Friction stir processing: A novel technique for fabrication of surface composite. *Mater. Sci. Eng. A* **2003**, *341*, 307–310. [CrossRef]
8. Naeini, M.F.; Shariat, M.H.; Eizadjou, M. On the chloride-induced pitting of ultra fine grains 5052 aluminum alloy produced by accumulative roll bonding process. *J. Alloys Compd.* **2011**, *509*, 4696–4700. [CrossRef]
9. Surekha, K.; Murty, B.S.; Rao, K.P. Effect of processing parameters on the corrosion behaviour of friction stir processed AA 2219 aluminum alloy. *Solid State Sci.* **2009**, *11*, 907–917. [CrossRef]
10. Surekha, K.; Murty, B.S.; Rao, K.P. Microstructural characterization and corrosion behavior of multipass friction stir processed AA2219 aluminium alloy. *Surf. Coat. Technol.* **2008**, *202*, 4057–4068. [CrossRef]
11. Eldeeb, M.; Khorshed, L.; Abdallah, S.; Gaafer, A.; Mahmoud, T. Effect of friction stir welding process parameters and post-weld heat treatment on the corrosion behaviour of AA6061-O aluminum alloys. *Egy. J. Chem.* **2019**, *62*, 1367–1375. [CrossRef]
12. Esmaily, M.; Mortazavi, N.; Osikowicz, W.; Hindsefelt, H.; Svensson, J.E.; Halvarsson, M.; Thompson, G.E.; Johansson, L.G. Influence of multi-pass friction stir processing on the corrosion behavior of an Al-Mg-Si alloy. *J. Electrochem. Soc.* **2016**, *163*, C124. [CrossRef]
13. Reddy, G.M.; Rao, K.S. Enhancement of wear and corrosion resistance of cast A356 aluminium alloy using friction stir processing. *Trans. Indian Inst. Met.* **2010**, *63*, 793–798. [CrossRef]
14. Moshwan, R.; Yusof, F.; Hassan, M.A.; Rahmat, S.M. Effect of tool rotational speed on force generation, microstructure and mechanical properties of friction stir welded Al-Mg-Cr-Mn (AA 5052-O) alloy. *Mater. Des.* **2015**, *66*, 118–128. [CrossRef]
15. Zhang, H.J.; Sun, S.L.; Liu, H.J.; Zhu, Z.; Wang, Y.L. Characteristic and mechanism of nugget performance evolution with rotation speed for high-rotation-speed friction stir welded 6061 aluminum alloy. *J. Manuf. Process.* **2020**, *60*, 544–552. [CrossRef]
16. Golezani, A.S.; Barenji, R.V.; Heidarzadeh, A.; Pouraliakbar, H. Elucidating of tool rotational speed in friction stir welding of 7020-T6 aluminum alloy. *Int. J. Adv. Manuf. Technol.* **2015**, *81*, 1155–1164. [CrossRef]

17. Sudhagar, S.; Sakthivel, M.; Mathew, P.J.; Daniel, S.A.A. A multi criteria decision making approach for process improvement in friction stir welding of aluminium alloy. *Measurement* **2017**, *108*, 1–8. [CrossRef]
18. Kadaganchi, R.; Gankidi, M.R.; Gokhale, H. Optimization of process parameters of aluminum alloy AA 2014-T6 friction stir welds by response surface methodology. *Def. Technol.* **2015**, *11*, 209–219. [CrossRef]
19. Banik, A.; Roy, B.S.; Barma, J.D.; Saha, S.C. An experimental investigation of torque and force generation for varying tool tilt angles and their effects on microstructure and mechanical properties: Friction stir welding of AA 6061-T6. *J. Manuf. Process.* **2018**, *31*, 395–404. [CrossRef]
20. *ASTM E112-13*; Standard Test Methods for Determining Average Grain Size. ASTM International: West Conshohocken, PA, USA, 2013. [CrossRef]
21. Kumar, R.; Singh, K.; Pandey, S. Process forces and heat input as function of process parameters in AA5083 friction stir welds. *Trans. Nonferrous Met. Soc. China* **2012**, *22*, 288–298. [CrossRef]
22. Forcellese, A.; Martarelli, M.; Simoncini, M. Effect of process parameters on vertical forces and temperatures developed during friction stir welding of magnesium alloys. *Int. J. Adv. Manuf. Technol.* **2016**, *85*, 595–604. [CrossRef]
23. Peel, M.J.; Steuwer, A.; Withers, P.; Dickerson, T.; Shi, Q.; Shercliff, H. Dissimilar friction stir welds in AA5083-AA6082. Part I: Process parameter effects on thermal history and weld properties. *Met. Mater. Trans. A* **2006**, *37*, 2183–2193. [CrossRef]
24. Cui, S.; Chen, Z.; Robson, J. A model relating tool torque and its associated power and specific energy to rotation and forward speeds during friction stir welding/processing. *Int. J. Mach. Tools Manuf.* **2010**, *50*, 1023–1030. [CrossRef]
25. Al-Badour, F.A.; Adesina, A.Y.; Ibrahim, A.B.; Suleiman, R.K.; Merah, N.; Sorour, A.A. Electrochemical Investigation of the Effect of Process Parameters on the Corrosion Behavior of Aluminum-Cladded Pressure Vessel Steel Using a Friction Stir Diffusion Cladding Process. *Metals* **2020**, *10*, 623. [CrossRef]
26. Azizieh, M.; Kokabi, A.; Abachi, P. Effect of rotational speed and probe profile on microstructure and hardness of AZ31/Al2O3 nanocomposites fabricated by friction stir processing. *Mater. Des.* **2011**, *32*, 2034–2041. [CrossRef]
27. Moghaddas, M.A.; Kashani-Bozorg, S.F. Effects of thermal conditions on microstructure in nanocomposite of Al/Si3N4 produced by friction stir processing. *Mater. Sci. Eng. A* **2013**, *559*, 187–193. [CrossRef]
28. Heidarzadeh, A.; Mironov, S.; Kaibyshev, R.; Çam, G.; Simar, A.; Gerlich, A.; Khodabakhshi, F.; Mostafaei, A.; Field, D.; Robson, J.; et al. Friction stir welding/processing of metals and alloys: A comprehensive review on microstructural evolution. *Prog. Mater. Sci.* **2020**, *117*, 100752. [CrossRef]
29. Nandan, R.; Debroy, T.; Bhadeshia, H. Recent advances in friction-stir welding—Process, weldment structure and properties. *Prog. Mater. Sci.* **2008**, *53*, 980–1023. [CrossRef]
30. Meng, X.; Huang, Y.; Cao, J.; Shen, J.; dos Santos, J.F. Recent progress on control strategies for inherent issues in friction stir welding. *Prog. Mater. Sci.* **2020**, *115*, 100706. [CrossRef]
31. Patle, H.; Dumpala, R.; Sunil, B.R. Machining Characteristics and Corrosion Behavior of Grain Refined AZ91 Mg Alloy Produced by Friction Stir Processing: Role of Tool Pin Profile. *Trans. Indian Inst. Met.* **2017**, *71*, 951–959. [CrossRef]
32. Pezeshkian, M.; Ebrahimzadeh, I.; Gharavi, F. Fabrication of Cu Surface Composite Reinforced by Ni Particles Via Friction Stir Processing: Microstructure and Tribology Behaviors. *J. Tribol.* **2017**, *140*, 011607. [CrossRef]
33. Sharma, C.; Dwivedi, D.K.; Kumar, P. Effect of welding parameters on microstructure and mechanical properties of friction stir welded joints of AA7039 aluminum alloy. *Mater. Des.* **2012**, *36*, 379–390. [CrossRef]
34. Chen, Y.; Ding, H.; Li, J.-Z.; Zhao, J.-W.; Fu, M.-J.; Li, X.-H. Effect of welding heat input and post-welded heat treatment on hardness of stir zone for friction stir-welded 2024-T3 aluminum alloy. *Trans. Nonferrous Met. Soc. China* **2015**, *25*, 2524–2532. [CrossRef]
35. Hansen, N. Hall−Petch relation and boundary strengthening. *Scr. Mater.* **2004**, *51*, 801–806. [CrossRef]
36. Maurya, R.; Kumar, B.; Ariharan, S.; Ramkumar, J.; Balani, K. Effect of carbonaceous reinforcements on the mechanical and tribological properties of friction stir processed Al6061 alloy. *Mater. Des.* **2016**, *98*, 155–166. [CrossRef]
37. Macdonald, D.D. Review of mechanistic analysis by electrochemical impedance spectroscopy. *Electrochim. Acta* **1990**, *35*, 1509–1525. [CrossRef]
38. McCafferty, E. *Introduction to Corrosion Science*; Springer Science and Business Media LLC: Berlin/Heidelberg, Germany, 2010. [CrossRef]
39. Prakashaiah, B.; Kumara, D.V.; Pandith, A.A.; Shetty, A.N.; Rani, B.A. Corrosion inhibition of 2024-T3 aluminum alloy in 3.5% NaCl by thiosemicarbazone derivatives. *Corros. Sci.* **2018**, *136*, 326–338. [CrossRef]
40. Scully, J.R.; Silverman, D.C.; Kendig, M.W. *Electrochemical Impedance: Analysis and Interpretation*; ASTM International: West Conshohocken, PA, USA, 1993. [CrossRef]
41. Chauhan, D.S.; Ansari, K.; Sorour, A.; Quraishi, M.; Lgaz, H.; Salghi, R. Thiosemicarbazide and thiocarbohydrazide functionalized chitosan as ecofriendly corrosion inhibitors for carbon steel in hydrochloric acid solution. *Int. J. Biol. Macromol.* **2018**, *107*, 1747–1757. [CrossRef]
42. Jinlong, L.; Hongyun, L.; Jinpeng, X. Experimental study of corrosion behavior for burnished aluminum alloy by EWF, EBSD, EIS and Raman spectra. *Appl. Surf. Sci.* **2013**, *273*, 192–198. [CrossRef]
43. Zheng, Z.; Gao, Y.; Gui, Y.; Zhu, M. Corrosion behaviour of nanocrystalline 304 stainless steel prepared by equal channel angular pressing. *Corros. Sci.* **2012**, *54*, 60–67. [CrossRef]

44. Trdan, U.; Grum, J. Evaluation of corrosion resistance of AA6082-T651 aluminium alloy after laser shock peening by means of cyclic polarisation and EIS methods. *Corros. Sci.* **2012**, *59*, 324–333. [CrossRef]
45. Heakal, F.E.-T.; Tantawy, N.; Shehta, O. Influence of chloride ion concentration on the corrosion behavior of Al-bearing TRIP steels. *Mater. Chem. Phys.* **2011**, *130*, 743–749. [CrossRef]
46. Yadav, M.; Kumar, S.; Sinha, R.; Bahadur, I.; Ebenso, E. New pyrimidine derivatives as efficient organic inhibitors on mild steel corrosion in acidic medium: Electrochemical, SEM, EDX, AFM and DFT studies. *J. Mol. Liq.* **2015**, *211*, 135–145. [CrossRef]
47. Mehrian, S.M.; Rahsepar, M.; Khodabakhshi, F.; Gerlich, A. Effects of friction stir processing on the microstructure, mechanical and corrosion behaviors of an aluminum-magnesium alloy. *Surf. Coat. Technol.* **2020**, *405*, 126647. [CrossRef]
48. Chen, M.-A.; Ou, Y.-C.; Fu, Y.-H.; Li, Z.-H.; Li, J.-M.; Liu, S.-D. Effect of friction stirred Al-Fe-Si particles in 6061 aluminum alloy on structure and corrosion performance of MAO coating. *Surf. Coat. Technol.* **2016**, *304*, 85–97. [CrossRef]
49. Milošev, I.; Kosec, T.; Strehblow, H.-H. XPS and EIS study of the passive film formed on orthopaedic Ti–6Al–7Nb alloy in Hank's physiological solution. *Electrochim. Acta* **2008**, *53*, 3547–3558. [CrossRef]
50. Hong, T.; Sun, Y.; Jepson, W. Study on corrosion inhibitor in large pipelines under multiphase flow using EIS. *Corros. Sci.* **2002**, *44*, 101–112. [CrossRef]
51. Pandey, V.; Singh, J.; Chattopadhyay, K.; Srinivas, N.S.; Singh, V. Optimization of USSP duration for enhanced corrosion resistance of AA7075. *Ultrasonics* **2018**, *91*, 180–192. [CrossRef]
52. Hsu, C.H.; Mansfeld, F. Technical Note: Concerning the Conversion of the Constant Phase Element Parameter Yo into a Capacitance. *Corrosion* **2001**, *57*, NACE-01090747. [CrossRef]
53. Ansari, K.; Quraishi, M.; Singh, A. Schiff's base of pyridyl substituted triazoles as new and effective corrosion inhibitors for mild steel in hydrochloric acid solution. *Corros. Sci.* **2014**, *79*, 5–15. [CrossRef]
54. Acuña, R.; Abreu, C.M.; Cristóbal, M.J.; Cabeza, M.; Nóvoa, X.R. Electrochemical study of the surface metal matrix composite developed on AA 2024-T351 by the friction stir process. *Corros. Eng. Sci. Technol.* **2019**, *54*, 715–725. [CrossRef]
55. Guitián, B.; Nóvoa, X.; Puga, B. Electrochemical Impedance Spectroscopy as a tool for materials selection: Water for haemodialysis. *Electrochim. Acta* **2011**, *56*, 7772–7779. [CrossRef]
56. Gray, J.; Orme, C. Electrochemical impedance spectroscopy study of the passive films of alloy 22 in low pH nitrate and chloride environments. *Electrochim. Acta* **2007**, *52*, 2370–2375. [CrossRef]
57. Liu, Y.; Meng, G.; Cheng, Y. Electronic structure and pitting behavior of 3003 aluminum alloy passivated under various conditions. *Electrochim. Acta* **2009**, *54*, 4155–4163. [CrossRef]
58. Navaser, M.; Atapour, M. Effect of Friction Stir Processing on Pitting Corrosion and Intergranular Attack of 7075 Aluminum Alloy. *J. Mater. Sci. Technol.* **2017**, *33*, 155–165. [CrossRef]
59. Viceré, A.; Roventi, G.; Paoletti, C.; Cabibbo, M.; Bellezze, T. Corrosion Behavior of AA6012 Aluminum Alloy Processed by ECAP and Cryogenic Treatment. *Metals* **2019**, *9*, 408. [CrossRef]
60. Rao, D.; Katkar, V.; Gunasekaran, G.; Deshmukh, V.; Prabhu, N.; Kashyap, B. Effect of multipass friction stir processing on corrosion resistance of hypereutectic Al-30Si alloy. *Corros. Sci.* **2014**, *83*, 198–208. [CrossRef]
61. Adesina, A.Y.; Gasem, Z.M.; Madhan Kumar, A. Corrosion Resistance Behavior of Single-Layer Cathodic Arc PVD Nitride-Base Coatings in 1 M HCl and 3.5 pct NaCl Solutions. *Metall. Mater. Trans. B* **2017**, *48*, 1321–1332. [CrossRef]
62. Ralston, K.D.; Birbilis, N.; Davies, C.H.J. Revealing the relationship between grain size and corrosion rate of metals. *Scr. Mater.* **2010**, *63*, 1201–1204. [CrossRef]
63. Ralston, K.D.; Birbilis, N. Effect of Grain Size on Corrosion: A Review. *Corrosion* **2010**, *66*, 075005. [CrossRef]
64. Chung, M.-K.; Choi, Y.-S.; Kim, J.-G.; Kim, Y.-M.; Lee, J.-C. Effect of the number of ECAP pass time on the electrochemical properties of 1050 Al alloys. *Mater. Sci. Eng. A* **2003**, *366*, 282–291. [CrossRef]
65. Jilani, O.; Njah, N.; Ponthiaux, P. Corrosion properties of anodized aluminum: Effects of equal channel angular pressing prior to anodization. *Corros. Sci.* **2014**, *89*, 163–170. [CrossRef]
66. Jinlong, L.; Tongxiang, L.; Chen, W.; Ting, G. The passive film characteristics of several plastic deformation 2099 Al–Li alloy. *J. Alloys Compd.* **2016**, *662*, 143–149. [CrossRef]
67. Vysotskiy, I.; Zhemchuzhnikova, D.; Malopheyev, S.; Mironov, S.; Kaibyshev, R. Microstructure evolution and strengthening mechanisms in friction-stir welded Al–Mg–Sc alloy. *Mater. Sci. Eng. A* **2019**, *770*, 138540. [CrossRef]
68. Charit, I.; Mishra, R. High strain rate superplasticity in a commercial 2024 Al alloy via friction stir processing. *Mater. Sci. Eng. A* **2003**, *359*, 290–296. [CrossRef]
69. Li, Y.; Guan, Y.; Liu, Y.; Zhai, J.; Hu, K.; Lin, J. Effect of processing parameters on the microstructure and tensile properties of a dual-phase Mg–Li alloy during friction stir processing. *J. Mater. Res. Technol.* **2022**, *17*, 2714–2724. [CrossRef]
70. Argade, G.; Kumar, N.; Mishra, R. Stress corrosion cracking susceptibility of ultrafine grained Al–Mg–Sc alloy. *Mater. Sci. Eng. A* **2012**, *565*, 80–89. [CrossRef]
71. Song, D.; Ma, A.-B.; Jiang, J.-H.; Lin, P.-H.; Yang, D.-H. Corrosion behavior of ultra-fine grained industrial pure Al fabricated by ECAP. *Trans. Nonferrous Met. Soc. China* **2009**, *19*, 1065–1070. [CrossRef]

Article

Preparation of One-Dimensional Polyaniline Nanotubes as Anticorrosion Coatings

Guangyuan Yang [1], Fuwei Liu [2], Ning Hou [1], Sanwen Peng [1,*], Chunqing He [2] and Pengfei Fang [2,*]

[1] China Tobacco Hubei Industrial Cigarette Materials, LLC, Wuhan 430051, China; yangguangyuan@whut.edu.cn (G.Y.); houning@hjl.hbtobacco.cn (N.H.)
[2] Key Laboratory of Artificial Micro- and Nano-Structures of Ministry of Education, Department of Physics, Wuhan University, Wuhan 430072, China; liufuwei168@163.com (F.L.); hecq@whu.edu.cn (C.H.)
* Correspondence: pengsanwen@126.com (S.P.); fangpf@whu.edu.cn (P.F.)

Abstract: Uniform polyaniline (PANI) nanotubes were synthesized by a self-assembly method under relatively dilute hydrochloric acid (HCl) solution. Scanning electron microscopy (SEM), Fourier transform infrared (FTIR) spectroscopy, and UV-Vis-NIR spectroscopy were employed to characterize the morphology and molecular structure of the PANI products. SEM images show that the PANI nanotubes have uniform morphology and form compact coating on the substrate surface. For comparison, aggregated PANI was also synthesized by conventional polymerization method. The performance of the PANI products on carbon steel was studied using eletrochemical measurement and immersion corrosion experiment in 3.5 wt% NaCl aqueous solution. The corrosion potentials of carbon steel samples increase by 0.196 V and 0.060 V after coated with PANI nanotubes and aggregated PANI, respectively, and the corrosion currents density decrease by about 76.32% and 36.64%, respectively. The 6-day immersion experiment showed that the carbon steel samples coated by PANI nanotubes showed more excellent anticorrosion performance, because the more compact coating formed by PANI nanotubes may inhibit the corrosion process between the anodic and cathodic.

Keywords: polyaniline; nanotubes; one-dimensional; electrochemical performance

1. Introduction

Conducting polymers have been studied for their excellent protective ability for metals in the last few years. Polyaniline is generally considered to be one of the best candidates for an excellent anticorrosion film, because of its ease of synthesis [1–5], interesting redox properties [6,7], and good environment stability [8,9]. Although enormous efforts have been devoted to the study of electrochemical polymerization [10–13], it is not easy to make large polyaniline films in this method and some corrosion-susceptive metals are often oxidized or dissolved in the potential domain of the polyaniline electro-polymerization. Another strategy explored was the use of the chemical polymerization method. However, most of the products in conventional polymerization method are aggregated PANI with poor solubility in conventional solvents, which is unfavorable to the formation of compact films.

In order to obtain large-area and compact coatings, specific PANI nanostructures synthesized by chemical polymerization may be an effective method to improve the solubility and dispersion stability [14–16]. It is generally considered that the compact PANI film itself can generate compact passive layers on the metal surface and inhibit the corrosion process between anodic and cathodic [17–20]. Therefore, some researchers have tried to explore novel PANI nanostructures, especially one-dimension nanostructures, for anticorrosion application. Yao et al. [15] synthesized polyaniline nanofibers with good solubility and dispersion stability in ethanol, and applied them to carbon steel surfaces as a suspension in ethanol and investigated the anticorrosion performance. The results showed that PANI nanofibers bring a better passive effect to carbon steel and provide more excellent corrosion protection than aggregated PANI. Yang et al. [16] synthesized polyaniline nanostructures in sulfuric acid solution using three different polymerization methods. They found that nanofibers synthesized by the direct mixed reaction method have highly uniform morphology which may improve the solubility and the formation of the compact coating thus effectively enhancing the corrosion protection property. The polyaniline nanotube is another one-dimension nanostructure which has attracted a great deal of attention because of its potentially interesting electrical and optical properties [21–23]. Moreover, PANI nanotubes with uniform morphology may have good solubility and dispersion stability which is necessary in the formation of compact coatings. On the other hand, it was found that pure PANI coating film had a good protective effect, but its corrosion protection was short term due to its poor barrier to water. It is necessary to develop a blended coating containing both conventional resin and PANI [19]. Although there are several previous reports of PANI nanotubes used as anticorrosion coating, the preparation of PANI nanotubes and design of high-performance anticorrosion PANI nanotube/polymer coatings are still a great challenge.

In this study, polyaniline nanotubes were prepared under a relatively dilute HCl solution. The obtained polyaniline products were not only directly applied on carbon steel surface but also as fillers dispersed in alkyd resin to investigate their anticorrosion performance. The results show that PANI nanotubes have more excellent protective properties than aggregated PANI. The protection mechanism was also discussed.

2. Materials and Methods

2.1. Materials

Aniline (analytical grade) was obtained from Tianjin Tianli Chemical Reagent Co. Ltd. and purified by distillation prior to use. Ammonium persulfate (APS), N-methyl-2-pyrrolidone (NMP), hydrochloric acid, and methanol were analytical grade and used without further purification. The alkyd resin we used was 3139 alkyd resin of Hubei Wuhan Lion Rock brand (viscosity: $\geq 300/25\ °C$, solid: $65 \pm 2\%$, acid value: ≤ 10 mgkoh/g). All aqueous solutions were prepared with deionized water by an ultrapure water treatment system.

2.2. Preparations

PANI nanotubes were prepared by a self-assembly method under relatively dilute HCl solution. One milliliter of aniline was dissolved in 10 mL of 1 mol/L HCl and 2.5 g APS was dissolved in 40 mL deionized water. The two solutions were then mixed and immediately shaken well for approximately 5 min. The reaction was carried out at room temperature for an extra 24 h. The reaction products were washed with deionized water and centrifuged to separate PANI. Separated PANI was then dispersed in deionized water and centrifuged. This treatment was repeated several times until the suspension reached a neutral pH and became colorless. The resulting polyaniline precipitate was centrifuged and repeatedly washed using methanol to remove the oligomer and finally dried in an oven at about 60 °C for 24 h.

In the conventional polymerization method, 1 mL of aniline was added to 10 mL of 1 mol/L HCl solution, and the mixed solution was transferred to an ice bath environment. Then 40 mL pre-cooled aqueous solution of APS (0.25 mol/L) was added dropwise to the pre-cooled aniline-acid mixed solution with constant stirring. The reaction was conducted at 5 ± 1 °C. After the addition, the stirring was continued for 1 h for ensuring complete polymerization. Then the reaction products were purified using deionized water and methanol. The operation is similar to the subsequent treatment of PANI nanotubes.

The rectangular (28 mm × 24 mm × 1 mm) carbon steel (the content of carbon is 0.22–0.45%) samples were polished by emery paper 1000 grit, and all steel working samples were treated in acetone and ethanol solution to degrease prior to coating. A certain quality of aggregated PANI or PANI nanotubes was dissolved in NMP to obtain a 10% suspension respectively.

The obtained PANI products were also used for the preparation of the aggregated PANI/alkyd and PANI nanotube/alkyd coatings. The coating systems were prepared separately by dispersing 2.0, 5.0, and 8.0 wt% of polymers (aggregated PANI and PANI nanotubes) in 20 wt% solution of alkyd in xylene. Additionally, the alkyd without PANI was prepared as well.

Both of the PANI and PANI/alkyd coatings were allowed to dry in air at room temperature for 144 h, then the coated samples were encapsulated by paraffin and a constant area of about 1 cm^2 was left for corrosion test.

2.3. Characterization

The thicknesses of the PANI alone and PANI/alkyd coatings were measured using a magnetic thickness gauge (QCC-A, Jiangdu Pearl Experimental Machinery Factory, Yangzhou, China). Morphologies of the PANI products and the coatings were investigated by a Sirion field-emission scanning electron microscopy (Hitachi S-4800, FEI Company, Hillsboro, OR, USA). The magnifications of PANI nanotubes and aggregated PANI in SEM (Figure 1) are both 50,000, and the magnifications of the PANI nanotubes and aggregated PANI coatings (Figure 2) are both 3000. The molecular structures of polyaniline nanostructure were studied by an attenuated total reflectance Fourier transform infrared spectrometer (ATR-FTIR, Nicolet iS10, Thermo Electron Corporation, Waltham, MA, USA) and a Cary 5000 UV-Vis-NIR spectrometer (Agilent Technologies, Ltd, Santa Clara, CA, USA). In ATR the frequency range was from 550 to 4000 cm^{-1}, number of scans was 32, and the resolution used was 6 with dataspacing 1.929 cm^{-1}. The electrochemical corrosion measurements were performed on a potentiostat/galvanostat (CHI 600C, Chenhua Company, Shanghai, China) in a three-electrode electrochemical cell using carbon steel samples as working electrodes. A Pt sheet was used as a counter electrode and all potentials were referred to the saturated calomel electrode (SCE). In Tafel plot, the scanning rate was 10 mV/s. The electrochemical impedance spectroscopy (EIS) measurements were taken in the frequency range of 100 k to 0.01 Hz, and the amplitude of the sinusoidal voltage signal was 10 mV. The EIS data were analyzed and fitted with ZSimpWin software. In both Tafel and EIS experiments, the corrosion environment was 3.5 wt% NaCl water solutions at room temperature.

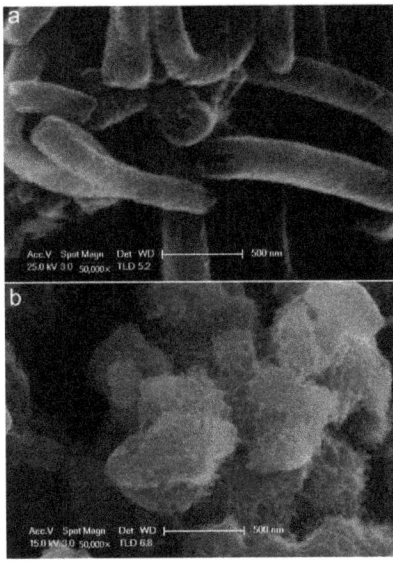

Figure 1. SEM images of PANI nanotubes (**a**), and aggregated (**b**).

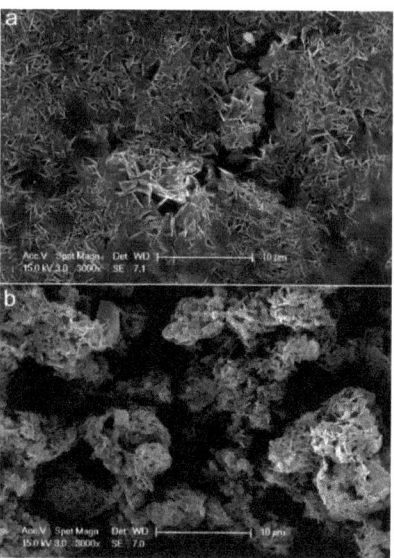

Figure 2. SEM images of the coatings of PANI nanotubes (**a**), and aggregated PANI (**b**).

3. Results and Discussion

Figure 1 shows the SEM images of PANI nanotubes and aggregated PANI. It is found that the PANI nanotubes have nearly uniform outer diameters of 180–250 nm, wall thicknesses of 40–80 nm, and lengths varying from 500 to 2000 nm (Figure 1a). However, the PANI synthesized by conventional polymerization method are irregularly shaped agglomerates containing varies of particulates (Figure 1b). In our experiments it is evident that a relatively dilute acidic condition was obtained when the APS solution was mixed with the aniline–acid solution directly, while a higher acidity was kept when the APS solution was added dropwise. The PANI nanotubes are proposed to produce within an intermediate acidity interval, and granular PANI are obtained at higher acidity [24]. Polymerization starting in mildly acidic conditions results in aniline oligomers, which are insoluble in water [25,26]. These aniline oligomers may be made of phenazine-like moieties oxidized from ortho-coupled aniline. They aggregate to constitute a template-like structure, which further dictates the directional growth of PANI, namely the production of PANI nanotubes. In addition, the final nanotubular productions are generated through the self-organized phenazine units and the stacking of aniline oligomers [24,25,27–32].

The thicknesses of aggregated PANI and PANI nanotubes coatings were measured at 8 ± 0.3 μm. The SEM morphologies of polyaniline coatings formed using PANI nanotubes and aggregated PANI are presented in Figure 2. It can be seen that the coating (Figure 2b) formed by aggregated PANI shows agglomerates and some cracks on the coating. In the presence of PANI nanotubes, the image (Figure 2a) gave the formation of a more uniform and compact coating. These different macroscopic properties of the polyaniline coatings could directly affect anticorrosion performance.

The FTIR spectra of the PANI nanotubes and aggregated PANI are given in Figure 3. The PANI nanotubes and aggregated PANI have similar FTIR spectra. The peaks at 1570 and 1487 cm^{-1} are the stretching mode of C=N and C=C for the quinoid and benzenoid rings. The peaks at 1294 and 1244 cm^{-1} are assigned to the C−N stretching mode of benzenoid ring, while the peaks at 1130 and 800 cm^{-1} are attributed to the aromatic C−H in-plane bending and the out-of-plane deformation of C−H in the 1,4-disubstituted benzene ring, respectively [33]. These values are similar to the previously reported infrared spectra for polyaniline systems [34,35]. Figure 4 shows the UV-Vis-NIR spectra of the polyaniline structures. Two distinctive absorption bands at 374.5 and 468 nm can be seen in the UV-Vis-

NIR spectra. The former is associated with the π → π* transition and the latter is caused by polaron band → π* transition [33,36]. FTIR and UV-Vis-NIR results show that the states of the nanostructures should be doped polyaniline in their emeraldine salt forms [16].

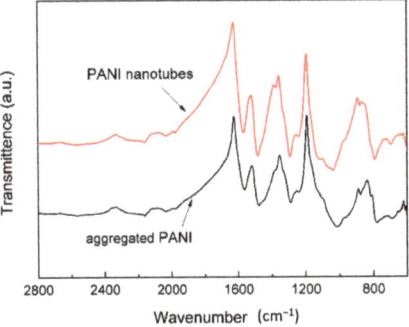

Figure 3. FT−IR spectra of PANI nanotubes and aggregated PANI.

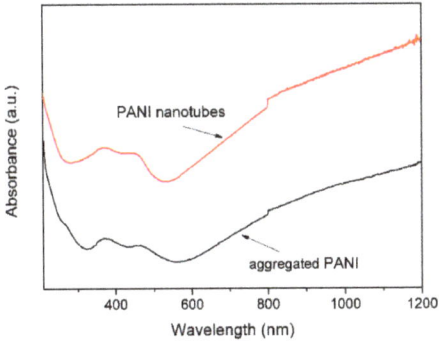

Figure 4. UV-Vis-NIR spectra of PANI nanotubes and aggregated PANI.

The potentiodynamic polarization curves for uncoated carbon steel (CS), carbon steel covered with aggregated PANI (CS-A-PANI), and carbon steel covered with PANI nanotubes (CS-N-PANI) in 3.5 wt% NaCl solution are shown in Figure 5. The values of corrosion potential (E_{corr}) and corrosion currents density (I_{corr}) obtained from the potentiodynamic polarization curves are presented in Table 1. Compared with uncoated carbon steel, the E_{corr} of CS-A-PANI and CS-N-PANI increased about 0.060 V and 0.196 V, while the I_{corr} of CS-A-PANI and CS-N-PANI decreased about 36.64% and 76.32%, respectively. When the carbon steel covered with aggregated PANI and nanotube PANI, the value of cathodic Tafel constants decrease from 95.1 mV/dec to 85.3 and 71 mV/dec, simultaneously the value of anodic Tafel constants decrease from 85.5 mV/dec to 70.9 and 66.5 mV/dec. These results indicate that PANI nanotubes have better corrosion protection to carbon steel than aggregated PANI. This is possibly due to PANI nanotubes possess favorable adhesion to carbon steel and guarantee the generation of the passive layer on the carbon steel surface [37].

Table 1. Tafel curves parameters for uncoated CS, CS-A-PANI, and CS-N-PANI in 3.5 wt% NaCl solution.

Electrodes	I_{corr} (µA cm^{-2})	E_{corr} (V)	b_a (mV/dec)	b_c (mV/dec)
Uncoated CS	9.514	−0.541	85.5	95.1
CS-A-PANI	6.028	−0.481	70.9	85.3
CS-N-PANI	2.253	−0.345	66.5	71

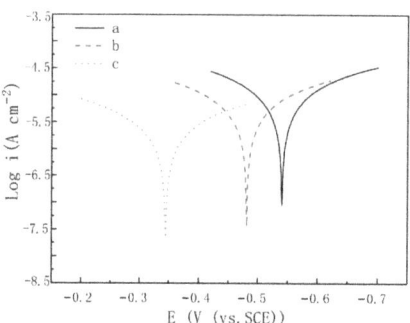

Figure 5. Tafel curves for uncoated CS (a), CS−A−PANI (b), and CS−N−PANI (c) in 3.5 wt% NaCl aqueous solution.

To further study the corrosion protection performance of PANI nanostructures, impedance spectra were employed to investigate the anticorrosion properties of the obtained PANI/alkyd coating system during immersion in 3.5 wt% NaCl solution. Figure 6 shows the impedance spectra of the alkyd coating containing aggregated PANI (the filler content is 0, 2, and 5 wt% and the obtained samples are abbreviated as Alkyd, Alkyd−A2, and Alkyd−A5, respectively). The obtained EIS spectra were fitted by two different equivalent circuits (depicted in Figure 6), in which R_s, R_c, and R_{ct} are the electrolyte solution resistance, coating resistance, and charge transfer resistance, respectively, while Q_c and Q_{dl} represent constant phase elements (CPE) associated with coating capacitance and double layer capacitance, respectively. As can be seen from Figure 6a, the impedance modulus of pure alkyd coating decreased with immersion time. After only two days of immersion, the impedance modulus at 0.01 Hz became less than 10^7 $\Omega \cdot cm^2$. Meanwhile, the second time constant can be also observed, which indicates that the electrochemical reactions at coating/metal interface took place. For the coating incorporated with PANI, it can be clearly seen (Figure 6b) that the addition of a small amount (2 wt%) of aggregated PANI significantly enhanced the anticorrosion properties of the alkyd coating, after 30-day immersion the impedance modulus also remains nearly 10^8 $\Omega \cdot cm^2$. However, with increasing aggregated PANI content (5 wt%), a dramatic decrease in impedance modulus is observed (Figure 6c). After 30 days of immersion, the time constant associated with the corrosion reactions at coating/metal interface also appears. Generally, the pure alkyd coating sample is partially heterogeneous and has lots of micro-defects, which affect the barrier properties of the polymer coating. The addition of a small amount of aggregated PANI may block the defects in some extent and thus form more compact coating structure, whereas an excess of aggregated PANI may agglomerate and affect the homogeneity of the coating, which result in the generation of some new micro-paths for corrosive media transportation [19].

For PANI nanotube/alkyd coating system (the samples containing 2, 5, and 8 wt% of PANI nanotubes are abbreviated as Alkyd−T2, Alkyd−T5, and Alkyd−T8, respectively), the variation of impedance spectra (as shown in Figure 7) shows a similar trend with that of aggregated PANI/alkyd coatings. The difference is that with the increase in nanotube content, the anticorrosion property does not decrease until the filler content is higher than 5 wt%. Additionally, after 30-day immersion, all the PANI nanotube/alkyd coatings remain relatively high impedance modulus, more than 10^8 $\Omega \cdot cm^2$. The values of Rc and Qc were also evaluated using the fits of experimental spectra and are shown in Table 2. Comparing the obtained data, it can be clearly observed that values of R_c and Q_c increase and decrease at first and then decrease and increase with the increase in filler (aggregated PANI or PANI nanotubes) content, respectively. The optimum content of aggregated PANI and PANI nanotubes are 2 wt% and 5 wt%, respectively. All these results are in agreement with those obtained from the impedance modulus, which indicates that PANI nanotubes are easier to disperse in alkyd resin forming more compact structures, and thus show more excellent protective properties than the aggregated PANI.

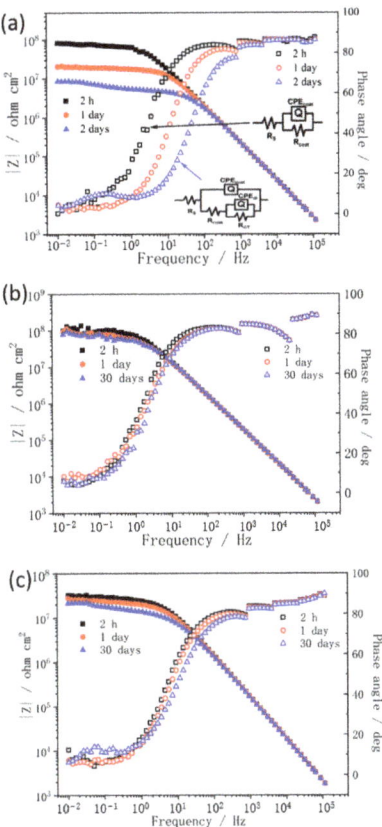

Figure 6. Impedance spectra for carbon steel coated with Alkyd (**a**), Alkyd−A2 (**b**), and Alkyd−A5 (**c**), in 3.5 wt% NaCl solution.

Table 2. Electrochemical parameters calculated from EIS spectra of alkyd coatings containing PANI nanostructures immersed in 3.5 wt% NaCl solution.

Samples	R_c ($\Omega \cdot cm^2$)		CPE_c-T ($F \cdot cm^{-2}$)		CPE_c-P	
	Value	Error (%)	Value	Error (%)	Value	Error (%)
Alkyd	1.48×10^7	1.54	2.55×10^{-9}	5.45	0.90	0.62
Alkyd−A2	7.90×10^7	5.31	1.69×10^{-9}	2.01	0.92	0.26
Alkyd−A5	3.01×10^7	2.88	2.43×10^{-9}	5.54	0.90	0.69
Alkyd−T2	7.38×10^8	9.14	1.26×10^{-9}	5.89	0.93	0.79
Alkyd−T5	1.32×10^9	4.87	1.08×10^{-9}	4.04	0.95	0.55
Alkyd−T8	5.30×10^7	1.42	2.31×10^{-9}	1.96	0.91	0.26

Based on above discussion, the protection mechanism of PANI nanotubes was schematically illustrated in Figure 8. It was widely reported that PANI prevents corrosion of steel in two ways [27,33,35]. On the one hand, it acts as a physical barrier preventing penetration of corrosive media across the film. On the other hand, it induces the formation of passive layer on the steel surface. As can be clearly seen, the PANI nanotubes with uniform diameters can be easily dispersed in alkyd resin, forming a more compact coating structure, and thus preventing the corrosive ions transport through the coating. In addition, the relatively high PANI nanotube content in the alkyd coating doubtlessly ensures the good physical contact between PANI and steel substrate, facilitating the formation of passive layer. All these factors guarantee the excellent protection properties of PANI nanotubes.

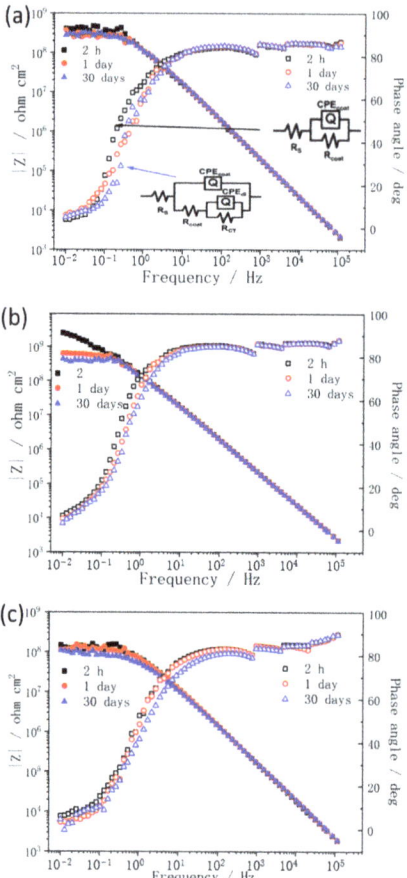

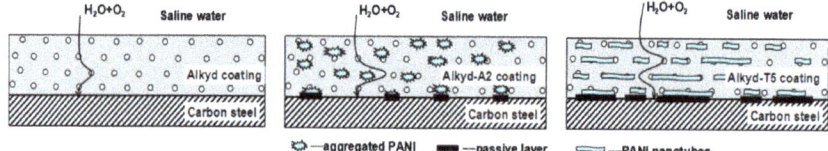

Figure 7. Impedance spectra for carbon steel coated with Alkyd−T2 (**a**), Alkyd−T5 (**b**), and Alkyd−T8 (**c**), in 3.5 wt% NaCl solution.

Figure 8. Mechanism of corrosion protection of alkyd containing PANI nanostructures.

4. Conclusions

Polyaniline nanotubes with uniform morphology were synthesized under relatively dilute acidic conditions. Phenazine-like moieties generated in mildly acidic conditions dictated the growth of PANI nanotubes. The obtained PANI products were not only directly applied on the carbon steel but also dispersed in alkyd resin as fillers to investigate their anticorrosion properties in the corrosive saline media. The results show that PANI nanotubes have more excellent protection properties than the aggregated PANI obtained by conventional polymerization method. The excellent anticorrosion properties of PANI nanotubes may be due to the formation of the compact coating on the carbon steel surface, guaranteeing the generation of the passive layer on the substrate surface. Moreover, a sufficient amount of PANI nanotubes ensures the homogeneity of the alkyd coating improving the barrier properties of the polymer film, as well as a good physical contact between PANI and steel substrate facilitating the formation of the passive layer.

Author Contributions: Conceptualization, S.P. and P.F.; methodology, F.L. and G.Y.; data curation, F.L.; writing—original draft preparation, F.L. and G.Y.; writing—review and editing, N.H. and C.H.; supervision, S.P. and P.F. All authors have read and agreed to the published version of the manuscript.

Funding: This work was financially supported by the National Natural Science Foundation of China (grant number 11875209).

Institutional Review Board Statement: Not applicable.

Informed Consent Statement: Not applicable.

Data Availability Statement: Raw data is available upon request.

Conflicts of Interest: The authors declare no conflict of interest.

References

1. Hatchett, D.W.; Josowicz, M.; Janata, J. Comparison of chemically and electrochemically synthesized polyaniline films. *J. Electrochem. Soc.* **2019**, *146*, 4535–4538. [CrossRef]
2. Viva, F.A.; Andrade, E.M.; Molina, F.V.; Florit, M.I. Electropolymerization of 2-methoxy aniline. Electrochemical and spectroscopical product characterization. *J. Electrochem. Soc.* **1999**, *471*, 180–189. [CrossRef]
3. MacDiarmid, A.G. Nobel Lecture: "Synthetic metals": A novel role for organic polymers. *Rev. Mod. Phys.* **2001**, *73*, 701–712. [CrossRef]
4. Pud, A.; Ogurtsov, N.; Korzhenko, A.; Shapoval, G. Some aspects of preparation methods and properties of polyaniline blends and composites with organic polymers. *Prog. Polym. Sci.* **2003**, *28*, 1701–1753. [CrossRef]
5. Sun, Q.; Park, M.-C.; Deng, Y. Dendritic superstructure formation of polyaniline prepared using a water-soluble polyelectrolyte copolymer as the support matrix. *Mater. lett.* **2007**, *61*, 3052–3055. [CrossRef]
6. Hugotlegoff, A.; Bernard, M.C. Protonation and oxidation processes in polyaniline thin-films studied by optical multichannel analysis and in-situ raman-spectroscopy. *Synth. Met.* **1993**, *60*, 115–131. [CrossRef]
7. Gospodinova, N.; Terlemezyan, L. Conducting polymers prepared by oxidative polymerization: Polyaniline. *Prog. Polym. Sci.* **1998**, *23*, 1443–1484. [CrossRef]
8. Neoh, K.G.; Kang, E.T.; Khor, S.H.; Tan, K.L. Stability studies of polyaniline. *Polym. Degrad. Stab.* **1990**, *27*, 107–117. [CrossRef]
9. Amano, K.; Ishikawa, H.; Kobayashi, A.; Satoh, M.; Hasegawa, E. Thermal-stability of chemically synthesized polyaniline. *Synth. Met.* **1994**, *62*, 229–232. [CrossRef]
10. Kraljić, M.; Mandić, Z.; Duić, L. Inhibition of steel corrosion by polyaniline coatings. *Corros. Sci.* **2003**, *45*, 181–198. [CrossRef]
11. Deberry, D.W. Modification of the electrochemical and corrosion behavior of stainless-steels with an electroactive coating. *J. Electrochem. Soc.* **1985**, *132*, 1022–1026. [CrossRef]
12. Bereket, G.; Hur, E.; Sahin, Y. Electrochemical synthesis and anti-corrosive properties of polyaniline, poly(2-anisidine), and poly(aniline-co-2-anisidine) films on stainless steel. *Prog. Org. Coat.* **2005**, *54*, 63–72. [CrossRef]
13. Karpakam, V.; Kamaraj, K.; Sathiyanarayanan, S. Electrosynthesis of PANI-nano TiO$_2$ composite coating on steel and its anti-corrosion performance. *J. Electrochem. Soc.* **2011**, *158*, C416. [CrossRef]
14. Abu, Y.M.; Aoki, K. Corrosion protection by polyaniline-coated latex microspheres. *J. Electrochem. Soc.* **2005**, *583*, 133–139. [CrossRef]
15. Yao, B.; Wang, G.C.; Ye, J.K.; Li, X.W. Corrosion inhibition of carbon steel by polyaniline nanofibers. *Mater. Lett.* **2008**, *62*, 1775–1778. [CrossRef]
16. Yang, X.G.; Li, B.; Wang, H.Z.; Hou, B.R. Anticorrosion performance of polyaniline nanostructures on mild steel. *Prog. Org. Coat.* **2010**, *69*, 267–271. [CrossRef]
17. Guo, R.; Wang, J.; Wang, H.H.; Fei, G.Q.; Wang, C.Y.; Sun, L.Y.; Wallace, G.G. Engineering the poly(vinyl alcohol)-polyaniline colloids for high-performance waterborne alkyd anticorrosion coating. *Appl. Surf. Sci.* **2019**, *481*, 960–971. [CrossRef]
18. Xiong, Y.; Hu, J.; Nie, X.; Wei, D.; Zhang, N.; Peng, S.; Dong, X.; Li, Y.; Fang, P. One-step firing of carbon fiber and ceramic precursors for high performance electro-thermal composite: Influence of graphene coating. *Mater. Des.* **2020**, *191*. [CrossRef]
19. Tian, Z.F.; Yu, H.J.; Wang, L.; Saleem, M.; Ren, F.J.; Ren, P.F.; Chen, Y.S.; Sun, R.L.; Sun, Y.B.; Huang, L. Recent progress in the preparation of polyaniline nanostructures and their applications in anticorrosive coatings. *RSC Adv.* **2014**, *4*, 28195–28208. [CrossRef]
20. Schauer, T.; Joos, A.; Dulog, L.; Eisenbach, C.D. Protection of iron against corrosion with polyaniline primers. *Prog. Org. Coat.* **1998**, *33*, 20–27. [CrossRef]

21. Olad, A.; Ramazani, Z. Preparation, characterization, and anticorrosive properties of polyaniline nanotubes. *Int. J. Polym. Mater.* **2012**, *61*, 949–962. [CrossRef]
22. Huang, J.; Virji, S.; Weiller, B.H.; Kaner, R.B. Nanostructured polyaniline sensors. *Chemistry* **2004**, *10*, 1314–1319. [CrossRef] [PubMed]
23. Mahmoudian, M.R.; Alias, Y.; Basirun, W.J. Effect of narrow diameter polyaniline nanotubes and nanofibers in polyvinyl butyral coating on corrosion protective performance of mild steel. *Prog. Org. Coat.* **2012**, *75*, 301–308. [CrossRef]
24. Trchova, M.; Sedenkova, I.; Konyushenko, E.N.; Stejskal, J.; Holler, P.; Ciric-Marjanovic, G. Evolution of polyaniline nanotubes: The oxidation of aniline in water. *J. Phys. Chem. B* **2006**, *110*, 9461–9468. [CrossRef]
25. Konyushenko, E.N.; Stejskal, J.; Šeděnková, I.; Trchová, M.; Sapurina, I.; Cieslar, M.; Prokeš, J. Polyaniline nanotubes: Conditions of formation. *Polym. Int.* **2006**, *55*, 31–39. [CrossRef]
26. Bhanvase, B.A.; Sonawane, S.H. New approach for simultaneous enhancement of anticorrosive and mechanical properties of coatings: Application of water repellent nano $CaCO_3$–PANI emulsion nanocomposite in alkyd resin. *Chem. Eng. J.* **2010**, *156*, 177–183. [CrossRef]
27. Wei, Z.X.; Zhang, Z.M.; Wan, M.X. Formation mechanism of self-assembled polyaniline micro/nanotubes. *Langmuir* **2002**, *18*, 917–921. [CrossRef]
28. Zhang, L.; Wan, M. Self-assembly of polyaniline—from nanotubes to hollow microspheres. *Adv. Funct. Mater.* **2003**, *13*, 815–820. [CrossRef]
29. Stejskal, J.; Sapurina, I.; Trchová, M.; Konyushenko, E.N.; Holler, P. The genesis of polyaniline nanotubes. *Polymer* **2006**, *47*, 8253–8262. [CrossRef]
30. Janosevic, A.; Ciric-Marjanovic, G.; Marjanovic, B.; Holler, P.; Trchova, M.; Stejskal, J. Synthesis and characterization of conducting polyaniline 5-sulfosalicylate nanotubes. *Nanotechnology* **2008**, *19*, 135606. [CrossRef]
31. Stejskal, J.; Sapurina, I.; Trchova, M.; Konyushenko, E.N. Oxidation of aniline: Polyaniline granules, nanotubes, and oligoaniline microspheres. *Macromolecules* **2008**, *41*, 3530–3536. [CrossRef]
32. Tran, H.D.; Li, D.; Kaner, R.B. One-dimensional conducting polymer nanostructures: Bulk synthesis and applications. *Adv. Mater.* **2009**, *21*, 1487–1499. [CrossRef]
33. Li, G.C.; Pang, S.P.; Xie, G.W.; Wang, Z.B.; Peng, H.R.; Zhang, Z.K. Synthesis of radially aligned polyaniline dendrites. *Polymer* **2006**, *47*, 1456–1459. [CrossRef]
34. Li, X.; Wang, G.; Li, X.; Lu, D. Surface properties of polyaniline/nano-TiO_2 composites. *Appl. Surf. Sci.* **2004**, *229*, 395–401. [CrossRef]
35. Gai, L.; Du, G.; Zuo, Z.; Wang, Y.; Liu, D.; Liu, H. Controlled synthesis of hydrogen titanate−polyaniline composite nanowires and their resistance−temperature characteristics. *J. Phys. Chem. C* **2009**, *113*, 7610–7615. [CrossRef]
36. Bian, C.; Yu, Y.; Xue, G. Synthesis of conducting polyaniline/TiO_2 composite nanofibres by one-step in situ polymerization method. *J. Appl. Polym. Sci.* **2007**, *104*, 21–26. [CrossRef]
37. Alam, J.; Riaz, U.; Ahmad, S. High performance corrosion resistant polyaniline/alkyd ecofriendly coatings. *Curr. Appl. Phys.* **2009**, *9*, 80–86. [CrossRef]

Article

In-Vitro Analysis of FeMn-Si Smart Biodegradable Alloy

Ana Maria Roman [1], Victor Geantă [2], Ramona Cimpoeșu [1,*], Corneliu Munteanu [3], Nicoleta Monica Lohan [1], Georgeta Zegan [4,*], Eduard Radu Cernei [4], Iulian Ioniță [1], Nicanor Cimpoeșu [1] and Nicoleta Ioanid [4]

1. Faculty of Materials Science and Engineering, Gh. Asachi Technical University from Iasi, 700050 Iasi, Romania; ana-maria.roman@academic.tuiasi.ro (A.M.R.); monica.lohan@yahoo.com (N.M.L.); iulian.ionita@academic.tuiasi.ro (I.I.); nicanor.cimpoesu@tuiasi.ro (N.C.)
2. Faculty of Materials Science and Engineering, University Politehn Bucuresti, Splaiul Independentei 313, 060042 Bucharest, Romania; victorgeanta@yahoo.com
3. Faculty of Mechanical, "Gh. Asachi" Technical University from Iasi, 700050 Iasi, Romania; cornelmun@gmail.com
4. Faculty of Dental Medicine, "Grigore T. Popa" University of Medicine and Pharmacy, 700050 Iasi, Romania; eduard-radu.cernei@umfiasi.ro (E.R.C.); nicole_ioanid@yahoo.com (N.I.)
* Correspondence: ramona.cimpoesu@tuiasi.ro (R.C.); georgeta.zegan@umfiasi.ro (G.Z.)

Abstract: Special materials are required in many applications to fulfill specific medical or industrial necessities. Biodegradable metallic materials present many attractive properties, especially mechanical ones correlated with good biocompatibility with vivant bodies. A biodegradable iron-based material was realized through electric arc-melting and induction furnace homogenization. The new chemical composition obtained presented a special property named SME (shape memory effect) based on the martensite transformation. Preliminary results about this special biodegradable material with a new chemical composition were realized for the chemical composition and structural and thermal characterization. Corrosion resistance was evaluated in Ringer's solution through immersion tests for 1, 3, and 7 days, the solution pH was measured in time for 3 days with values for each minute, and electro-corrosion was measured using a potentiostat and a three electrode cell. The mass loss of the samples during immersion and electro-corrosion was evaluated and the surface condition was studied by scanning electron microscopy (SEM) and energy dispersive spectroscopy (EDS). SME was highlighted with differential scanning calorimetry (DSC). The results confirm the possibility of a memory effect of the materials in the wrought case and a generalized corrosion (Tafel and cyclic potentiometry and EIS) with the formation of iron oxides and a corrosion rate favorable for applications that require a longer implantation period.

Keywords: iron based biodegradable alloy

Citation: Roman, A.M.; Geantă, V.; Cimpoeșu, R.; Munteanu, C.; Lohan, N.M.; Zegan, G.; Cernei, E.R.; Ioniță, I.; Cimpoeșu, N.; Ioanid, N. In-Vitro Analysis of FeMn-Si Smart Biodegradable Alloy. *Materials* 2022, 15, 568. https://doi.org/10.3390/ma15020568

Academic Editor: Lili Tan

Received: 10 November 2021
Accepted: 10 January 2022
Published: 12 January 2022

Publisher's Note: MDPI stays neutral with regard to jurisdictional claims in published maps and institutional affiliations.

Copyright: © 2022 by the authors. Licensee MDPI, Basel, Switzerland. This article is an open access article distributed under the terms and conditions of the Creative Commons Attribution (CC BY) license (https://creativecommons.org/licenses/by/4.0/).

1. Introduction

A special class of degradable biomaterials is intended for temporary implants whose presence is necessary to heal diseased tissue [1]. These types of implants work based on the same principle, but applications for different physiological environments differentiate them. For example, in temporary cardiovascular applications, coronary stents must open a narrowed artery and keep it open until the blood vessel is healed by replacing old tissues with newly formed ones [2,3]. In the case of orthopedic applications, implants of this type heal a fractured bone, keep it sustained until a healthy bone tissue is formed to replace the implant, which should then degrade. For use in cardiovascular applications such as stents, these biodegradable metals have shown adequate properties and special purity of the metal following the process of obtaining them from metallurgy and electrodeposition. [4]. The classical methods of thermal and thermomechanical treatments also play an essential role in obtaining properties with a specific destination for medical applications.

To fulfill their function in the healing process and to be successful in application, the biodegradable materials used for coronary stents must have a balance between the mechanical properties and the degradation process [5]. The speed with which the degradation occurs is crucial to allow the stent to maintain its mechanical resistance long enough to be able to heal the diseased arterial vein. The healing period in this case can be between 6 and 12 months [6,7]. The speed with which the material degrades must be optimal to allow the waste resulting from the degradation to be eliminated from the body. Waste accumulation around the implant can be harmful and can cause other unwanted injuries. Some studies indicate a favorable period for complete degradation of a stent to be between 12–24 months after implantation [8].

Studies on biodegradable metals present different methods in material development and improvement of mechanical properties. In vitro and in vivo studies are performed to obtain an optimal rate of degradation. Data were recorded on Fe-Mn alloy coronary stents [9], WZ21 Mg gastrointestinal implant [10], and Mg implant for laryngeal microsurgery [11].

Fe plays an important role in the breakdown of lipids, proteins, and DNA damage by producing reactive species following the Fenton reaction [12–14].

Following in vitro research, pure Fe has had a positive effect on the prevention of restenosis [12,15]. Another alloying element that can be associated with Fe is Mn. Excess Mn in the body has not been shown to be toxic. The alloying of Fe with Mn led to the production of new austenitic and some antiferromagnetic alloys, compatible with the magnetic field of MRI [16]. Fe-Mn alloys are influenced in the degradation process by increasing the corrosion rate given by Mn from the oxide layer formed. Zhang et al. [17] showed that the additional corrosion of the substrate is due to the Mn oxide present on the metal surface. Dargusch et al. [16] confirmed this in his paper, noting the uniform distribution of Mn oxide on the corrosion layer of the Fe-Mn alloy. Another important factor that could influence the rate of degradation is deformation. In the study by Heiden et al. [18], it was concluded that the rate of degradation of cold-rolled Fe-20Mn alloy was slower than that of the same cast alloy due to the more protective oxide layer formed on the metal surface. Hermawan et al. [19–22] has numerous studies on biodegradable Fe-Mn alloys, thus giving encouraging prospects for future studies in the design of new alloys based on Fe-Mn alloys. Several classes of new materials have been proposed, such as Fe-Mn-Pd alloys [23], Fe-Mn-(Co, C, Al, etc.) [24] as having a good degradation behavior and mechanical properties suitable for these types of implants. Specialist studies have shown the mechanism of Fe-Mn alloy degradation during dynamic degradation tests in the solution modified by Hank's solution. Following in vitro tests, Fe-Mn alloys showed a better degradation rate, 220–240 μm/year, than in the case of pure Fe [25].

The basic property of SMA (shape memory alloys) is that when thermally or mechanically activated they have the SME and the pseudo-elastic effect. If we add to these materials with special properties noted above, good resistance to corrosion and bending, and compatibility with magnetic and biological resonance, we obtain special materials that will undoubtedly be the best candidates in choosing materials for different medical applications. The addition of Si in the alloys of the Fe-Mn system leads to the appearance of the SME, as previously demonstrated [26,27]. B. Liu et al. in 2010 obtained favorable results following studies conducted on FeMn-Si alloys as candidates for biodegradable alloys. The important aspects that raise the issue of investigations on these alloys for applications in biodegradable implants are related to the microstructure, mechanical properties and SME, biocompatibility, and good degradation rate. The SME of FeMn-Si alloys formed by ε-martensite and γ-austenite phases was clearly shown. Also, with the addition of Si, an increase in the content of the γ-austenite phase was observed [28].

In this article, the behavior of the corrosive environment in the cast and wrought state of FeMnSi alloy was presented with emphasis on degradation properties of the materials. SEM, EDS, and EIS results are given to highlight the surface state of the samples after contact with Ringer's solution. The enhancement of corrosion rate with addition of Si to an FeMn alloy is expected and differences of corrosion rate between cast and wrought samples were observed.

2. Materials and Methods

An experimental alloy, FeMn-Si was obtained from high purity materials in a vacuum Arc Melting Facility MRF ABJ 900 (University Politehnica Bucharest, Bucharest, Romania), which ensured the melting of metallic materials in Ar-controlled atmosphere after preemptive working chamber up to 10^{-5} mbar by using a non-consumable throttle tungsten mobile electrode. The re-melting process, repeated five times, occurred in an induction furnace (Inductro, Bucharest, Romania) in ceramic crucible at "Gheorghe Asachi" Technical University in Iasi. The ingot was wrought until it was a 1 mm sheet, using a hot rolling equipment with the sample heated to 1100 °C and 5 reduction passes. The samples analyzed in this article were in cast and wrought states (C and W), both heat-treated through solution water quenching (heated to 1100 °C, maintained for 5 min for temperature homogenization and cooled in water + ice). For experimental tests, the specimens were mechanically polished with Al_2O_3 suspension solution (2–5 μm) after metallographic grinding with paper disks with 160–4000 MPi granulation. The cleaning of the surface was done with ethyl-alcohol for 30 min and the microstructure was highlighted by chemical etching using Nital 2% solution [29].

Two sample fragments were cut, weighing less than 50 mg, for DSC experiments (Partner digital balance). A differential scanning calorimeter type DSC 200 F3 Maya (NETZSCH, Selb, Deutschland) was used, with sensitivity: <1 W, temperature accuracy of 0.1 K, and enthalpy accuracy generally <1%. The calibration was done according to the standards with Hg, Bi, In, Sn, and Zn. Temperature scans were performed with the following temperature program: cooling from room temperature to −50 °C, heating from −50 °C to 200 °C, and cooling to room temperature. The cooling and heating rate was 10 K/min. All experiments were performed under an Ar protective atmosphere. NETZSCH's Proteus software version 4.8.5 was used to evaluate the DSC thermograms resulting from cooling and heating using the tangent method for determining critical temperatures and a rectilinear baseline for dissipated/absorbed heat.

The samples were subject to immersion tests in Ringer's solution (one of the first laboratory solutions of salts in water shown to greatly prolong the survival time of excised tissue; the solution contains calcium, potassium, and sodium chlorides, and sodium bicarbonate in the concentrations in which they occur in body fluids) at 37 °C, for 1, 3, and 7 days to analyze the interaction between the metallic materials and an electrolyte solution. The samples were weighed using a AS220 Partner analytical balance (Partner Co., Bucharest, Romania), before immersion, after immersion, and after an ultrasound cleaning stage (ultrasonic bath, 60 min in technical alcohol). The solution pH values were recorded with an Arduino set-up each minute and the variations were analyzed to establish the chemical reactions that occurred during the contact of the samples with the solution. Chemical composition of the surface was established with an EDS detector, Bruker X-flash, Mannheim, Germany Scanning electron microscopy (SEM, VegaTescan LMH II, SE detector, 30 kV, Brno—Kohoutovice, Czech Republic) was used to analyze the experimental materials structure and the state of the surface after immersion tests and electro-corrosion. X-ray diffraction tests were made with an Expert PRO-MPD system, (XRD, Panalytical, Almelo, The Netherlands type, Cu-X-ray tube (Kα-1.54°)).

The corrosion behavior of the samples was studied by comparing the method applied in the case of Fe-based alloys, using Ringer's solution with standard composition (chemical composition for Ringer's solution: 1000 mL contains: sodium chloride 8.6 g, calcium chloride $\times$ $6H_2O$ 0.5 g, potassium chloride 0.3 g, distilled water up to 1000 mL) as the liquid medium. The VoltaLab-21 potentiometer (Radiometer, Copenhagen, Denmark) was used to determine the corrosion resistance by analyzing linear and cyclic curves in Ringer's electrolyte solution, and the acquisition and processing of data was done with the Volta Master 4 package, version 6.0. A cell was used to expose the sample (working electrode) to Ringer's solution, with an auxiliary Pt electrode and one saturated with calomel. The samples were isolated with Teflon, so only one area was exposed to the electrolyte, an area of 0.78 cm^2. The solution was aerated permanently with a magnetic stirrer to remove gas bubbles from the metal surface following the removal of hydrogen. The potential-dynamic polarization test recorded data on electrode behavior. Through the polarization mechanisms of the direct current, approximate information was obtained about the corrosion speed of the working electrode (Fe-based samples), the type of surface corrosion (generalized or pitting), formation and stability of the passivation layer, anodic reactions (oxidation), or cathodic reactions (reduction).

The authors chose the following coordinates corresponding to the function: for the current density [mA/cm^2] the potential [V], a variation that allows the accentuation of the corrosion potential (E_{corr}) and the corrosion current (J_{corr}). The temperature of the experiments was room temperature ($\pm 24\ ^\circ$C) and the potential was recorded (line graphs were recorded at a scan rate of 1 mV/s and cycle graphs at a scan rate of 10 mV/s). For the accuracy of the results the experiments were repeated four times. Corrosion current values helped to obtain the instantaneous corrosion rate: V_{corr} (μm/year) [30].

3. Results

Microstructural, chemical, and thermal characteristics of the new chemical composition SMA were characterized using scanning electron microscopy (SEM), energy dispersive spectroscopy (EDS), and differential scanning calorimetry (DSC). The corrosion behavior of the alloy was evaluated through pH variation of electrolyte solution, immersion tests, linear and cyclic potentiometry, and EIS experiments.

3.1. Experimental FeMnSi Materials Analysis

The experimental materials were mechanically ground to remove oxides from the surface and cleaned in an ultrasound bath in technical alcohol. The chemical composition of the samples (cast and wrought state) was determined in five different areas of the surface, and the average values are shown in Table 1. Standard deviations of the elements show a homogeneous chemical composition of the material that will confirm the same properties of the material for the entire volume. Good chemical and structural homogeneity are essential for biodegradable materials and for SMA properties [31].

Table 1. Experimental results after chemical composition analysis for the initial cast sample and wrought state (average values after five determinations on different 1 mm^2 areas).

Elements/State	Fe		Mn		Si	
	wt%	at%	wt%	at%	wt%	at%
Cast	82.29	79.11	13.88	13.56	3.83	7.32
Wrought	82.16	79.33	14.47	14.2	3.37	6.48
EDS error %	0.06		0.02		0.03	

Standard deviation (StDev): Fe ± 0.15, Mn: ± 0.1 and Si: ± 0.05.

The chemical composition of the FeMnSi samples that was obtained leads to the appearance of the SME [32]. The main alloying element required by these FeMnSi alloys is manganese, with two roles in the thermodynamic stability of the phases. First, manganese stabilizes γ-austenite with the FCC structure being a phase that occurs at high temperatures in the case of iron. At normal pressure, manganese plays the role of stabilizing the ε phase with the HCP type structure, the thermodynamically stable phase only at high pressure for pure iron. The thermodynamic equilibrium temperature between phases γ- and ε- is close to room temperature, too low for the diffusion of atoms; phase ε is formed as martensite under cooling or loading. In Figure 1, XRD peaks of FeMnSi in cast, analyzed in [29] and wrought state, are given.

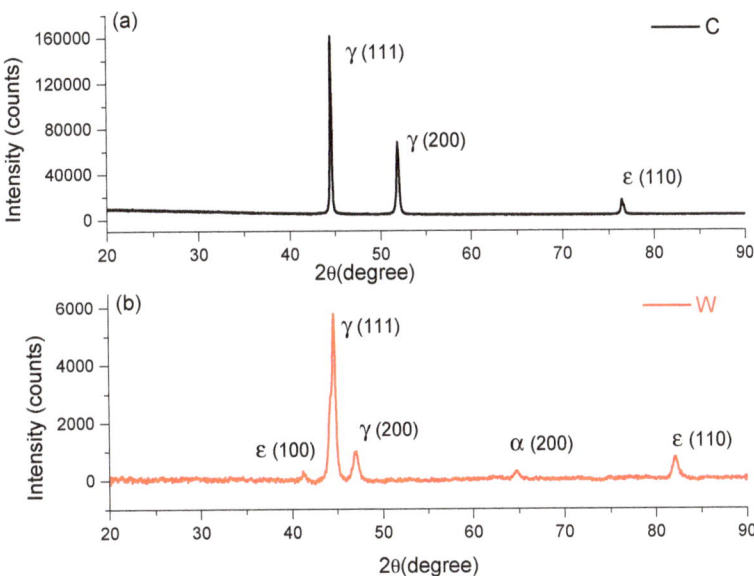

Figure 1. XRD spectra of FeMnSi: (**a**) cast and (**b**) wrought.

On the rolled sample we observed and identify five main phases (see Figure 1), at the following angles: 41.23619°, 44.65729°, 46,984.93°, 64.69378°, and 82.12412°. It is known that iron has several allotropic forms, specifically (at normal pressure): α-Fe with a cube structure with centered volume, stable up to 912 °C; γ-Fe with a cube-type structure with centered faces, stable up to at 1394 °C; Pand δ-Fe with a cube structure with centered volume, stable up to 1538 °C (melting temperature). Iron-based alloys will have phases in the structure with similar structures, but depending on the alloying elements, the field of these phases is modified. In addition to these phases, new phases such as intermetallic compounds or carbides may appear in the iron alloy systems, which will lead to the appearance of new metallographic constituents. Phase diagrams give indications of the phases that can occur in various alloy systems, especially binary or ternary. The higher the number of alloying elements and the more diverse the chemical composition, the more phases can occur in the alloy system.

Given the chemical composition of the analyzed alloys (Mn is the second alloying element as a percentage), the analysis of the equilibrium diagram of the Fe-Mn binary system shows that at the mass concentration of this element, at room temperature, there are an α-Fe phase and a γ-Fe phase, solid manganese iron solutions that may have similar structures but with different network parameters. Manganese is an alloying element that increases the range of the γ-Fe phase. Instead, both silicon and aluminium are alloying elements that increase the range of the α-Fe phase. From the analysis of the phase diagram and of the considerations stated above, it can be considered as a working hypothesis that the two alloys will have in the structure either the α-Fe type phase or both α-Fe and γ-Fe type phases. In this case, after the rolling deformation process, the peak of ε (110) at 82° presented an increase in the wrought state compared to cast state and, likewise, the appearance of the peak ε (100) at 41.23619°.

The transformation γ → ε depends on SME, so pretensioning, annealing treatment, thermomechanical training, and deformation temperature influence the FeMnSi SME [33]. Sato et al. studied Fe-30Mn-1Si alloy single crystals that showed a large shape recovery strain. [34]. In the FeMnSi system, the most indicated concentration ranges are 14–33% mass for Mn and 4–6% mass for Si, respectively. The composition is chosen for the beginning of the martensitic transformation, Ms, temperatures close to room temperature. For

SMA, different chemical compositions based on FeMnSi have been proposed by partially replacing Mn with the elements Cr, Ni, Cu, Al, etc. [35–38]. The ideal proportion of the alloy components is made according to the temperature Ms of the transformation $\gamma \to \varepsilon$ and the stability of austenite compared to á-martensite. Studies show that a small number of interstitial elements, such as C and N, strongly stabilize the γ-austenite phase and reduce the concentration of Mn when they are dissolved in the γ phase. [39]. The properties of the material can be improved by alloying with new elements. The new properties obtained refer to aspects related to the increase in corrosion resistance, mechanical resistance, the formability, and the decrease in the production cost. To conduct the phase transformation and to obtain the SME, the appropriate concentrations of the alloy can be calculated with the equations for the Gibbs free energy difference between the phases γ- and ε- and the equations for the temperature Ms of the transformation $\gamma \to \varepsilon$ [40]. Alloying with Si strengthens the matrix to suppress the dislocation slip and helps the martensitic transformation $\gamma \to \varepsilon$ by decreasing the energy of the stacking defect [41,42]. An important property such as that related to magnetism affects the phase transformation in its transformation into ε-martensite. Silicon lowers the magnetic transition temperature of FeMnSi-based alloys to sub-zero temperatures and leads to the martensitic transformation $\gamma \to \varepsilon$ at room temperature. Another influence that Si has is that it can lead to the short-term ordering of atoms to improve the martensitic transformation $\varepsilon \to \gamma$ [43]. Another factor that raises the reversibility of the inverse transformation is the improved coherence between the γ and ε networks with the addition of silicon. An interesting aspect would be that despite such complicated factors, the optimal amount of Si needed to reach the best shape recovery strain is always between 4 and 6% by mass.

The SEM electron microscope was used to investigate the microstructures of alloys in both states (cast and rolled) at high amplification. A detailed analysis of the microstructure did not show the presence of á martensite, which is usually present in the lenticular form, respectively, Figure 2a,b.

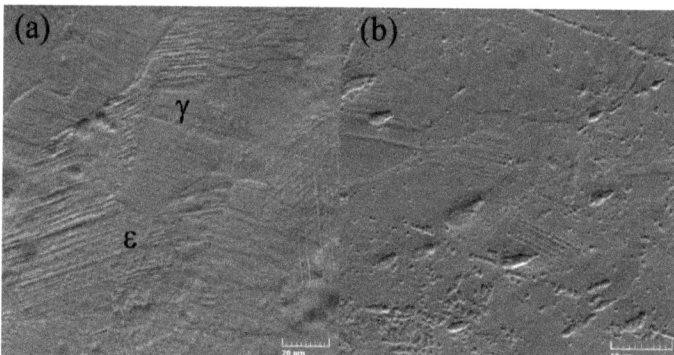

Figure 2. SEM micrographs of FeMnSi SMA (**a**) cast and (**b**) wrought.

The γ/ε interfaces do not have the normal direction to the interface, present in the thermo-elastic martensitic transformations, for example, in the case of SMA based on Ti-Ni. In contrast, ε-martensite increases in the direction parallel to the γ/ε interfaces, and thickening occurs due to the coalescence of nearby thin ε plates. Microstructure analysis showed that ε-martensite has fine lamellar structures involving thin remaining layers and/or high probabilities of stacking defects [44]. Stacking failure inside ε-martensite is a thin plate with a thickness of two atomic layers. The appearance of the microstructure can be associated with the distribution of nucleation sites and the growth of martensite crystals; this increase in martensite occurs due to the displacement of partial Shockley dislocations [45]. The SME consists in the deformed state in which ε-martensite is induced as shown in Figure 3, which subsequently returns to the original γ—austenite shown in

Figure 2a by the inverse transformation to heating. Even so, the shape recovery strain in the binary FeMn alloys is tiny [45]. To achieve an optimal SME, a second necessary element should be added, specifically, silicon.

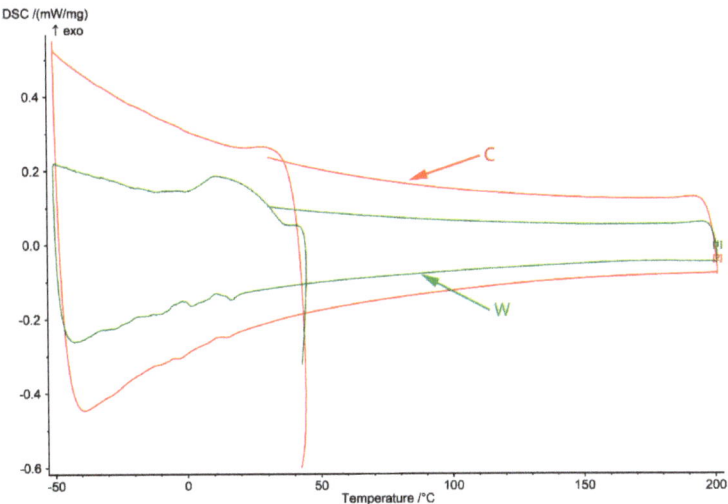

Figure 3. DSC diagram of C and W samples in −50 to 200 °C domain (heating from 25 to 200 °C and cooling to −50 and back to 25 °C) and γ-ε forward transformation through cooling and ε-γ reverse transformation through heating.

In Figure 3, DSC diagram of C and W samples in the −50 to 200 °C domain (cooling from room temperature to −50 °C, heating to 200 °C and cooling to 25 °C) are presented. In the case of the cast sample, no variation of the thermal flux is observed except for two small variations on cooling around 10 and −10 °C that appear after heating of the cast sample to 200 °C.

The SME of these FeMn-Si-based alloys is closely related to the martensitic transformation induced by deformation of austenite with cubic structure with centered faces in martensite with closed hexagonal structure, the reverse of the phenomenon being possible at subsequent heating. In the wrought sample, Ms temperature is at 35 °C and Mf at 1.5 °C with dissipated heat ($\Delta H/m$) [kJ/kg] of 7.6 associated with the $\gamma \rightarrow \varepsilon$ transition. The SME will be evaluated in a different article through tensile and bending tests to establish the application potential of this material as an SMA element.

The plastic deformation of the FCC and HCP structures is achieved by different sliding modes with extended dislocation (a), mechanical γ-twinning (b), and martensitic transformation $\gamma \rightarrow \varepsilon$ (c), because of the expansion of stacking defects and their regular or irregular duplication, depending on the relative stability γ/ε. [46]. An important role in determining the mechanisms of plasticity is played by the energy of stacking defects. The DSC result of the wrought state confirms the appearance of the martensitic peak ε (100) at 41.23619° on XRD result.

3.2. Electrolyte Solution pH Variation in Contact with Metallic Sample Analysis

Corrosion initially began when the samples were immersed in Ringer's solution. The oxidation reaction has randomly occurred in several areas of the anodic outer surface, at the grain boundaries, and at the interface between phases; see Equations (1) and (2). The cathodic reaction of the reduction of water that consumed the released electrons fol-lowed (see Equation (3)). Further, the layers of insoluble hydroxides (metal oxides) were formed from free metal ions, which reacted with hydroxide ions (OH−) (see Equations (4) and (5)).

In the subsequent visual observations, these hydroxides appeared as a red-brown (Fe_2O_3) layer on the top and a black (Fe_3O_4 and FeO) layer on the bottom.

Based on the fact that the general reaction consumes H^+ and produces OH^-, the electrolyte solution pH will increase, enhancing the formation of an $Fe(OH)_2$ thin layer on the experimental alloy surface (precipitation reaction). This process characterizes both samples, melted and wrought (see Figure 4), in the first 16–17 h of contact. The porous penetrable layer, in which most of the compounds are oxides, formed on the surface play a protective role for substrate in this time, slowing down new corrosion processes.

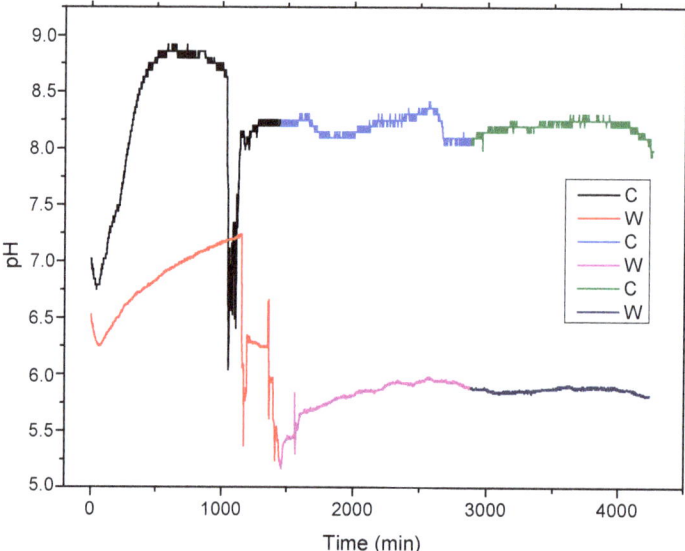

Figure 4. The pH variation during 72 h of immersion at 37 °C of melted and wrought sample.

Chloride ions compensated for the growth of metal ions under the hydroxide layer by penetrating the metal substrate. The metal chloride formed was then hydrolyzed into hydroxide and free acid; see Equation (6). This decreased the pH value in the pitting pits, and the solution remained neutral. An in vitro static and dynamic degradation was performed [4,5]. The decrease in the solution's pH was observed after immersion for 16–17 h for both C and W samples, with different rates, approximately 60 to 90 min for melted sample and 500 min, with pH variations based on different areas of breakthrough, for wrought sample. It can be observed that the cast sample of FeMnSi presents a faster corrosion rate than the wrought material, due to a bigger grain structure and main structural defects obtained from melting.

Initial corrosion reaction (a):

$$Fe \rightarrow Fe^{2+} + 2e^- \qquad (1)$$

$$Mn \rightarrow Mn^{2+} + 2e^- \qquad (2)$$

$$2H_2O + O_2 + 4e^- \rightarrow 4OH^- \qquad (3)$$

Formation of hydroxide layers (b):

$$2Fe^{2+} + 4OH^- \rightarrow 2Fe(OH)_2 \text{ or } 2FeO \cdot 2H_2O \qquad (4)$$

$$4Fe(OH)_2 + O_2 + 2H_2O \rightarrow 4Fe(OH)_3 \text{ or } 2Fe_2O_3 \cdot 6H_2O \qquad (5)$$

Pitting formation (c):

$$Fe^{2+} + 2Cl^- \rightarrow FeCl_2 + H_2O \rightarrow Fe(OH)_2 + HCl \tag{6}$$

Further pH variations can be observed in the subsequent two days (2880 min) for both samples based on passivation and repassivation of the surface in the electrolyte solution [47].

3.3. Immersion Experiments Analyses

Immersion experiments present a clearly loose of mass after one or seven days in Ringer's solution at 37 °C. Mass variation of the samples was recorded for cast and wrought samples, see Table 2, and present an increase in mass after one day based on the compounds formed after the interaction of the material with Ringer's solution (generally oxides) with a bigger value for cast sample (enhanced corrosion).

Table 2. The results of the masses of the experimental samples after 1, 3, and 7 days of immersion in Ringer's solution (five repetitive mass determinations were done on the same sample) and the corresponding degradation rate.

Sample	1 Day		3 Days		7 Days	
	Cast Sample (C)	Wrought Sample (W)	Cast Sample (C)	Wrought Sample (W)	Cast Sample (C)	Wrought Sample (W)
Initial mass (mg)	3182.7	563.1	2756.6	471.9	2989.7	513.8
Mass after immersion (mg)	3184.3 (+1.6)	563.3 (+0.2)	2756.3 (−0.3)	465.7 (−6.2)	2984.3 (−5.4)	511.5 (−2.3)
Mass after ultrasonic cleaning (mg)	3182.0 (−0.7)	562.6(−0.5)	2752.8 (−3.8)	463.9 (−8.0)	2983.9 (−5.8)	510.5 (−3.3)
DR (mm/year)	0.088	0.084	0.159	0.451	0.104	0.080

Standard deviation: ±0.1 mg.

The degradation rate presented in Table 2 resulted according to the formula [48]:

$$DR = \frac{8.76 \times 10^4 W}{At\rho} \tag{7}$$

where: DR = degradation rate (mm/year); W = mass loss (g); A = sample area (cm^2); t = time of immersion (h); and ρ = metal density (g/cm^3). This confirms the higher value of corrosion of the cast sample compared with the wrought one.

The sample mass presents a decrease after ultrasound cleaning (in all cases) based on the removal of the compounds formed on the surface through immersion. The stability of the compounds formed on the surface is low even after one day of reactions. After 3 and 7 days before and after ultrasound cleaning, we observed a decrease in the sample mass, more exfoliation of the surface being done during this period, and the corrosion compounds passed in Ringer's solution. A bigger mass loss is observed in case of the cast sample, meaning a higher corrosion than the wrought sample (around two times higher for the cast sample).

Chemical composition insights of the surface after immersion in Ringer's solution before and after ultrasound cleaning are given in Table 3 after EDS determination on a 4 mm^2 surface. In addition to the alloy main elements, respectively, Fe, Mn, and Si, new elements were identified on the surface after the interaction with Ringer's solution, generally O, C, Cl, and Na. The difference between the chemical composition of the surface before and after ultrasound cleaning are substantial, indicating that most of the compounds are unstable on the surface from the first day of contact with the electrolyte solution. However, after seven days a lower loss of compounds is observed, indicating that the stability of the compounds is better, the interaction with the substrate is higher, and the loss of material occur in larger quantities.

Table 3. The chemical composition of the FeMnSi alloy after 1 day and 7 days of immersion in Ringer's solution and after ultrasonic cleaning after each period for both cast and wrought samples.

El./Samples			Fe		Mn		Si		O		C		Cl		Na	
			wt%	at%	wt%	at%	wt%	at%	wt%	at%	wt%	at%	wt%	at%	wt%	at%
1 day (I)	C	I	56.16	29.39	8.75	4.66	1.86	1.93	22.02	40.23	8.39	20.42	0.49	0.41	2.33	2.96
		I + UC	67.55	42.43	12.3	7.85	2.78	3.47	5.89	12.9	11.37	33.2	0.05	0.04	0.07	0.1
	W	I	48.77	22.01	5.74	2.63	1.23	1.1	33.53	52.83	9.82	20.62	0.51	0.37	0.4	0.44
		I + UC	53.11	25.8	8.38	4.14	1.77	1.71	25.06	42.49	11.32	25.57	0.32	0.25	0.05	0.06
7 days (I)	C	I	57.92	29.92	10.29	5.4	2.32	2.39	13.54	24.41	15.65	37.6	0.15	0.12	0.13	0.17
		I + UC	56.85	28.53	9.78	4.99	1.94	1.94	14.56	25.5	16.59	38.7	-	-	0.28	0.34
	W	I	45.57	19.99	7.1	3.17	1.57	1.37	32.53	49.83	11.87	24.23	0.12	0.08	1.24	1.32
		I + UC	52.81	25.16	7.7	3.73	2.05	1.94	21.73	36.13	14.13	31.3	0.21	0.16	1.23	1.42
EDS detector error %			1.54		0.31		0.17		4.04		4.23		0.07		0.18	

C: cast, W: wrought, I: after immersion, I + UC: after immersion and ultrasound cleaning. StDev: Fe: ±0.9, Mn: ±0.5, Si: ±0.22, O: ±0.2, C: ±0.1, Cl: ±0.1, Na: ±0.1.

Structural aspects of the surface after immersion tests were taken using scanning electron microscopy, see Figure 5, and present the main aspects of the compounds formed on the surface. A more stable anchoring of the compounds after 7 days is observed from Figure 5e–h, confirming the observations made from chemical composition analyses. After initial immersion (Figure 5a–d), the entire surface is covered by reaction compounds, generally iron and manganese oxides (see Table 3), which are mainly removed from the surface after cleaning. Figure 5b shows the compounds formed at the micro scale as well as the corroded surface of the sample. On the wrought sample a reduced quantity of compounds is observed (see Figure 5g,h) before and after sonication confirming the mass variation quantities given in Table 2.

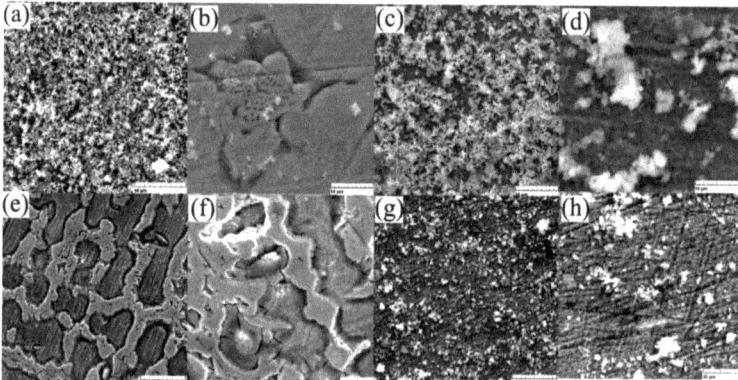

Figure 5. SEM images of the surface: (**a**) C sample after 1 day immersion (1 kx) and (**b**) after ultrasound cleaning (5 kx), (**c**) W sample after 1 day immersion (1 kx) and (**d**) after ultrasound cleaning (5 kx), (**e**) C sample after 7 days immersion (1 kx), and (**f**) after ultrasound cleaning (2 kx), (**g**) W sample after 7 days immersion (1 kx) and (**h**) after ultrasound cleaning (2 kx).

For the cast state, the corrosion begins at the surface on the first day without the formation of many compounds and more an attack of the structure (Figure 5b,f), and after seven days of immersion a thick layer of compounds is observed on the wrought sample (Figure 5g,h). Considering a low corrosion rate of Fe-based biodegradable materials compared to Mg-based, we can note an advantage of the higher corrosion rate of the cast material.

In Figure 6, the state of the experimental materials surface is presented through elemental distributions. In the initial state (C and W) no visible differences of chemical elements distribution is observed (Figure 6a,f) so the phase transformations occur with very small modifications of the chemical composition (γ to ε or the reverse).

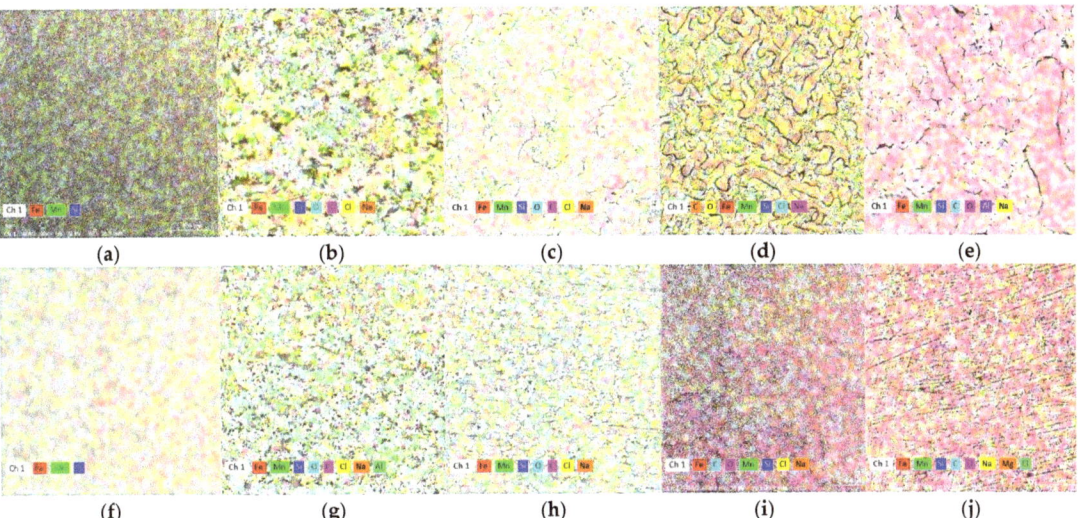

Figure 6. X-ray mapping of chemical elements on the surface. The initial samples, cast (**a**) and wrought (**f**), (**b**); C sample after 1 day immersion and (**c**) after ultrasound cleaning; (**d**) C sample after 7 days immersion and (**e**) after ultrasound cleaning; (**g**) W sample after 1 day immersion and (**h**) after ultrasound cleaning; (**i**) W sample after 7 days immersion and (**j**) after ultrasound cleaning.

By addition to FeMn system, silicon replaces iron, contributing to an increase of the Mn:Fe report in the general system composition in the ternary alloy. Silicon is not homogeneously spread in the microstructure and enriched in inter-dendritic regions [49], and will promote a higher degradation rate.

3.4. Electro-Chemical Corrosion Resistance

Complementary data from other sources or sensors must be obtained for a more complete picture of the corrosion process. These complementary data are extracted at the same time as the purchase from the corrosion sensor. Corrosion monitoring is done most accurately by highly sensitive methods that provide an instantaneous signal when the corrosion rate changes. Changes in corrosion potential can give indications of active/passive behavior in alloys. Moreover, the corrosion potential can give fundamental indications on the possibility of corrosion from a thermodynamic perspective (thermodynamic probability of corrosion. The corrosion potential, E_{corr}, is a measure of the corrosion tendency of an alloy immersed in an electrolytic environment. Corrosion is evaluated indirectly, from the linear polarization curves, using a Tafel diagram (see Figure 7). The intersection of the linear portions of the anodic and cathodic branches of the polarization curve gives the potential the value of the corrosion potential, E_{corr}. Electrochemical properties of the samples C and W were determined by Tafel polarization and EIS measurements. Values of the corrosion parameters, such as corrosion current density (i_{corr}), corrosion potential (E_{corr}), and protection efficiency (V_{corr}), were extracted from the Tafel curves and are presented in Table 4. The (E_{corr}) value of Sample W is −678.9 mV and for sample T is −930.7 mV. The shift of the E_{corr} toward negative values implies a higher corrosion resistance for the wrought sample.

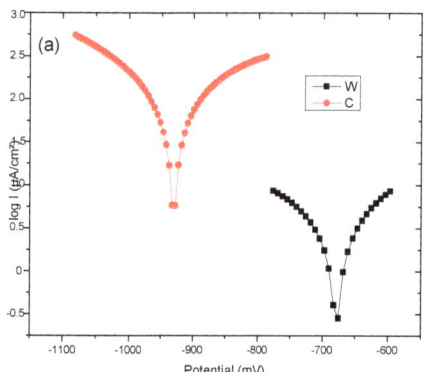 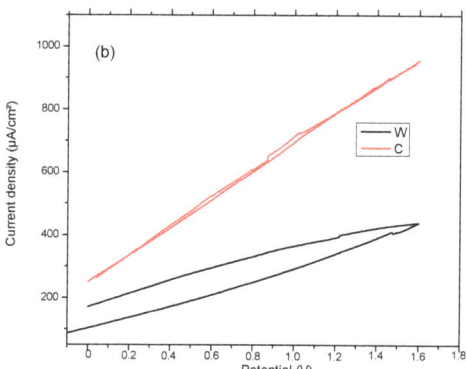

Figure 7. (**a**) Tafel and (**b**) cyclic diagrams for C and W samples.

Table 4. Linear Tafel parameters for C and W samples.

Sample	E_{corr} mV	b_a mV	b_c mV	R_p ohm·cm^2	J_{corr} mA/cm^2	V_{corr} mm/Year
C	−930.7	324.8	−224.3	388.57	2.17	132.9
W	−678.9	95.2	−156.7	1000.42	0.11	24.47

The corrosion current density j_{corr} value for W, which is directly proportional to the corrosion rate, is lower than the value for C. This can be attributed to the surface defects on the sample C. Based on the analysis of cyclic polarization diagrams, it can be observed that there are not any significant differences between the two samples. This presents uniform (generalized) corrosion when polarized anodically. Hydrogen is also released at lower potentials, but in tiny quantities for Fe-based alloys.

The FeMnSi system (C and W) exhibits a tendency of active dissolution under anodic polarization conditions, compared to the Fe-Mn alloy [50]. The presence of Si appears to slow down the kinetics of the anodic reaction, a fact suggested by the slope change of the binary alloy from 70 mV/decade to 100 mV/decade in the case of the ternary alloy. A transformation to SiO_2 may occur due to the release of Mn and Fe ions from the surface, which will cover the surface and thus reduce the dissolution rate, as shown in the diagram of the potential pH of Si-H2O [50]. The process of dissolving Mn and Fe leads to a more alkaline pH and destabilization of surface oxides [50]. Before switching to a controlled mass transfer regime, the slope suddenly increases to over 600 mV from the SCE in the case of the polarization curve of FeMnSi. The increase of degradation rate by Si adding to FeMn system is given by the irregular spread of Si at the grains limits and promotion of areas with different corrosion potential (similar to micro-piles formation).

3.5. Electro-Impedance Spectroscopy (EIS) Experiments

The data obtained show that, for these systems, the behavior in solution can be described using a circuit equivalent to a single time constant: a Randles-type circuit, which indicates that the corrosion process occurs on the entire surface of the sample (generalized corrosion) by a single chemical reaction (most likely corrosion of iron with the formation of soluble and insoluble products).

A constant phase element (CPE) was used instead of the Randles circuit capacitance, for a better explanation of the deviation of the Nyquist diagrams from the ideal behavior (a semicircle on the abscissa axis, Zr) due to the change in capacitances with frequency. The impedance of CPE is expressed mathematically with the help of the following relation [51–53]:

$$Z_{CPE} = \frac{1}{Q(j\omega)^n} \tag{8}$$

where: Q is a constant proportional to the active area; $<Q> = \Omega^{-1}\,s^n/cm^2 \equiv S\cdot s^n/cm^2$, ω is the angular frequency ($\omega = 2\pi f$, f is the frequency of the applied alternating current), and j is the imaginary number; $j = (-1)^{\frac{1}{2}}$.

Because of this relationship, the phase angle of the CPE is independent of frequency and has a value of $(90°)^n$, which is also the reason why it is called a constant phase element.

The circuit elements have the following meanings: R_s—electrolyte resistance between the working electrode and the reference electrode, R_{ct}—resistance to charge transfer through the double-electric layer (ct → charge transfer) thus controlling the speed of the corrosion process, and CPE—element of constant phase, which, in theory, would represent the capacity of the double-electric layer (C_{dl}), but here it has the meaning of an imperfect capacitor ($n < 1$). Imperfections can be mechanical (rough surface) and/or chemical (non-uniform chemical composition).

The effective value of the double-layer capacity can be calculated using the Brug relation [50]:

$$C_{dl} = \left[Q \left(\frac{1}{R_s} + \frac{1}{R_{ct}} \right)^{n-1} \right]^{\frac{1}{n}} \tag{9}$$

Using this relationship and the data in Table 5, we obtain: $C_{dl} = 4.9155 \times 10^{-3}$ F/cm^2. In Table 5, ε represents the percentage error of evaluation of each circuit element.

From the viewpoint of the values of χ^2, the circuits R (QR) (QR) and R (CR) (QR) seem much more appropriate than the simple circuit R (QR) ($\chi^2 \sim 10^{-3}$), but from the viewpoint of the individual evaluation errors of the circuit elements seems to be more appropriate is the last circuit, in which the percentage errors are insignificant.

The R (CR) (QR) circuit fits the experimental curve very well. The admission of this circuit indicates that the complex layer on the sample surface is forcibly divided into two layers: SDE (electric double layer where the reaction occurs) and another layer. To see if this is so, for the circuit R (QR) (QR) we calculated the global resistance series: R= $R_1 + R_2$ and Q global series: $Q = (Q_1 \cdot Q_2)/(Q_1 + Q_2)$ (because Q has character capacitor), and we found R = 165.14 ohm·cm^2, respectively Q = 7.838 10^{-3} S·s^n/cm^2, which are close to the values R = 199.8 ohm·cm^2, respectively Q = 9.424 10^{-3} S·s^n/cm^2 found for circuit R (Q). The differences could be attributed to large errors in the evaluation of the parameters with the R (QR) (QR) circuit.

In the Nyquist diagram, the experimental curve shows a negative loop in the low frequency range. This distortion is most often attributed to an inductive behavior of the electrochemical system due to the process of adsorption of intermediates on the surface of the electrode (sample). Here, it would most likely be adsorption of Fe or Fe_2O_3. For data simulation, it is necessary to use an equivalent circuit with inductance (Figure 8).

Table 5. The values of the equivalent circuit for sample C.

R(QR)

R_s (ohm·cm^2)	$10^3 \cdot Q$ (S·s^n/cm^2)	n	R_{ct} (ohm·cm^2)	$10^3 \cdot \chi^2$
29.8	9.424	0.684	199.8	1.33
ε (%): 0.6998	2.135	1.705	3.828	

R(QR)(QR)

R_s (ohm·cm^2)	Q_1 (S·s^n/cm^2)	n_1	R_1 (Ω·cm^2)	Q_2 (S·s^n/cm^2)	n_2	R_2 (Ω·cm^2)	$10^3 \cdot \chi^2$
29.26	0.01193	0.610	73.63	0.02281	0.9851	91.51	0.440
ε: 0.6998	13.06	3.691	129.8	100.2	17.1	90.7	

R(CR)(QR)

R_s (ohm·cm^2)	C (F/cm^2)	R_1 (Ω·cm^2)	Q (S·s^n/cm^2)	n	R_{ct} (Ω·cm^2)	$10^3 \cdot \chi^2$
29.27	0.02468	85.33	0.01175	0.612	80.47	0.438
ε: 0.6437	13.51	13.55	6.191	3.33	21.8	

The authors tried three circuits that seem to be suitable for interpretation, with the values presented in Table 5.

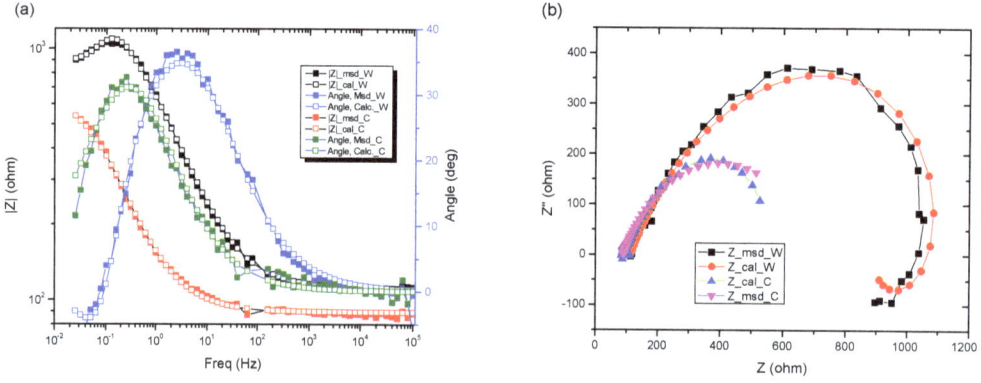

Figure 8. Electrochemical measurements of tested samples for (a) Bode plots, (b) Nyquist plots.

In the circuit used in Table 6, R_s, Q, and R_{ct} have the same meaning as in the previous circuits; Q and R_{ct} characterize the double-electric layer that controls the corrosion rate, in the Nyquist diagram being represented by the capacitive loop in the dolmen of high frequencies. In the same circuit W represents an inductance and R_L is the total resistance of the adsorbed particles on the surface unit of the sample.

Table 6. The values of the equivalent circuit for sample C.

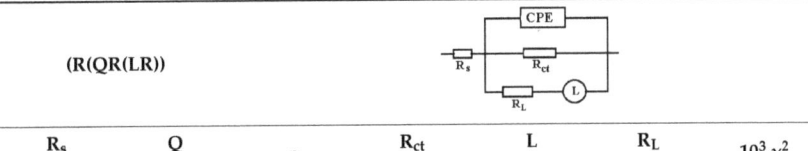

(R(QR(LR))							
R_s (ohm·cm^2)	Q (S·s^n/cm^2)	n	R_{ct} (ohm·cm^2)	L (Henry·cm^2)	R_L (ohm·cm^2)	$10^3 \cdot \chi^2$	
36.29	0.001161	0.665	424.1	1881	655.8	1.30	
ε%: 1.737	4.379	1.829	4.237	12.94	7.04		

Note that the adsorbed intermediate does not form a compact or porous layer but has the appearance of islands on the surface of the sample, a structure that favors a relaxation or reduction of sample capacity in the corrosion process (the effective active surface of the sample is reduced).

In this case, the effective capacity of the electric double layer is calculated using the modified Brug relation:

$$C_{dl} = \left[Q \left(\frac{1}{R_s} + \frac{1}{R_p} \right)^{n-1} \right]^{\frac{1}{n}} \tag{10}$$

where R_p is the polarization resistance: $R_p = ((R_{ct} \cdot R_L)/R_{ct} + R_L)$

This circuit describes a system in which a single uniform corrosion reaction takes place, without the involvement of other types of processes such as diffusion or adsorption.

Moreover, both the Nyquist and Bode curves indicate that the experimental data are suitable for a circuit with a single time constant: the R (CR) circuit.

4. Conclusions

An iron-based SMA was realized thorough a classical melting method with a new chemical composition for possible applications in the medical field as biodegradable material. The analysis of the experimental results led to the following conclusions:

- A smart Fe-based biodegradable alloy can be applied to the medical field with a proper thermo-mechanical treatment to modify the transformation temperatures;
- Reactions between the alloy and electrolyte modify the pH of the environment;
- The cast sample presents a higher corrosion rate than the wrought one based on the microstructure arrangement, microstructural defects, or grain dimension being suitable for certain medical applications;
- The corrosion current density j_{corr} value for W, which is directly proportional to the corrosion rate, is lower than the value compared for C. This can be attributed to the surface defects on the sample C;
- Based on the analysis of cyclic polarization diagrams, it can be observed that there is not any significant difference between the two samples. They present uniform (generalized) corrosion when polarized anodically;
- The data obtained from EIS experiments show that, for these systems, the behavior in solution can be described by using a circuit equivalent to a single time constant; a Randles-type circuit, which indicates that the corrosion occurs on the entire surface of the sample (generalized corrosion) by a single chemical reaction (most likely corrosion of iron with the formation of soluble and insoluble products).

Author Contributions: Conceptualization, R.C., G.Z. and N.C.; data curation, N.I.; formal analysis, A.M.R., V.G., R.C., C.M. and N.C.; funding acquisition, I.I. and N.C.; investigation, A.M.R., V.G., C.M., N.M.L., G.Z., E.R.C. and N.I.; methodology, R.C., N.M.L., G.Z. and E.R.C.; project administration, N.C.; resources, C.M. and I.I.; software, N.M.L.; supervision, I.I.; validation, N.M.L., I.I. and N.I.; visualization, E.R.C.; writing—original draft, A.M.R., R.C., G.Z. and N.C.; writing—review and editing, R.C. and N.C. All authors have read and agreed to the published version of the manuscript.

Funding: This research was funded by publication grant of the TUIASI, project number GI/P2/2021.

Institutional Review Board Statement: Not applicable.

Informed Consent Statement: Not applicable.

Data Availability Statement: All data presented in this study are contained within the article.

Acknowledgments: Not applicable.

Conflicts of Interest: The authors declare no conflict of interest.

References

1. Hermawan, H.; Mantovani, D. Degradable metallic biomaterials: The concept, current developments and future directions. *Minerva Biotecnol.* **2009**, *21*, 207–216.
2. Datta, M.K.; Chou, D.T.; Hong, D.; Saha, P.; Chung, S.J.; Lee, B.; Sirinterlikci, A.; Ramanathan, M.; Roy, A.; Kumta, P.N. Structure and Thermal Stability of Bio-Degradable Mg–Zn–Ca Based Amorphous Alloys Synthesized by Mechanical Alloying. *Mater. Sci. Eng. B* **2011**, *176*, 1637–1643. [CrossRef]
3. Wegener, B.; Sievers, B.; Utzschneider, S.; Müller, P.; Jansson, V.; Rößler, S.; Nies, B.; Stephani, G.; Kieback, B.; Quadbeck, P. Microstructure, Cytotoxicity and Corrosion of Powdermetallurgical Iron Alloys for Biodegradable Bone Replacement Materials. *Mater. Sci. Eng. B* **2011**, *176*, 1789–1796. [CrossRef]
4. Moravej, M.; Prima, F.; Fiset, M.; Mantovani, D. Electroformed iron as new biomaterial for degradable stents: Development process and structure-properties relationship. *Acta Biomater.* **2010**, *6*, 1726–1735. [CrossRef] [PubMed]
5. Hermawan, H.; Dube, D.; Mantovani, D. Developments in Metallic Biodegradable Stents. *Acta Biomater.* **2010**, *6*, 1693–1697. [CrossRef]
6. El-Omar, M.M.; Dangas, G.; Iakovou, I.; Mehran, R. Update on in-stent restenosis. *Curr. Intervent. Cardiol. Rep.* **2001**, *3*, 296–305.
7. Schomig, A.; Kastrati, A.; Mudra, H.; Blasini, R.; Schuhlen, H.; Klauss, V.; Richardt, G.; Neumann, F.J. Four-year experience with Palmaz-Schatz stenting in coronary angioplasty complicated by dissection with threatened or present vessel closure. *Circulation* **1994**, *90*, 2716–2724. [CrossRef]
8. Serruys, P.W.; Kutryk, M.J.; Ong, A.T. Coronary-artery stents. *N. Engl. J. Med.* **2006**, *354*, 483–495. [CrossRef]
9. Hermawan, H.; Mantovani, D. New generation of medical implants: Metallic biodegradable coronary stent. In Proceedings of the International Conference on Instrumentation, Communications, Information Technology, and Biomedical Engineering (ICICI-BME), Bandung, Indonesia, 8–9 November 2011; pp. 399–402.
10. Hänzi, A.C.; Metlar, A.; Schinhammer, M.; Aguib, H.; Lüth, T.C.; Löffler, J.F.; Uggowitzer, P.J. Biodegradable wound-closing devices for gastrointestinal interventions: Degradation performance of the magnesium tip. *Mater. Sci. Eng. C* **2011**, *3*, 1098–1103. [CrossRef]
11. Chng, C.B.; Lau, D.P.; Choo, J.Q.; Chui, C.K. A bio-absorbable micro-clip for laryngeal microsurgery—Design and evaluation. *Acta Biomater.* **2012**, *8*, 2835–2844. [CrossRef] [PubMed]
12. Mueller, P.P.; May, T.; Perz, A.; Hauser, H.; Peuster, M. Control of smooth muscle cell proliferation by ferrous iron. *Biomaterials* **2006**, *27*, 2193–2200. [CrossRef]
13. Fontcave, M.; Pierre, J.L. Iron: Metabolism, toxicity and therapy. *Biochimie* **1993**, *75*, 767–773. [CrossRef]
14. Zhu, S.; Huang, N.; Xu, L.; Zhang, Y.; Liu, H.; Sun, H.; Leng, Y. Biocompatibility of pure iron: In vitro assessment of degradation kinetics and cytotoxicity on endothelial cells. *Mater. Sci. Eng. C* **2009**, *29*, 1589–1592. [CrossRef]
15. Peuster, M.; Wohlsein, P.; Brugmann, M.; Ehlerding, M.; Seidler, K.; Fink, C.; Brauer, H.; Fischer, A.; Hausdorf, G. A novel approach to temporary stenting: Degradable cardiovascular stents produced from corrodible metal-results 6–18 months after implantation into New Zealand white rabbits. *Heart* **2001**, *86*, 563–569. [CrossRef]
16. Dargusch, M.S.; Dehghan-Manshadi, A.; Shahbazi, M.; Venezuela, J.; Tran, X.; Song, J.; Liu, N.; Xu, C.; Ye, Q.; Wen, C. Exploring the Role of Manganese on the Microstructure, Mechanical Properties, Biodegradability, and Biocompatibility of Porous Iron-Based Scaffolds. *ACS Biomater. Sci. Eng.* **2019**, *5*, 1686–1702. [CrossRef]
17. Zhang, Y.S.; Zhu, X.M.; Zhong, S.H. Effect of alloying elements on the electrochemical polarization behavior and passive film of Fe–Mn base alloys in various aqueous solutions. *Corros. Sci.* **2004**, *46*, 853–876. [CrossRef]
18. Heiden, M.; Walker, E.; Nauman, E.; Stanciu, L. Evolution of novel bioresorbable iron-manganese implant surfaces and their degradation behaviors in vitro. *J. Biomed. Mater. Res. A* **2015**, *103*, 185–193. [CrossRef] [PubMed]
19. Hermawan, H.; Moravej, M.; Dube, D.; Mantovani, D. Degradation behaviour of metallic biomaterials for degradable stents. *Adv. Mater. Res.* **2007**, *15*, 113–118.
20. Hermawan, H.; Purnama, A.; Dube, D.; Couet, J.; Mantovani, D. Fe–Mn alloys for metallic biodegradable stents: Degradation and cell viability studies. *Acta Biomater.* **2010**, *6*, 1852–1860. [CrossRef] [PubMed]
21. Hermawan, H.; Dubé, D.; Mantovani, D. Degradable metallic biomaterials: Design and development of Fe-Mn alloys for stents. *J. Biomed. Mater. Res. A* **2010**, *93*, 1–11. [CrossRef]
22. Hermawan, H.; Mantovani, D. Process of prototyping coronary stents from biodegradable Fe–Mn alloys. *Acta Biomater.* **2013**, *9*, 8585–8592. [CrossRef]

23. Schinhammer, M.; Hänzi, A.C.; Löffler, J.F.; Uggowitzer, P.J. Design Strategy for Biodegradable Fe-Based Alloys for Medical Applications. *Acta Biomater.* **2010**, *6*, 1705–1713. [CrossRef]
24. Liu, B.; Zheng, Y.F. Effects of Alloying Elements (Mn, Co., Al, W, Sn, B, C and S) on Biodegradability and In Vitro Biocompatibility of Pure Iron. *Acta Biomater.* **2011**, *7*, 1407–1420. [CrossRef]
25. Hermawan, H. *Biodegradable Metals—From Concept to Applications*; Springer: Berlin, Germany, 2012.
26. Wada, M.; Naoi, H.; Yasuda, H.; Maruyama, T. Shape recovery characteristics of biaxially prestrained Fe–Mn–Si-based shape memory alloy. *Mater. Sci. Eng. A* **2008**, *481*, 178–182. [CrossRef]
27. Wang, Y.; Venezuela, J.; Dargusch, M. Biodegradable shape memory alloys: Progress and prospects. *Biomaterials* **2021**, *279*, 121215. [CrossRef] [PubMed]
28. Liu, B.; Zheng, Y.F.; Ruan, L. In vitro investigation of Fe30Mn6Si shape memory alloy as potential biodegradable metallic material. *Mater. Lett.* **2011**, *65*, 540–543. [CrossRef]
29. Cimpoesu, N.; Săndulache, F.; Istrate, B.; Cimpoeșu, R.; Zegan, G. Electrochemical Behavior of Biodegradable FeMnSi–MgCa Alloy. *Metals* **2018**, *8*, 541. [CrossRef]
30. Burduhos-Nergis, D.P.; Vasilescu, G.D.; Burduhos-Nergis, D.D.; Cimpoesu, R.; Bejinariu, C. Phosphate Coatings: EIS and SEM Applied to Evaluate the Corrosion Behavior of Steel in Fire Extinguishing Solution. *Appl. Sci.* **2021**, *11*, 7802. [CrossRef]
31. Cimpoesu, N.; Mihalache, E.; Lohan, N.M.; Suru, M.G.; Comaneci, R.I.; Ozkal, B.; Bujoreanu, L.G.; Pricop, B. Structural-morphological fluctuations induced by thermomechanical treatment in a Fe—Mn—Si Shape memory alloy. *Met. Sci. Heat Treat.* **2018**, *60*, 471–477. [CrossRef]
32. Rahman, R.A.; Juhre, D.; Halle, T.; Mehmood, S.; Asghar, W. Types, DSC Thermal Characterization of Fe-Mn-Si based Shape Memory Smart Materials and their Feasibility for Human Body in Terms of Austenitic Start Temperatures. *J. Eng. Technol.* **2019**, *8*, 185–206.
33. Li, H.; Dunne, D.; Kennon, N. Factors influencing shape memory effect and phase transformation behaviour of Fe–Mn–Si based shape memory alloys. *Mater. Sci. Eng. A* **1999**, *273*, 517–523. [CrossRef]
34. Donner, P.; Hornbogen, E.; Sade, M. Shape memory effects in melt-spun Fe-Mn-Si alloys. *J. Mater. Sci. Lett.* **1989**, *8*, 37–40. [CrossRef]
35. Forsberg, A.; Ågren, J. Thermodynamic evaluation of the Fe-Mn-Si system and the γ/ε martensitic transformation. *J. Phase Equilib.* **1993**, *14*, 354–363. [CrossRef]
36. Stanford, N.; Dunne, D.P. Martensite/particle interactions and the shape memory effect in an Fe–Mn–Si-based alloy. *J. Mater. Sci.* **2007**, *42*, 4334–4343. [CrossRef]
37. Dogan, A.; Arslan, H. Effect of ball-milling conditions on microstructure during production of Fe–20Mn–6Si–9Cr shape memory alloy powders by mechanical alloying. *J. Therm. Anal. Calorim.* **2012**, *109*, 933–938. [CrossRef]
38. Ariapour, A.; Perovic, D.D.; Yakubtsov, I. Shape-memory effect and strengthening mechanism in an Nb and N-doped Fe-Mn-Si-based alloy. *Metall. Mater. Trans. A* **2001**, *32*, 1621–1628. [CrossRef]
39. Jiang, B.H.; Qi, X.A.; Zhou, W.M.; Xi, Z.L.; Hsu, T.Y. The effect of nitrogen on shape memory effect in Fe-Mn-Si alloys. *Scr. Mater.* **1996**, *34*, 1437–1441. [CrossRef]
40. Eyercioglu, O.; Kanca, E.; Pala, M.; Ozbay, E. Prediction of martensite and austenite start temperatures of the Fe-based shape memory alloys by artificial neural networks. *J. Mater. Proc. Technol.* **2008**, *200*, 146–152. [CrossRef]
41. Tomota, Y.; Nakagawara, W.; Tsuzaki, K.; Maki, T. Reversion of stress-induced ε martensite and two-way shape memory in Fe-24Mn and Fe-24Mn-6Si alloys. *Scr. Metall. Mater.* **1992**, *26*, 1571–1574. [CrossRef]
42. Tsuzaki, K.; Murakami, Y.; Natsume, Y.; Maki, T. Role of pre-straining of austenite in the improvement of shape memory effect in an Fe-33Mn-6Si alloy. In Proceedings of the Symposium K: Environment Conscious Materials of the 3rd IUMRS International Conference on Advanced Materials, Ikebukuro, Tokyo, Japan, 31 August–4 September 1994; pp. 961–964.
43. Gavriljuk, V.G.; Bliznuk, V.V.; Shanina, B.D.; Kolesnik, S.P. Effect of silicon on atomic distribution and shape memory in Fe–Mn base alloys. *Mater. Sci. Eng. A* **2005**, *406*, 1–10. [CrossRef]
44. Kikuchi, T.; Kajiwara, S.; Tomota, Y. Microscopic Studies on Stress-induced Martensite Transformation and Its Reversion in an Fe–Mn–Si–Cr–Ni Shape Memory Alloy. *Mater. Trans. JIM* **1995**, *36*, 719–728. [CrossRef]
45. Sawaguchi, T.; Maruyama, T.; Otsuka, H.; Kushibe, A.; Inoue, Y.; Tsuzaki, K. Design concept and applications of Fe-Mn-Si-based alloys-from shape-memory to seismic response control. *Mater. Trans.* **2016**, *57*, 283–293. [CrossRef]
46. Remy, L.; Pineau, A. Twinning and strain-induced F.C.C. → H.C.P. transformation in the Fe Mn Cr C system. *Mater. Sci. Eng.* **1977**, *28*, 99–107. [CrossRef]
47. Cimpoesu, N.; Trinca, L.C.; Bulai, G.; Stanciu, S.; Gurlui, S.; Mareci, D. Electrochemical Characterization of a New Biodegradable FeMnSi Alloy Coated with Hydroxyapatite-Zirconia by PLD Technique. *J. Chem. N. Y.* **2016**, *2016*, 9520972. [CrossRef]
48. Mouzou, E.; Paternoster, C.; Tolouei, R.; Purnama, A.; Chevallier, P.; Dubé, D.; Prima, F.; Mantovani, D. In vitro degradation behavior of Fe–20Mn–1.2C alloy in three different pseudo-physiological solutions. *Mater. Sci. Eng. C* **2016**, *61*, 564–573. [CrossRef]
49. Babacan, N.; Kochta, F.; Hoffmann, V.; Gemming, T.; Kühn, U.; Giebeler, L.; Gebert, A.; Hufenbach, J. Effect of silver additions on the microstructure, mechanical properties and corrosion behavior of biodegradable Fe-30Mn-6Si. *Mat. Tod. Comm.* **2021**, *28*, 102689. [CrossRef]
50. Wolff, U.; Schneider, F.; Mummert, K.; Schultz, L. Stability and electrochemical properties of passive layers on Fe-Si alloys. *Corrosion* **2000**, *56*, 1195–1201. [CrossRef]

51. Mansfield, F.; Bertocci, U. *Electrochemical Corrosion Testing*; ASTM STP 1981: Philadelphia, PA, USA, 1981; p. 727.
52. Conway, B.E.; Bockris, J.O.M.; White, R.E. *Modern Aspects of Electrochemistry*; Kluwer Academic/Plenum Publishers: New York, NY, USA, 1999; Volume 32, p. 143.
53. Baciu, E.R.; Cimpoesu, R.; Vitalariu, A.; Baciu, C.; Cimpoesu, N.; Sodor, A.; Zegan, G.; Murariu, A. Surface Analysis of 3D (SLM) Co-Cr-W Dental Metallic Materials. *Appl. Sci.* **2021**, *11*, 255. [CrossRef]

Article

Corrosion Resistance of Electrochemically Synthesized Modified Zaccagnaite LDH-Type Films on Steel Substrates

Michael Kahl and Teresa D. Golden *

Department of Chemistry, University of North Texas, Denton, TX 76203, USA; mkahl87@gmail.com
* Correspondence: tgolden@unt.edu

Abstract: Modified zaccagnaite layered double hydroxide (LDH) type films were synthesized on steel substrates by pulsed electrochemical deposition from aqueous solutions. The resulting films were characterized by X-ray diffraction, scanning electron microscopy/X-ray dispersive spectroscopy, and Fourier transform infrared spectroscopy. Structural characterization indicated a pure layered double hydroxide phase; however, elemental analysis revealed that the surface of the films contained Zn:Al ratios outside the typical ranges of layered double hydroxides. Layer thickness for the deposited films ranged from approximately 0.4 to 3.0 μm. The corrosion resistance of the film was determined using potentiodynamic polarization experiments in 3.5 wt.% NaCl solution. The corrosion current density for the coatings was reduced by 82% and the corrosion potential was shifted 126 mV more positive when 5 layers of modified LDH coatings were deposited onto the steel substrates. A mechanism was proposed for the corroding reactions at the coating.

Keywords: zaccagnaite; LDH; corrosion protection; ceramic matrix composites; electrochemical deposition

Citation: Kahl, M.; Golden, T.D. Corrosion Resistance of Electrochemically Synthesized Modified Zaccagnaite LDH-Type Films on Steel Substrates. *Materials* 2021, 14, 7389. https://doi.org/10.3390/ma14237389

Academic Editors: Costica Bejinariu and Nicanor Cimpoesu

Received: 14 October 2021
Accepted: 30 November 2021
Published: 2 December 2021

Publisher's Note: MDPI stays neutral with regard to jurisdictional claims in published maps and institutional affiliations.

Copyright: © 2021 by the authors. Licensee MDPI, Basel, Switzerland. This article is an open access article distributed under the terms and conditions of the Creative Commons Attribution (CC BY) license (https://creativecommons.org/licenses/by/4.0/).

1. Introduction

Steels are utilized in many applications for a variety of industries, including architectural/civil engineering, medical equipment, oil and gas, food and drink processing/storage, water treatment/transport, automotive, and pharmaceutical [1–7]. Stainless steel has good corrosion resistance in various corrosive environments, with resistance derived from its chromium component. A minimum of 12% chromium allows for the formation of a protective chromium oxide layer. The oxide layer is self-repairing in oxygen-rich environments [8]. However, stainless steels are susceptible to localized corrosion in various environments [9,10]. Stainless steels are often coated to prolong lifetime when utilized in a chloride environment [11,12]. Many types of coatings have been developed and recently there has been a shift towards more environmentally friendly coatings. There have been several studies showing the potential of using layered double hydroxides (LDHs) which are anionic clays as coatings, for metal or alloy substrates [13–16].

LDHs are a class of layered anionic clays derived from the natural clay hydrotalcite. They are comprised of metal hydroxide layers with anions and water in the interlayer regions between the metal sheets. LDH is represented by the formula $[M^{2+}_{1-x}M^{3+}_{x}(OH)_2][A^{n-}]_{x/n}\cdot zH_2O$, where M^{2+} and M^{3+} are divalent and trivalent metal cations, respectively. A^{n-} is an anion such as CO_3^{2-} or NO_3^{-}, x is the $M^{3+}/(M^{2+} + M^{3+})$ ratio, and z is the number of associated water molecules [17]. The positive charge is derived from the substitution of divalent ions with trivalent ions in brucite-like metal hydroxide. This positive charge is balanced by interlayer anions which can be exchanged. A divalent—to-trivalent ratio ($M^{2+}:M^{3+}$) between 4:1 and 2:1 is considered the range for a material to be reliably classified as an LDH, although there are exceptions [17–19].

LDHs are thermally stable, inexpensive and eco-friendly and are known for their high anion exchange capacities and adsorption properties. Because of their anion exchange properties, LDHs can be used in corrosion resistant coatings. The coating can undergo anion exchange to trap chloride ions and prevent them from attacking the substrate [20]. Tedim et al. showed that an LDH-NO$_3$ coating can lower the permeability of chlorides to the substrate, delaying coating degradation and improving corrosion resistance. The LDH coating acted as a 'nanotrap' for chloride ions [20]. LDH coatings also exhibit barrier properties, especially when the thickness is in the micron range [21]. LDH films can be spin-coated to produce thicker films improving corrosion protection by blocking penetration of the chloride ions [22]. Hydrotalcite films have also been prepared to provide barrier (passive) protection against chloride attack due to a decrease in pinhole type defects [23]. The growth mechanism along with the corrosion mechanism for the Mg-Al hydrotalcite films have been studied on magnesium alloys [24–26]. Mg-Al-CO$_3$ LDH films were synthesized by a combined co-precipitation and hydrothermal process on AZ31 alloy [27]. These films increased the corrosion protection of the substrate but required a 48 h synthesis, 12 h aging process and a 24–48 h heat treatment in an autoclave. Films containing both crystalline Mg(OH)$_2$ and Mg-Al-CO$_3$ LDH were generated by a steam coating method on magnesium alloy AZ31 at temperatures up to 453 K [28]. These films displayed excellent corrosion resistance in 5 wt.% NaCl solution. Films grown by in situ crystallization have exhibited self-healing properties in 3.5 wt.% NaCl [29].

Electrochemical deposition is another technique for the in situ generation of films on various substrates [30]. Electrodeposition of thin films is an attractive technique because of the low cost, simple setup, short synthesis time, and ability to deposit on large or unconventional substrate shapes [11]. Furthermore, there is greater control over film properties and deposition rate by changing the deposition parameters [31–33]. Zaccagnaite is a hexagonal Zn-Al-LDH, a substituted variant of hydrotalcite, and represented by the formula Zn$_4$Al$_2$(OH)$_{12}$[CO$_3$] 3H$_2$O [19]. It has a metal ratio of 2:1 in natural mineral formations and various ratios have been synthetically prepared. There has been very little work done on the direct electrodeposition of LDH-type coatings for corrosion protection. Yarger et al. electrodeposited Zn-Al-NO$_3$ films onto gold-coated glass substrates with a nitrate solution containing Zn^{2+} and Al^{3+} ions [34]. An optically transparent Li-Al-CO$_3$ LDH was electrochemically deposited onto AZ31 substrate from a Li$^+$/Al^{3+} aqueous solution [21]. The coating provided excellent corrosion protection to the substrate but synthesis of the electrolyte solution required many steps. Wu et al. deposited Zn-Al-NO$_3$ films onto AZ91D Mg alloy substrate in a Zn^{2+}/Al^{3+} aqueous solution [35]. The LDH coating showed great corrosion resistance and improved adhesion to the substrate. For LDH electrodeposition, the electrochemical generation of OH$^-$ species via a reduction reaction of the nitrate ions is important since the increase in local pH facilitates the formation of the LDH films at the substrate surface [34]. However, for Zn-Al LDH deposition, other Zn phases can interfere with the LDH deposition if the deposition conditions are not precisely controlled. If the deposition potential is too negative, Zn and Zn hydroxide species may form and if the deposition potential is too positive Zn oxide species may deposit [36]. Additionally, if the OH$^-$ production rate slows, Al(OH)$_3$ precipitate can compete with LDH deposition. If the OH$^-$ production rate is kept high enough, then LDH directly forms [37]. Since the M(II) and M(III) ions are rapidly consumed near the electrode surface, we therefore propose a pulse deposition method to produce the LDH type coatings. Pulsing the deposition conditions resets the deposition process allowing regeneration of the OH$^-$ species with each pulse step and time for diffusion of the metal species. This could improve protection capacity of the resulting LDH-type coatings. In this study, a facile pulsing method was developed to electrochemically deposit modified zaccagnaite films onto a steel substrate at room temperature. These films are modified zaccagnaite materials for two reasons. First, the elemental ratio of Zn:Al is outside the typical range for LDH and secondly the carbonate group has been exchanged with nitrate. A designed step potential method was used to synthesize the films. A short pulse deposition duration followed by drying of the film was repeated up to five times to mitigate the fracturing of the film which commonly occurs in hydroxide and oxide films prepared from aqueous solutions. These films were characterized and tested for their corrosion resistance in 3.5 wt.% NaCl aqueous solution.

2. Materials and Methods

2.1. Film Synthesis

The substrates were stainless-steel (430) discs from Ted Pella, Inc (Redding, CA, USA). with a diameter of 10 mm, a thickness of 0.76 mm and an area of 1.77 cm^2. The substrates contained Fe, <0.12% C, 16–18% Cr, <0.75% Ni, <1.0% Mn, <1.0% Si, <0.040% P, and <0.030% S by weight. The discs were ground with SiC 320 grit paper (LECO, St. Joseph, MI, USA) and then degreased by sonication in acetone, and then attached to coiled copper wire leads with conductive silver epoxy (EPO-TEK, Billerica, MA, USA). Once dry, the substrate was mounted in epoxy utilizing molds. After 24 h curing, the mounted electrodes were ground successively with 320, 400, and 600 grits of SiC paper then polished with 1 μm diamond suspension on a felt polishing pad followed by ultrasonication in ethanol. The resulting substrate surface had a mirror finish with an embedded uniform, flat, even surface with the epoxy.

The electrolytic solution was prepared by dissolving a 2:1 molar ratio of Zn^{2+} to Al^{3+} ions in DI water. Aluminum nitrate nonahydrate ($Al(NO_3)_3 \cdot 9H_2O$, Alfa Aesar, Tewksbury, MA, USA) was the aluminum source, zinc nitrate hexahydrate ($Zn(NO_3)_2 \cdot 6H_2O$, Alfa Aesar, Tewksbury, MA, USA) was the source of zinc, and potassium nitrate (KNO_3, Fisher Scientific, Walham, MA, USA) was used as the electrolyte to help facilitate the formation of modified zaccagnaite film at the electrode surface. The electrolytic concentration of each compound was 0.02 M $Zn(NO_3)_2 \cdot 6H_2O$, 0.01 M $Al(NO_3)_3 \cdot 9H_2O$, and 0.2 M KNO_3. The final pH of the deposition solution was 3.8. An EG&G Princeton Applied Research (PAR) Model 273A potentiostat/galvanostat (Oak Ridge, TN, USA) was used to electrochemically deposit films. The depositions were performed at room temperature utilizing a three-electrode configuration. Due to the fact that carbonate ions (carbon dioxide dissolution) have an exceptionally high affinity to the LDHs [18], the deposition solution was purged with nitrogen for 20 min prior to film fabrication in order to minimize carbonate contamination and allow nitrate insertion in the interlayers. The working electrode was the polished stainless-steel disc, a platinum mesh was used as the counter electrode and the reference electrode was a saturated calomel electrode (SCE) (Fisher Scientific, Walham, MA, USA). Films were deposited in multiple steps. A step potential method was used for film deposition. The applied potential started at −1.5 V vs. SCE for 5 sec and then stepped to −1.0 V vs. SCE for 20 sec; this cycling was done for a total of 50 sec and represented deposition of one layer. Each layer was electrodeposited and allowed to dry undisturbed before the next layer was added. A total of 1, 2, and 5 layers were deposited, (1L, 2L, 5L).

2.2. Characterization

The structure and phase composition of the modified zaccagnaite films were identified by X-ray diffraction (XRD) using a Siemens D500 diffractometer (KSA, Aubrey, TX, USA) with Cu Kα radiation (λ = 1.5405 Å) as the source in a standard Bragg-Brentano configuration. The x-ray tube was operated at 35 kV and 24 mA. Each sample was scanned from 2 to 40° (2θ), at a step size of 0.05° and a dwell time of 1.0 s. The surface morphology of the films was characterized by scanning electron microscopy (SEM) with an X-ray dispersive spectroscopy (EDX) attachment (FEI Quanta 200 ESEM, Hillsboro, OR, USA). A spot size of 3.0 and an accelerating voltage of 25 kV were used for SEM. Film thickness measurements were performed with a Veeco Dektak 150 stylus profilometer (Plainview, NY, USA). A Perkin Elmer Spectrum One FT-IR Spectrophotometer (Waltham, MA, USA) was used to analyze the composition of the films. The films were scraped off with a blade and ground up further before being placed onto an ATR stage. Each sample was scanned 16 times at a wavenumber range of 4000–450 cm^{-1}.

2.3. Immersion Test and Corrosion Measurements

Potentiodynamic polarization measurements and immersion tests were performed in 3.5 wt.% NaCl aqueous solutions at room temperature. Electrochemical measurements were conducted with an EG&G Princeton Applied Research (PAR) Model 273A potentiostat/galvanostat. The coated film and a SCE electrode were used as the working and reference electrodes, respectively. Two graphite rods were used as the counter electrodes for polarization measurements. Each sample was immersed in the NaCl solution for 30 min before polarization curves were measured with respect to open circuit potential (OCP) at a scan rate of 1 mV/s. The polarization resistance, R_p, was the slope value obtained from linear polarization (LPR) scanning ±150 mV with respect to E_{corr}. E_{corr} was determined as the point of intersection of the anodic and cathodic polarization branches. To obtain a more accurate estimation of i_{corr}, the cathodic polarization region was used since the anodic region contained current density oscillations. In the tafel plot, a horizontal line was drawn at the E_{corr} value and another horizontal line was drawn 100 mV cathodic from E_{corr}. A slope line was drawn from the 100 mV meeting point on the cathodic branch to intersect with the E_{corr} line. The point of intersection was taken as the value of i_{corr} [38–40]. Immersion tests were performed in 3.5 wt.% NaCl solution at room temperature for up to 168 h to examine the long-term corrosion resistance of the zaccagnaite-coated films.

3. Results and Discussion

3.1. Film Deposition Mechanism

The electrochemical synthesis of hydroxides can occur by electrogeneration of base via a nitrate reduction at the working electrode interface (Equation (1)) [41].

$$NO_3^- + H_2O + 2e^- \rightarrow NO_2^- + 2OH^- \qquad (1)$$

The deposition solution contains nitrates from the supporting electrolyte (0.2 M KNO_3) and also as the salt of the divalent/trivalent cations. By applying a cathodic potential to the electrode, the pH at the electrode surface increases because of the consumption of H^+ and generation of OH^-. The higher the applied cathodic potential, the faster the pH increases; this increase in local pH leads to metal hydroxide precipitation at the electrode surface. Precipitation of metal hydroxides at the electrode consumes OH^- and lowers the pH (Equation (2)) [36,37,42].

$$Zn^{2+} + Al^{3+} + nOH^- + NO_3^- \rightarrow Zn\text{-}Al\ LDH \qquad (2)$$

Characterization of the surface helps identify the structural and chemical properties of the electrodeposited coating.

3.2. Structural Characterization

The x-ray diffraction pattern of the modified zaccagnaite film deposited on the substrate is displayed in Figure 1. The film was grown during one continuous deposition in order to obtain enough material for characterization. The peaks at 9.89° and 20.00° (2θ) represent the characteristic (003) and (006) reflections for LDH. The peak at 44.61° corresponds to the substrate. A basal spacing of 0.89 nm was calculated from the most intense peak at 9.89° according to Bragg's equation. This value is in agreement with previous studies [34,35,42]. The absence of non-basal reflections is evidence that the film is composed of highly oriented platelets [22,43]. The peaks may be of low intensity due to the thickness of the films (0.43 to 2.8 μm) and transparency of the lighter weight elements to X-rays, as well as some slight amorphous nature. Furthermore, the diffraction peaks are slightly broad due to poor crystallinity as also seen with previous electrosynthesized LDH films [37].

The FT-IR spectrum for the deposited film is displayed in Figure 2. The spectrum confirms the presence of hydroxides with only water and nitrate ions in the interlayer region. The broad peak at 3385 cm^{-1} corresponds to O-H stretching of hydroxide and water in the interlayer region. The peak at 1638 cm^{-1} shows the bending vibration of the interlayer water molecules. The peak at 1353 cm^{-1} represents the asymmetric stretching of nitrate ions in the interlayer region [34]. The peaks at 947, 760 and 542 cm^{-1} are associated with Al-O and Zn-O stretching modes [44–47]. The peaks from 2600 to 1800 cm^{-1} are from the diamond ATR surface [48].

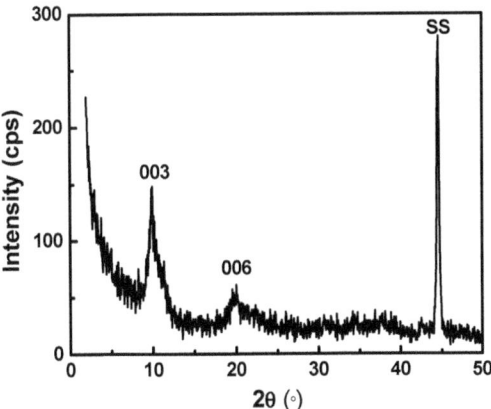

Figure 1. X-ray diffraction pattern of an electrodeposited modified zaccagnaite coating on stainless steel (SS) substrate.

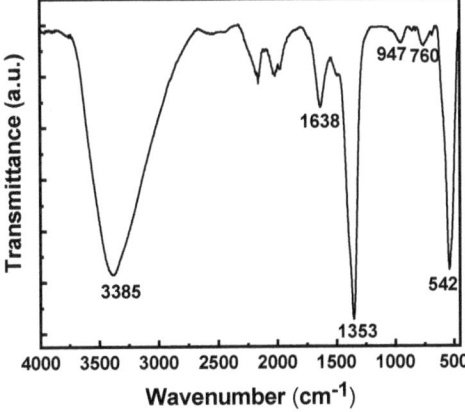

Figure 2. The FTIR spectrum of an electrodeposited modified zaccagnaite film.

Figure 3 shows the surface morphology of modified zaccagnaite films from a one layer (1L), two layer (2L), and five layer (5L) successive depositions. The surface of the 1L film (Figure 3a) shows a flattened disorganized coating with uneven coverage. The elemental ratio of Zn:Al measured by EDX on the surface of the film is approximately 1:1.2. Table 1 lists the elemental ratios and film thicknesses for the coatings. The SEM image of the 2L film (Figure 3b) shows a fairly homogenous surface without the flaws observed in the 1L films (Figure 3a). The double deposition reduced the uneven coverage. The elemental ratio of Zn:Al on the surface of the 2L film is approximately 1:2.8. The 5L film (Figure 3c) exhibits a surface which is different from both the 1L and 2L films. The elemental ratio of the 5L film's surface is approximately 1:5, while this film does not have the uneven coverage in material as seen in the 1L film, there are cracks observed. This cracking and shrinkage of the LDH coatings has been observed by other researchers and is typically attributed to dehydration of the coating in atmosphere and shown to be spontaneous upon drying (Equation (3)) [34,49–51]. Lu et al. showed that a dehydration/rehydration process was reversible by monitoring the LDH gallery spacings with XRD for Ni-Fe LDH coatings [52]. Gualandi et al. measured the gallery d-spacings of several LDH coatings with XRD and noted that Zn-Al LDH had the highest hydration levels (larger d-spacing) compared to Ni-Al LDH and Co-Al LDH coatings [37] and this cracking could even be observed in the SEM cross-sections [53]. Others have shown that hydrogen gas evolution during the deposition can occur at too high of cathodic potentials [35]. Even with the cracking process, adhesion was still good for the coatings [54].

$$[Zn^{2+}_{1-x}Al^{3+}_{x}(OH)_2][NO_3^-]_x \cdot zH_2O \rightarrow [Zn^{2+}_{1-x}Al^{3+}_{x}(OH)_2][NO_3^-]_x + zH_2O \quad (3)$$

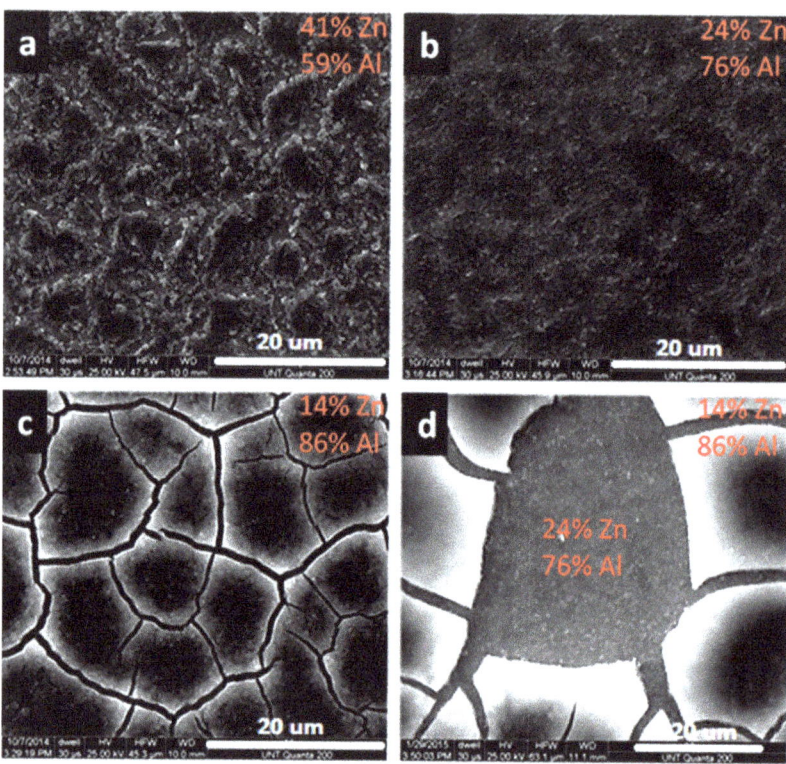

Figure 3. SEM images of modified zaccagnaite films for (**a**) 1L, (**b**) 2L, (**c**) 5L, and (**d**) 5L film with a portion of top phase removed (L = layer). (Image bars are 20 μm).

Table 1. Elemental composition measured by EDS and thickness of the electrodeposited films (n = 3).

Number of Layers	Zinc Atomic %	Aluminum Atomic %	Thickness (nm)
1	41 ± 4	59 ± 4	431 ± 50
2	24 ± 2	76 ± 2	612 ± 46
5	14 ± 3	86 ± 3	2814 ± 45

In Figure 3d, it is apparent that these cracks do not penetrate all the way to the substrate. Two separate phases can be observed, a fractured phase upon a continuous underlying phase. The surface of the underlying phase in Figure 3d has slightly more zinc (1:4 Zn:Al ratio) than the 5L film. Previous research has shown that LDH with an acceptable divalent:trivalent cation ratio is only formed during a certain time frame, dependent on the deposition potential, formation pH of the divalent cation hydroxide and divalent:trivalent cation ratio. At longer synthesis times, an aluminum dominated hydroxide phase is formed [48]. These electrochemical deposited coatings show the same trend, with increasing Al composition corresponding to increasing deposition times.

Film thickness measurements for 1L, 2L, and 5L films are reported in Table 1. The film thickness does not increase proportionally with the number of deposition layers. The lack of linearity may be due to deposition defects caused by dehydration of previous layers. Furthermore, the rate of film growth may change due to the differences in initial growth on the substrate versus growth on the previously deposited film. The formation of two sequential phases may also cause irregular film growth rate. The conductivity variation in each layer and the non-conducting nature of the zinc and aluminum hydroxide film will also contribute to the nonlinear growth rate.

Figure 4 shows a schematic representing the changing deposition mechanism where initially hydroxides are generated at the electrode surface. When the pH increases sufficiently at the electrode, metal hydroxides began to precipitate and any hydroxides not consumed move into the bulk of the solution. The diffusion of hydroxides away from the electrode increases the pH of the solution resulting in the precipitation of aluminum hydroxide. Aluminum hydroxide forms at approximately pH 4 and zinc aluminum LDH precipitates at approximately pH 6 [35,55]. The aluminum hydroxide precipitate coats the mixed hydroxide phase. This deposition process results in the film structure depicted in Figure 5, which is comprised of a mostly aluminum hydroxide phase on top of a mixed hydroxide phase. The zinc content of the mixed hydroxide phase is higher at the substrate interface and decreases as the film thickness increases. Zinc hydroxide is slightly more soluble ($K_{sp} = 3 \times 10^{-17}$) than aluminum hydroxide ($K_{sp} = 3 \times 10^{-34}$), also leading to a higher aluminum hydroxide solid on the surface.

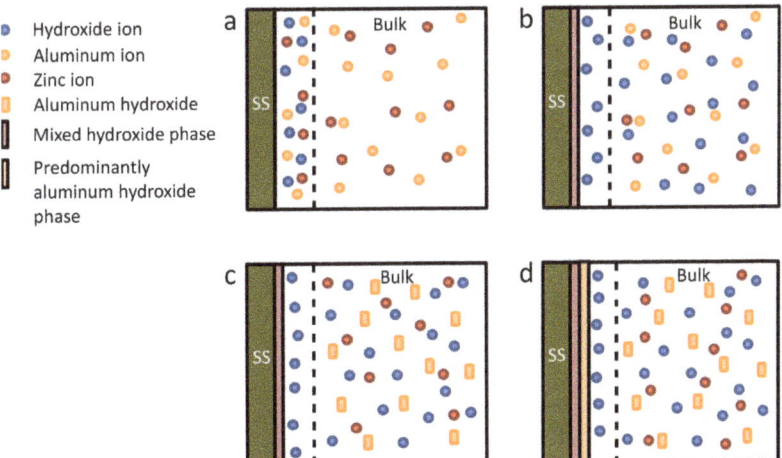

Figure 4. Film formation process of the mixed hydroxide phase and top aluminum hydroxide layer for the electrodeposited zaccagnaite coating on stainless-steel substrate, (**a**) generation of hydroxides at the electrode, (**b**) deposition of metal hydroxides and diffusion of unconsumed hydroxides away from the electrode surface, (**c**) pH increases in the bulk solution resulting in aluminum hydroxide precipitation, and (**d**) aluminum hydroxide precipitate formed in bulk deposits onto mixed hydroxide layer.

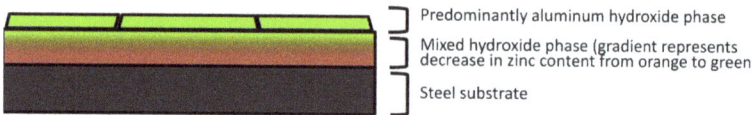

Figure 5. Changing film composition from the substrate to the outer top layer for the electrodeposited zaccagnaite coating.

3.3. Corrosion Resistance

The potentiodynamic polarization curves of the bare substrate, 1L, 2L, and 5L as-deposited films measured in 3.5 wt.% NaCl solution are shown in Figure 6. The corrosion potential (E_{corr}), corrosion current density (i_{corr}), and polarization resistance (R_p) data for each sample are listed in Table 2. Deposition of the modified zaccagnaite film resulted in a positive shift in the E_{corr} and a decrease in the i_{corr} compared to the bare substrate (better corrosion resistance). The 1L film had the greatest effect on E_{corr}, shifting almost 100 mV in the positive direction while the 2L and 5L films had a continued shift in E_{corr} by smaller increments of approximately 30 and 3 mV, respectively. The i_{corr} decreased from 0.66 µA/cm^2 to 0.61, 0.38, and 0.12 µA/cm^2 for the 1L, 2L, and 5L films, respectively. Polarization resistance, R_p, increased with the number of deposited layers resulting in the 5L film having a R_p an order of magnitude larger than the substrate R_p. The beginning of a pitting potential can be observed for the bare substrate sample at the end of the polarization curve around −50 mV. For each deposition layer, an increasing barrier property is observed in the anodic branch of potentiodynamic curves. These polarization measurements indicate that the film provides a barrier to the transport of aqueous species to the substrate so that the ability of the chloride ions to attack the substrate is reduced. This barrier increases and defects are minimized as more layers are deposited. Some current density oscillation is observed in the anodic branch of the polarization curves. The current density oscillation becomes greater as the number of deposited layers increases. This may be due to the dissolution of hydroxides from the coating surface or even some interlayer exchange of species between the film and the solution.

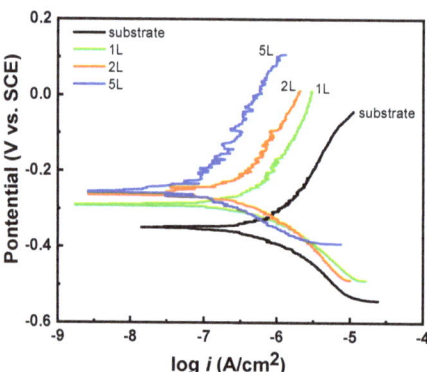

Figure 6. Potentiodynamic polarization curves of the substrate, and 1L, 2L, 5L modified zaccagnaite films measured in a 3.5 wt.% NaCl solution.

The stability of the modified zaccagnaite film in a corrosive marine environment was simulated with immersion testing. A 5L film was immersed at increasing durations up to 168 h in a 3.5 wt.% NaCl solution. Figure 7 shows the polarization curves of the film at four different immersion times. Table 3 lists the E_{corr}, i_{corr}, and R_p as the average of 3 measurements for the 5L film at various times. The E_{corr} remained relatively unchanged at approximately −0.240 V vs. SCE from 1 to 72 h. At 168 h the E_{corr} shifted positively to

−0.178 V vs. SCE and the corrosion resistance potential, R_p, began to decrease indicating beginning of corrosion damage at the longer emersion time. The i_{corr} was also stable starting at 0.12 µA/cm² for 1 h and ending at 0.18 µA/cm² at 168 h. Only the film immersed for 168 h exhibits a pitting potential at the end of the anodic region beginning around 0.0 V suggesting that the film was weakened by the longer exposure to the chloride environment. Current density oscillation is visible again beginning around the corrosion potential and throughout the anodic region.

Table 2. Corrosion potential (E_{corr}), corrosion current densities (i_{corr}), and LPR polarization resistance (R_p) measured for all samples in a 3.5 wt.% NaCl solution (n = 3).

Sample	E_{corr} (V vs. SCE)	i_{corr} (µA/cm²)	R_p (kΩ·cm²)
substrate	−0.363 ± 0.027	0.66 ± 0.01	19 ± 6
1L	−0.269 ± 0.020	0.61 ± 0.19	57 ± 27
2L	−0.240 ± 0.008	0.38 ± 0.11	137 ± 40
5L	−0.237 ± 0.013	0.12 ± 0.01	230 ± 54

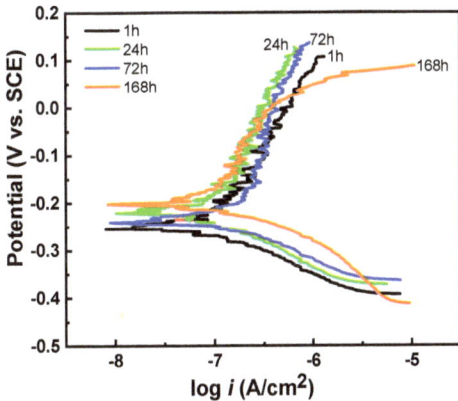

Figure 7. Potentiodynamic polarization curves of 5L films immersed in a 3.5 wt.% NaCl solution for 1, 24, 72, and 168 h.

Table 3. Corrosion potential (E_{corr}), corrosion current densities (i_{corr}), and LPR polarization resistance (R_p) measured for a 5L film in 3.5 wt.% NaCl at various immersion times (n = 3).

Immersion Time (h)	E_{corr} (V vs. SCE)	i_{corr} (µA/cm²)	R_p (kΩ·cm²)
1	−0.239 ± 0.017	0.12 ± 0.02	218 ± 59
24	−0.245 ± 0.031	0.08 ± 0.02	232 ± 37
72	−0.235 ± 0.020	0.18 ± 0.03	309 ± 138
168	−0.178 ± 0.013	0.18 ± 0.02	233 ± 72

SEM images were taken of the 5L film before (Figure 8a) and after immersion in 3.5 wt.% NaCl solution up to 168 h. Figure 8b shows that the aluminum dominated layer on the surface of the film forms pits after immersion of 168 h. These defects range in size from submicron to a couple of microns. These flaws may be evidence of dissociation of the film into hydroxide ions. These defects only affect the top phase and do not appear to penetrate to the substrate because of the stability observed during immersion testing. The top phase may behave as a sacrificial barrier for the underlying coating and possibly releases hydroxide ions. Figure 8d shows a section of the 5L film with the top phase detached after immersion in 3.5 wt.% NaCl solution up to 168 h. While the exposed bottom phase is not free of blemishes, it does not exhibit the pitting that is observed on the top phase.

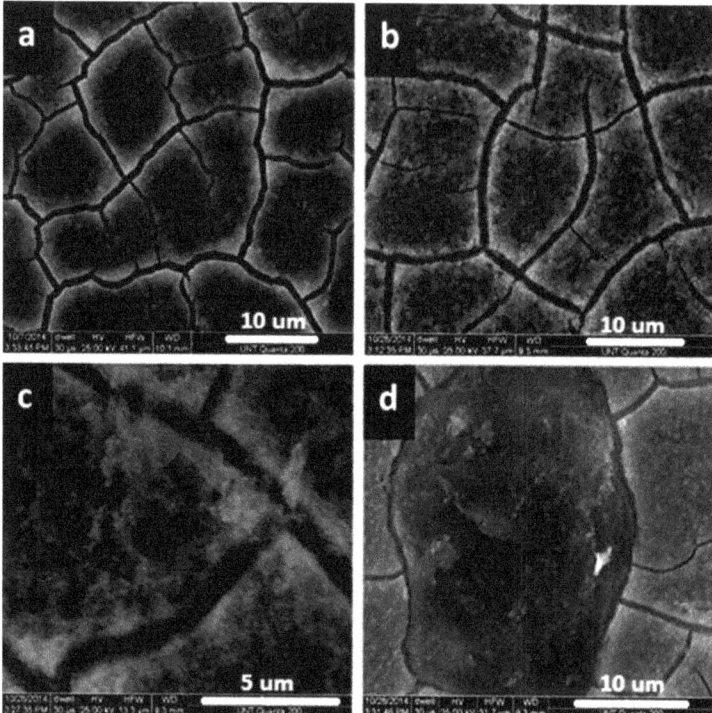

Figure 8. SEM images of 5L film (**a**) before and (**b**) after immersion in 3.5 wt.% NaCl solution for 168 h as well as (**c**) an enlarged image of a defect in the immersed film from (**b**), and (**d**) top section of 5L film removed after 168 h.

The anticorrosion performance of LDH films has been attributed to a multitude of factors. One factor is the high anion exchange capacity of LDH. The nitrate ions in the film have a lower affinity for intercalation than chloride ions. Surface anion exchange with chloride ions traps them and prevents their migration to the substrate. This anion exchange phenomenon has been observed in previous studies [20,21,35]. Another explanation is that the deterioration of the film may release hydroxide ions into the local environment increasing the pH [21]. The release of hydroxide ions can slow down the occurrence of pitting corrosion by reducing the rate of chloride migration to the pit and neutralizing the local solution environment. Previous research has also shown that layered double hydroxides can undergo dissolution/recrystallization or self-healing reactions during the corrosion process [56]. Hydrotalcite has been shown to form a protective amorphous aluminum hydroxide layer to prevent dissolution in mildly acidic solutions [57]. The mostly aluminum hydroxide top phase behaves as a protective coating for the mixed hydroxide phase preventing its dissolution. Figure 9 depicts a proposed corrosion mechanism for the coating after immersion of the film in corrosive media, depicting the sacrificial protection of the aluminum dominate phase and possible crystallization of aluminum hydroxide in pits on the surface. Other explanations for the corrosion resistance of the film is its barrier property. The film is dense and thick enough to prevent the penetration of chloride ions to the substrate surface [21,23]. This can be seen as successive layers are deposited onto the substrate. The corrosion resistance increases from 1 to 5 layers. As the thickness increases (~0.4 to 3 μm), the chloride ions are blocked against attack. Additionally, the film is insulating, resulting in a decrease in the rate of any electrochemical reactions including those involving corrosion. Further study can help determine a comprehensive mechanism of corrosion resistance for the electrodeposited modified zaccagnaite films.

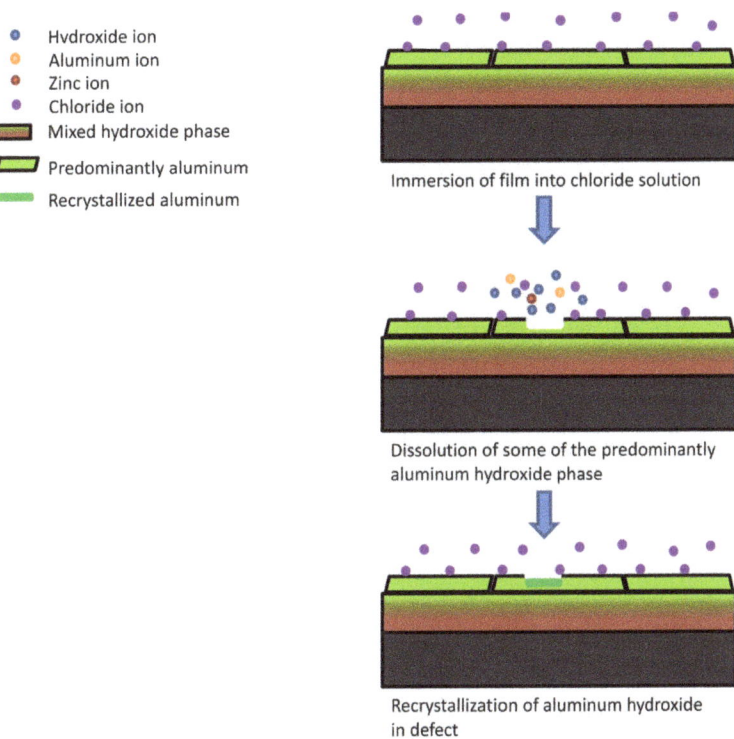

Figure 9. Postulated corrosion mechanism for immersed electrodeposited zaccagnaite film.

4. Conclusions

In this study a facile method was developed to electrochemically deposit modified zaccagnaite films onto a stainless-steel substrate. The electrodeposition occurred in multiple layers in order to minimize defects generated during deposition and drying of the film. XRD and FTIR studies showed the presence of an LDH phase; however, the elemental ratios of Zn:Al were outside typical limits. The metal substrate coated with modified zaccagnaite exhibited higher corrosion resistance than the bare substrate in 3.5 wt.% NaCl solution. Corrosion protection increased with the number of layers deposited. Aluminum concentration in the coating also increased with the number of layers. For the 5-layer coating, E_{corr} shifted positively ~100 mV and the corrosion current density was reduced by 82% when compared to the bare substrate in 3.5 wt.% NaCl. The 5-layered film also maintained its corrosion resistance while immersed for 168 h in 3.5 wt.% NaCl demonstrating that electrochemically generated modified zaccagnaite is an effective material to reduce corrosion on the substrate surface.

Author Contributions: Conceptualization, T.D.G. and M.K.; methodology, T.D.G. and M.K.; resources, T.D.G.; data curation, M.K.; writing—Original draft preparation, M.K.; writing—Review and editing, T.D.G.; visualization, T.D.G. and M.K.; supervision, T.D.G.; project administration, T.D.G.; funding acquisition, T.D.G. All authors have read and agreed to the published version of the manuscript.

Funding: This research was supported by Brewer Scientific and Tauen Analytical.

Institutional Review Board Statement: Not applicable.

Informed Consent Statement: Not applicable.

Data Availability Statement: The data presented in this study are available on reasonable request from the corresponding author.

Acknowledgments: The author acknowledges the Center for Advance Research and Technology (CART) at the University of North Texas for access to the SEM facilities used for this study.

Conflicts of Interest: The authors declare no conflict of interest.

References

1. Baddoo, N.R. Stainless steel in construction: A review of research, applications, challenges and opportunities. *J. Constr. Steel Res.* **2008**, *64*, 1199–1206. [CrossRef]
2. Bekmurzayeva, A.; Duncanson, W.J.; Azevedo, H.S.; Kanayeva, D. Surface modification of stainless steel for biomedical applications: Revisiting a century-old material. *Mater. Sci. Eng. C* **2018**, *93*, 1073–1089. [CrossRef]
3. Huynh, V.; Ngo, N.; Golden, T.D. Review article: Surface activation and pretreatments for biocompatible metals and alloys used in biomedical applications. *Int. J. Biomater.* **2019**, *2019*, 3806504. [CrossRef] [PubMed]
4. Mesquita, T.J.; Chauveau, E.; Mantel, M.; Bouvier, N.; Koschel, D. Corrosion and metallurgical investigation of two supermartensitic stainless steels for oil and gas environment. *Corros. Sci.* **2014**, *81*, 152–161. [CrossRef]
5. Herting, G.; Wallinder, I.O.; Leygraf, C. Corrosion-induced release of chromium and iron from ferritic stainless steel grade AISI 430 in simulated food contact. *J. Food Eng.* **2008**, *87*, 291–300.
6. Bitondo, C.; Bossio, A.; Monetta, T.; Curioni, M.; Bellucci, F. The effect of annealing on the corrosion behavior of 444 stainless steel for drinking water applications. *Corros. Sci.* **2014**, *87*, 6–10. [CrossRef]
7. Sugimoto, K.; Hojo, T.; Srivastava, A.K. Review: Low and medium carbon advanced high-strength forging steels for automotive applications. *Metals* **2019**, *9*, 1263. [CrossRef]
8. Streicher, M.A.; Grubb, J.F. *Austenitic and Ferritic Stainless Steels, in Uhlig's Corrosion Handbook*, 3rd ed.; Revie, R.W., Ed.; John Wiley & Sons, Inc.: Hoboken, NJ, USA, 2011; pp. 657–693.
9. Hu, Q.; Zhang, G.; Qiu, Y.; Guo, X. The crevice corrosion behaviour of stainless steel in sodium chloride solution. *Corros. Sci.* **2011**, *53*, 4065–4072. [CrossRef]
10. Zaffora, A.; Di Franco, F.; Santamaria, M. Corrosion of stainless steel in food and pharmaceutical industry. *Curr. Opin. Electrochem.* **2021**, *29*, 100760.
11. Ngo, N.; Shao, S.; Conrad, H.; Sanders, S.F.; D'Souza, F.; Golden, T.D. Synthesis, characterization, and the effects of organo-grafted nanoparticles in nickel coatings for enhanced corrosion protection. *Mater. Today Commun.* **2020**, *25*, 101628. [CrossRef]
12. González, M.B.; Saidman, S.B. Electrodeposition of bilayered polypyrrole on 316 L stainless steel for corrosion prevention. *Prog. Org. Coat.* **2015**, *78*, 21–27. [CrossRef]
13. Yasakau, K.A.; Kuznetsova, A.; Kallip, S.; Starykevich, M.; Tedim, J.; Ferreira, M.G.S.; Zheludkevich, M.L. A novel bilayer system comprising LDH conversion layer and sol-gel coating for active corrosion protection of AA2024. *Corros. Sci.* **2018**, *143*, 299–313. [CrossRef]
14. Daugherty, R.E.; Zumbach, M.M.; Golden, T.D. Design challenges in electrodepositing metal-anionic clay nanocomposites: Synthesis, characterization, and corrosion resistance of nickel-LDH nanocomposite coatings. *Surf. Coat. Technol.* **2018**, *349*, 773–782. [CrossRef]
15. Iqbal, M.A.; Sun, L.; Barrett, A.T.; Fedel, M. Layered Double Hydroxide protective films developed on aluminum and aluminum alloys: Synthetic methods and anti-corrosion mechanisms. *Coatings* **2020**, *10*, 428. [CrossRef]
16. Kaseem, M.; Ramachandraiah, K.; Hossain, S.; Dikici, B. A review on LDH-smart functionalization of anodic films of Mg alloys. *Nanomaterials* **2021**, *11*, 536. [CrossRef] [PubMed]
17. Evans, D.G.; Slade, R.C.T. Structural aspects of layered double hydroxides. *Struct. Bond.* **2006**, *119*, 1–87.
18. Salak, A.N.; Lisenkov, A.D.; Zheludkevich, M.L.; Ferreira, M.G.S. Carbonate-free Zn-Al (1:1) layered double hydroxide film directly grown on zinc-aluminum alloy coating. *ECS Electrochem. Lett.* **2014**, *3*, C9–C11. [CrossRef]
19. Mills, S.J.; Christy, A.G.; Génin, J.M.R.; Kameda, T.; Colombo, F. Nomenclature of the hydrotalcite supergroup: Natural layered double hydroxides. *Mineral. Mag.* **2012**, *76*, 1289–1336. [CrossRef]
20. Tedim, J.; Kuznetsova, A.; Salak, A.N.; Montemor, F.; Snihirova, D.; Pilz, M.; Zheludkevich, M.L.; Ferreira, M.G.S. Zn–Al layered double hydroxides as chloride nanotraps in active protective coatings. *Corros. Sci.* **2012**, *55*, 1–4. [CrossRef]
21. Syu, J.; Uan, J.; Lin, M.; Lin, Z. Optically transparent Li-Al-CO$_3$ layered double hydroxide thin films on an AZ31 Mg alloy formed by electrochemical deposition and their corrosion resistance in a dilute chloride environment. *Corros. Sci.* **2013**, *68*, 238–248. [CrossRef]
22. Zhang, F.Z.; Sun, M.; Xu, S.L.; Zhao, L.L.; Zhang, B.W. Fabrication of oriented layered double hydroxide films by spin coating and their use in corrosion protection. *Chem. Eng. J.* **2008**, *141*, 362–367. [CrossRef]
23. Zhang, W.; Buchheit, R.G. Hydrotalcite coating formation on Al-Cu-Mg alloys from oxidizing bath chemistries. *Corrosion* **2002**, *58*, 591–600. [CrossRef]

24. Chen, J.; Song, Y.W.; Shan, D.Y.; Han, E.H. Study of the in situ growth mechanism of Mg–Al hydrotalcite conversion film on AZ31 magnesium alloy. *Corros. Sci.* **2012**, *63*, 148–158. [CrossRef]
25. Chen, J.; Song, Y.W.; Shan, D.Y.; Han, E.H. Study of the corrosion mechanism of the in situ grown Mg–Al–CO_3^{2-} hydrotalcite film on AZ31 alloy. *Corros. Sci.* **2012**, *65*, 268–277. [CrossRef]
26. Huang, M.; Lu, G.; Pu, J.; Qiang, Y. Superhydrophobic and smart MgAl-LDH anti-corrosion coating on AZ31 Mg surface. *J. Ind. Eng. Chem.* **2021**, *103*, 154–164. [CrossRef]
27. Zhang, F.; Liua, Z.; Zeng, R.; Li, S.; Cui, H.; Song, L.; Han, E. Corrosion resistance of Mg–Al-LDH coating on magnesium alloy AZ31. *Surf. Coat. Technol.* **2014**, *258*, 1152–1158. [CrossRef]
28. Ishizaki, T.; Chiba, S.; Watanabe, K.; Suzuki, H. Corrosion resistance of Mg–Al layered double hydroxide container-containing magnesium hydroxide films formed directly on magnesium alloy by chemical-free steam coating. *J. Mater. Chem. A* **2013**, *1*, 8968–8977. [CrossRef]
29. Wang, Y.; Zhou, X.; Yin, M.; Pu, J.; Yuan, N.; Ding, J. Superhydrophobic and self-healing Mg-Al Layered Double Hydroxide/Silane composite coatings on the Mg alloy surface with a long-term anti-corrosion lifetime. *Langmuir* **2021**, *37*, 8129–8138. [CrossRef] [PubMed]
30. Kahl, M.; Golden, T.D. Electrochemical determination of phenolic acids at a Zn/Al layered double hydroxide film modified glassy carbon electrode. *Electroanalysis* **2014**, *26*, 1664–1670. [CrossRef]
31. Wang, A.Q.; Golden, T.D. Electrochemical formation of Cerium oxide/Layered Silicate nanocomposite films. *J. Nanotechnol.* **2016**, *2016*, 8459374. [CrossRef]
32. DeLeon, V.; Golden, T.D. Effect of electrochemical parameters on the morphology and Ca/P ratios of deposited apatite coatings on metal and alloy substrates. *ECS Trans.* **2011**, *33*, 43–50. [CrossRef]
33. Singh, A.P.; Roccapriore, K.; Algarni, Z.; Salloom, R.; Golden, T.D.; Philipose, U. Structure and electronic properties of InSb nanowires grown in flexible polycarbonate membranes. *Nanomaterials* **2019**, *9*, 1260. [CrossRef] [PubMed]
34. Yarger, M.S.; Steinmiller, E.M.P.; Choi, K.S. Electrochemical synthesis of Zn-Al layered double hydroxide (LDHs) films. *Inorg. Chem.* **2008**, *47*, 5859–5865. [CrossRef] [PubMed]
35. Wu, F.; Liang, J.; Peng, Z.; Liu, B. Electrochemical deposition and characterization of Zn-Al layered double hydroxides (LDHs) films on magnesium alloy. *Appl. Surf. Sci.* **2014**, *313*, 834–840. [CrossRef]
36. He, Q.-Q.; Zhou, M.-J.; Hu, J.-M. Electrodeposited Zn-Al layered double hydroxide films for corrosion protection of aluminum alloys. *Electrochim. Acta* **2020**, *355*, 136796. [CrossRef]
37. Gualandi, I.; Monti, M.; Scavetta, E.; Tonelli, D.; Prevot, V.; Mousty, C. Electrodeposition of layered double hydroxides on platinum: Insights into the reactions sequence. *Electrochim. Acta* **2015**, *152*, 75–83. [CrossRef]
38. Jayaraj, J.; Raj, S.A.; Srinivasan, A.; Ananthakumar, S.; Pillai, U.T.S.; Dhaipule, N.G.K.; Mudali, U.K. Composite magnesium phosphate coatings for improved corrosionresistance of magnesium AZ31 alloy. *Corros. Sci.* **2016**, *113*, 104–115. [CrossRef]
39. Kaseem, M.; Hussain, T.; Baek, S.H.; Ko, Y.G. Formation of stable coral reef-like structures via self-assembly of functionalized polyvinyl alcohol for superior corrosion performance of AZ31 Mg alloy. *Mater. Des.* **2020**, *193*, 108823. [CrossRef]
40. Fischer, D.A.; Vargas, I.T.; Pizarro, G.E.; Armijo, F.; Walczak, M. The effect of scan rate on the precision of determining corrosion current by Tafel extrapolation: A numerical study on the example of pure Cu in chloride containing medium. *Electrochim. Acta* **2019**, *313*, 457–467. [CrossRef]
41. Therese, G.H.A.; Kamath, P.V. Electrochemical synthesis of metal oxides and hydroxides. *Chem. Mater.* **2000**, *12*, 1195–1204. [CrossRef]
42. De Roy, A.; Forano, C.; Besse, J.P. *Layered Double Hydroxides: Synthesis and Post-Synthesis Modification*; Rives, V., Ed.; Layered Double Hydroxides: Present and Future; Nova Science Publishers, Inc.: New York, NY, USA, 2001; pp. 1–39.
43. Lei, X.; Wang, L.; Zhao, X.; Chang, Z.; Jiang, M.; Yan, D.; Sun, X. Oriented CuZnAl ternary layered double hydroxide films: In situ hydrothermal growth and anticorrosion properties. *Ind. Eng. Chem. Res.* **2013**, *52*, 17934–17940. [CrossRef]
44. Gupta, V.; Gupta, S.; Miura, N. Potentiostatically deposited nanostructured Co_xNi_{1-x} layered double hydroxides as electrode materials for redox-supercapacitors. *J. Power Source* **2008**, *175*, 680–685. [CrossRef]
45. Kloprogge, J.T.; Frost, R.L.; Hickey, L. FT-Raman and FT-IR spectroscopic study of synthetic Mg/Zn/Al-hydrotalcites. *J. Raman Spectrosc.* **2004**, *35*, 967–974. [CrossRef]
46. Lin, Y.; Wang, J.; Evans, D.G.; Li, D. Layered and intercalated hydrotalcite-like materials as thermal stabilizers in PVC resin. *J. Phys. Chem. Solids* **2006**, *67*, 998–1001. [CrossRef]
47. Li, P.; Xu, Z.P.; Hampton, M.A.; Vu, D.T.; Huang, L.; Rudolph, V.; Nguyen, A.V. Control preparation of zinc hydroxide nitrate nanocrystals and examination of the chemical and structural stability. *J. Phys. Chem. C* **2012**, *116*, 10325–10332. [CrossRef]
48. Hu, G.; Wang, N.; O'Hare, D.; Davis, J. Synthesis of magnesium aluminium layered double hydroxides in reverse microemulsions. *J. Mater. Chem.* **2007**, *17*, 2257–2266. [CrossRef]
49. Rhee, S.W.; Kang, M.J. Kinetics on dehydration reaction during thermal treatment of MgAl-CO_3-LDHs. *Korean J. Chem. Eng.* **2002**, *19*, 653–657. [CrossRef]
50. Martin, J.; Jack, M.; Hakimian, A.; Vaillancourt, N.; Villemure, G. Electrodeposition of Ni-Al layered double hydroxide thin films having an inversed opal structure: Application as electrochromic coatings. *J. Electroanal. Chem.* **2016**, *780*, 217–224. [CrossRef]
51. Uan, J.-Y.; Lin, J.-K.; Tung, Y.-S. Direct growth of oriented Mg–Al layered double hydroxide film on Mg alloy in aqueous HCO_3^-/CO_3^{2-} solution. *J. Mater. Chem.* **2010**, *20*, 761–766. [CrossRef]

52. Lu, Z.; Qian, L.; Xu, W.; Tian, Y.; Jiang, M.; Li, Y.; Sun, X.; Duan, X. Dehydrated layered double hydroxides: Alcohothermal synthesis and oxygen evolution activity. *Nano Res.* **2016**, *9*, 3152–3161. [CrossRef]
53. Nakamura, K.; Shimada, Y.; Miyashita, T.; Serizawa, A.; Ishizaki, T. Effect of vapor pressure during the steam coating treatment on structure and corrosion resistance of the $Mg(OH)_2$/Mg-Al LDH composite film formed on Mg alloy AZ61. *MDPI Mater.* **2018**, *11*, 1659. [CrossRef] [PubMed]
54. Xue, L.; Cheng, Y.; Sun, X.; Zhou, Z.; Xiao, X.; Hu, Z.; Liu, X. The formation mechanism and photocatalytic activity of hierarchical NiAl–LDH films on an Al substrate prepared under acidic conditions. *Chem. Commun.* **2014**, *50*, 2301–2303. [CrossRef] [PubMed]
55. Monti, M.; Benito, P.; Basile, F.; Fornasari, G.; Gazzano, M.; Scavetta, E.; Tonelli, D.; Vaccari, A. Electrosynthesis of Ni/Al and Mg/Al layered double hydroxides on Pt and FeCrAlloy supports: Study and control of the pH near the electrode surface. *Electrochim. Acta* **2013**, *108*, 596–604. [CrossRef]
56. Yan, T.; Xu, S.; Peng, Q.; Zhao, L.; Zhao, X.; Lei, X.; Zhang, F. Self-healing of layered double hydroxide film by dissolution/recrystallization for corrosion protection of aluminum. *J. Electrochem. Soc.* **2013**, *160*, C480–C486. [CrossRef]
57. Jobbágy, M.; Regazzoni, A.E. Dissolution of nano-size Mg-Al-Cl hydrotalcite in aqueous media. *Appl. Clay Sci.* **2011**, *51*, 366–369. [CrossRef]

Article

New Zn3Mg-xY Alloys: Characteristics, Microstructural Evolution and Corrosion Behavior

Catalin Panaghie [1], Ramona Cimpoeșu [1,*], Bogdan Istrate [2], Nicanor Cimpoeșu [1], Mihai-Adrian Bernevig [1], Georgeta Zegan [3,*], Ana-Maria Roman [1], Romeu Chelariu [1] and Alina Sodor [3]

1. Faculty of Materials Science and Engineering, "Gh. Asachi" Technical University from Iasi, 700050 Iasi, Romania; catalin.panaghie@student.tuiasi.ro (C.P.); nicanor.cimpoesu@tuiasi.ro (N.C.); mihaibernevig@gmail.com (M.-A.B.); ana-maria.roman@academic.tuiasi.ro (A.-M.R.); chelariu@tuiasi.ro (R.C.)
2. Faculty of Mechanical Engineering, "Gh. Asachi" Technical University from Iasi, 700050 Iasi, Romania; bogdan_istrate1@yahoo.com
3. Faculty of Dental Medicine, "Grigore T. Popa" University of Medicine and Pharmacy, 700115 Iasi, Romania; alinasodor@yahoo.com
* Correspondence: ramona.cimpoesu@tuiasi.ro (R.C.); georgeta.zegan@umfiasi.ro (G.Z.)

Citation: Panaghie, C.; Cimpoeșu, R.; Istrate, B.; Cimpoeșu, N.; Bernevig, M.-A.; Zegan, G.; Roman, A.-M.; Chelariu, R.; Sodor, A. New Zn3Mg-xY Alloys: Characteristics, Microstructural Evolution and Corrosion Behavior. Materials **2021**, 14, 2505. https://doi.org/10.3390/ma14102505

Academic Editors: Guang-Ling Song, Mikhail Zheludkevich, Frank Czerwinski and Raman Singh

Received: 12 April 2021
Accepted: 11 May 2021
Published: 12 May 2021

Publisher's Note: MDPI stays neutral with regard to jurisdictional claims in published maps and institutional affiliations.

Copyright: © 2021 by the authors. Licensee MDPI, Basel, Switzerland. This article is an open access article distributed under the terms and conditions of the Creative Commons Attribution (CC BY) license (https://creativecommons.org/licenses/by/4.0/).

Abstract: Zinc biodegradable alloys attracted an increased interest in the last few years in the medical field among Mg and Fe-based materials. Knowing that the Mg element has a strengthening influence on Zn alloys, we analyze the effect of the third element, namely, Y with expected results in mechanical properties improvement. Ternary ZnMgY samples were obtained through induction melting in Argon atmosphere from high purity (Zn, Mg, and Y) materials and MgY (70/30 wt%) master alloys with different percentages of Y and keeping the same percentage of Mg (3 wt%). The corrosion resistance and microhardness of ZnMgY alloys were compared with those of pure Zn and ZnMg binary alloy. Materials were characterized using scanning electron microscopy (SEM), energy dispersive spectroscopy (EDS), X-ray diffraction (XRD), linear and cyclic potentiometry, and immersion tests. All samples present generalized corrosion after immersion and electro-corrosion experiments in Dulbecco solution. The experimental results show an increase in microhardness and indentation Young Modulus following the addition of Y. The formation of YZn12 intermetallic phase elements with a more noble potential than pure Zinc is established. A correlation is obtained between the appearance of new Y phases and aggressive galvanic corrosion.

Keywords: biodegradable alloy; ZnMgY; corrosion; immersion test; 10xDPBS; SEM; XRD; EDS

1. Introduction

Certain metals have begun to attract great interest over the years and started to be used in various medical applications due to their good mechanical properties, formability, and wear resistance in time [1–5]. These metals can be characterized by three words: 'biodegradable', 'bioabsorbable', and 'bioresorbable'. Materials such as magnesium (Mg), iron (Fe), and zinc (Zn) were studied and accepted as good implant material [6–8]. From these three materials, zinc (Zn) is the last studied [2] and is considered promising for more reasons. Researchers established that some issues related to Mg and Fe can be solved using Zn [9,10].

Zinc is an ideal candidate for its good degradation rate and acceptable biocompatibility [11]. Moreover, zinc is the second most abundant trace element in the human body and influences important metabolic processes, and it regulates the cell cycle [12]. Zinc plays an important role in the prevention of heart disease, for example, maintaining the integrity of endothelial cells [13], simulating the proliferation of endothelial cells through increasing the levels of endogenous basic fibroblast growth factor [14], and can prevent further damage caused by ischemia and infarction [15]. Pure zinc stents have been tested in vivo on rabbits and the results have been promising in terms of biodegradability, with no major signs of inflammation or thrombosis formation.

However, the mechanical properties of zinc are below the standards required for metals used in medical applications such as vascular stents. To improve the mechanical properties of pure zinc, it is necessary to add alloying elements with properties that should compensate for the disadvantages of zinc. Alloying elements such as Ca, Mg, Mn, Ge, and Cu is standard practice, and they have been used and studied for developing biodegradable Zn-based alloys with mechanical properties significantly improved [16–20].

The Zn-Mg biodegradable alloy system was studied by many researchers. Mg has 0.1 wt% solubility in Zn or less at 364 °C. As cast Zn-Mg alloys contain Zn primary dendrites and lamellar eutectic mixture of Zn and Mg2Zn11 confirmed by phase diagram and metallographic observations [21]. By adding the Y element in different mass percentages of the Zn-Mg system, the first days of immersion improvement of corrosion resistance and mechanical properties are considered.

In this article, preliminary results obtained on a new alloy (ZnMgY) were analyzed to appreciate the corrosion resistance and hardness variation compared with pure Zn and Zn3Mg materials. Zn alloys are, nowadays, appreciated as a promising material for medical biodegradable applications since their corrosion rate is between magnesium (too big corrosion rate-degradation) and iron (small corrosion rate-degradation), besides other benefic biological reactions of Zn [22–26].

2. Materials and Methods

2.1. Materials

The samples were obtained using high purity Zn and Mg and master alloy MgY (70/30 wt%) acquisitioned from Hunan China Co., Hunan, China [27], maintain for 10 min at T = 450 °C in a classical induction furnace with Ar atmosphere (~0.75 atm), Inductro, Bucharest, Romania. The experimental set consists of five samples, respectively, pure Zn, ZnMg, and three samples based on ZnMgxY with x = 0.4, 0.5, and 0.6. Experimental ingots (100 g) were obtained from the next material quantities: for alloy Zn3Mg0.4Y we use 96.6g pure Zn, 1.35 g MgY, and 2.05 pure Mg; for Zn3Mg0.5Y we use 96.5 g pure Zn, 1.68 g MgY, and 1.825 pure Mg; for Zn3Mg0.6Y we use 96.0 g pure Zn, 2.6 g MgY, and 1.6 pure Mg. Zinc loss by volatilization was achieved by keeping a reduced temperature of overheating of the metal bath and dilution of the alloying elements. The samples were five times re-melted to obtain proper chemical and structural homogeneity and to reduce the voids and micro-cracks from the melting process. To establish the re-melting effect, we perform surface state analyzes (nondestructive test: NDT) using fluorescent penetrant liquids.

All samples were subjected to nondestructive testing. The samples were cleaned in an ultrasound machine Geti (Tipa, Sadovacity, Czech Republic) before penetrant testing. Fluorescent penetrant testing provides a means of detecting surface-opening discontinuities such as porosity, cracks, or inclusions. The method used is hydrophilic post-emulsification. In this scope, we used an ultra-high sensitivity level 4 penetrant solution and a hydrophilic emulsifier in 5% concentration. A dry developer was used for a better contrast that amplifies the location of pores. Steps used for each stage were as follows: penetrant dwells time 20 min, emulsifier time 2 min, and developer time 10 min. The parts were inspected under UV light with an intensity of 3800 $\mu W/cm^2$ measured at 38 cm distance from the lamp's bulb.

2.2. Corrosion Behavior Analysis

The corrosion behavior was evaluated through immersion and electro-corrosion tests. In all chemical experiments, we use a Dulbecco's Phosphate Buffered Saline solution (SAFC Bioscience LTD, Hampshire, UK.) (without calcium or magnesium, code: 17-515Q): 10xDPBS (chemical composition: KCl:0.2; KH_2PO_4: 0.2; NaCl: 8.0 and $NaHPO_4$(anhydrous): 1.15).

Biodegradable materials can be used in melted, heat-treated or plastic deformed state, based on the application requirements; in this article, we analyze the melted state behavior of the samples. Immersion experiments (24, 48, and 72 h) were realized in a thermal-controlled equipment at 37 °C for different periods with regulation of the pH value at 24 h. The mass variation was determined by a Partner analytical balance, Partner Co., Bucharest, Romania, after the immersion tests and after cleaning in an ultrasound bath with an alcohol –based solution.

For electro-corrosion analyze a VoltaLab-21 potentiostat (Radiometer, Copenhagen, Denmark) was used to investigate the linear and cyclic potentiometry in 10xDPBS electrolyte solution. Tests were made using a three-electrode cell. A calomel-saturated electrode was used as a reference electrode and as an auxiliary electrode a platinum one. For the working electrode, all samples were inserted in Teflon support to expose a 0.80 cm^2 area to the electrolyte. The electrolyte was continuously mixed with a magnetic stirrer to eliminate the bubbles due to H$_2$ elimination even that in the case of Zn-alloys the release of H$_2$ is much lower than Mg-alloys. To obtain the corrosion potential (Ecorr) and corrosion current (Jcorr) characteristics for each sample, we plot the results as current density (mA/cm^2) function of potential (V). The corrosion current density, Jcorr, was determined using two equations by extrapolation of the Tafel lines: a complete polarization curve consisting of a cathodic part and an anodic part. The cathodic portion of the polarization curve contains information concerning the kinetics of the reduction reactions occurring for a particular system. The corrosion potential is deduced from the intersection of the anodic and cathodic Tafel slopes. The tests were made at room temperature (RT) (23 °C), and the potential records registered (linear plots were registered at a scan rate of 1 mV/s and the cyclic plots at a scanning rate of 10 mV/s). A scan rate of 1mV/s and a potential scan of approximately ±350 mV about Ecorr is generally required to made reasonably accurate extrapolation. The tests were repeated five times to achieve proper repeatability of the results.

Surface and microstructure were investigated using scanning electron microscopy, VegaTescan LMH II SEM (VegaTescan, Brno—Kohoutovice, Czech Republic), Secondary Electrons (SE) detector (VegaTescan, Brno—Kohoutovice, Czech Republic), electron gun supply: 30 kV, high pressure, 15.5 mm distance between the electron gun and the sample. Chemical composition insights were taken with an EDS detector, Bruker X-Flash, Mannheim, Germany using automatic and element list mode, in Point analysis a surface of 0.05 μm^2 is investigated. XRD experiments were realized on Expert PRO MPD equipment, Panalytical (XRD, Panalytical, Almelo, The Netherlands model, with copper—X-ray tube (Kα-1.54°). In processing the obtained data, diffractogram patterns were analyzed using the Highscore Plus software (2.2, Panalytical, Almelo, The Netherlands). Samples with area of 50 mm^2 were scanned in transmission mode as polished discs under the following parameters: 2 theta: 10°–1100°, step size: 0.13°, time/step: 51 s, and a scan speed of 0.065651°/s.

2.3. Microhardness Investigation

Microhardness experiments were realized using laboratory equipment CETR UMT-2 Tribometer (Universal Micro-Materials Tester), from the Tribology laboratory of the Mechanical Engineering Faculty from Iasi, Romania. The microindentation evaluation. For tests, we prepared rectangular samples with plan parallel faces with (L:w:t) 10 cm × 2 cm × 1 cm dimensions. The indentations (neatness of the results was supported by five determinations for each sample at 0.25 mm distance one of the others) were made with Rockwell diamond tip (120° opening angle). Dimensions of Rockwell's tip indenter are a radius of 200 ± 5 μm; angle 120° ± 0.30° and a standard deviation from the median line of ±2 μm. Using this equipment, we managed to evaluate and compare the values of indentation Young modulus (GPa), hardness (Gpa), and contact stiffness (N/μm) based on the determination of the maximum load, (N), maximum displacement, (μm), contact depth, (μm) and contact area, (μm^2). The method is based on a gradual increase in the indentation load from 0 to 10 N and release to the starting value. The electro-electronic part based on a capacitive sensor along with the load sensor permits the realization of a typical indentation diagram (load-depth). UTM Viewer program (CETR, Campbell, CA, USA) translates data files produced with the UTM testing software (2.16) (CETR, Campbell, CA, USA) into a graphical display for analysis [28].

3. Results

Preliminary results from the obtaining and characterization of a new alloy (ZnMgY) with possible medical applications are presented.

3.1. Preliminary Results of the Samples

Experimental materials were NDT tested to establish de effects of the melting operation. The sample surface presents important differences before and after five times re-melting of the alloys (Figure 1), fact that strongly recommends this operation if the classical melting process is used to obtain Zn-alloy. In Figure 1a, the penetrant liquids reveal some surface defects like pores, micro-cracks, or inclusions that will influence alloy properties like corrosion resistance. After a re-melting operation, we obtain a better quality of material surface with less and smaller defects, as can be observed in Figure 1b.

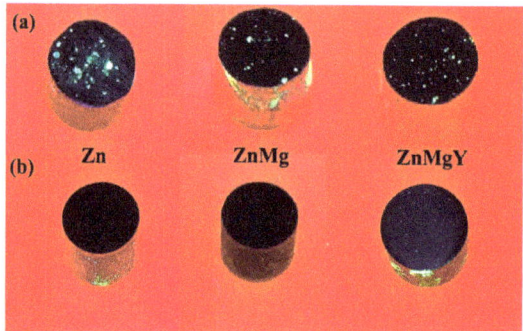

Figure 1. NDT analysis of melted Zn, ZnMg, and ZnMgY alloy (**a**) first melt sand (**b**) after five re-melts.

In this article, we will focus on analysis of the ZnMgY alloy, a less researched system than binary ZnMg material in comparison with pure Zn and Zn3Mg, also cast samples. Figure 2 shows the qualitative results of chemical analysis of ZnMgY alloy through the energies of dispersive characteristic X-ray of the new material with the identification of main components: Zn, Mg and Y. A simple distribution of the elements (Figure 2b–e) shows that different chemical composition compounds are present in the alloy. Beside a compound with more Y, located on the right corner of the investigated area, the elements appear in a homogeneous spread.

From elemental distributions (Figure 2c–e), we observe that Y is spread on the entire area, with different intensities, and magnesium participates at different phases with areas with a low or at all contributions.

Analyzing binary diagrams, Mg–Zn, Mg–Y and Y–Zn published in ASM Handbook, vol. 3, Alloys and Phase Diagrams [29], we identified the possibility of formation of different binary compounds such as Mg_2Zn_{11} (for 93.7 wt% Zn), $MgZn_2$ (for 84–84,6 wt%), Mg_2Zn_3 (for 80.1 wt% Zn), $MgZn$ (for 74 wt% Zn) or Mg_7Zn_3 (for 53.6 wt% Zn). No compounds Mg–Y with more than 30 wt% Y were identified on our analyses. From Y–Zn binary diagram, the formation of the following compounds was identified: α and β YZn_2 (for 59.6 wt% Zn), YZn_3 (for 69 wt% Zn), Y_3Zn_{11} (for 73 wt% Zn), $Y_{13}Zn_{58}$ (for 76,7 wt% Zn), YZn_5 (for 76.6 wt% Zn), Y_2Zn_{17} (for 86.2 wt% Zn), and YZn_{12} (for 89.8 wt% Zn).

Using a ternary diagram of Mg–Zn–Y predicted from the energy's formation [30], eleven stable compounds were determined in this alloy: $Mg_{21}Zn_{25}$, $MgZn_2$, Mg_2Zn_{11},

$Mg_{24}Y_5$, Mg_2Y, MgY, $Zn_{12}Y$, $Zn17Y2$, $Zn3Y$, ZnY, and $MgZnY$. Part of these compounds, based on thermodynamic conditions and chemical composition, formed in alloy and were identified by XRD analysis.

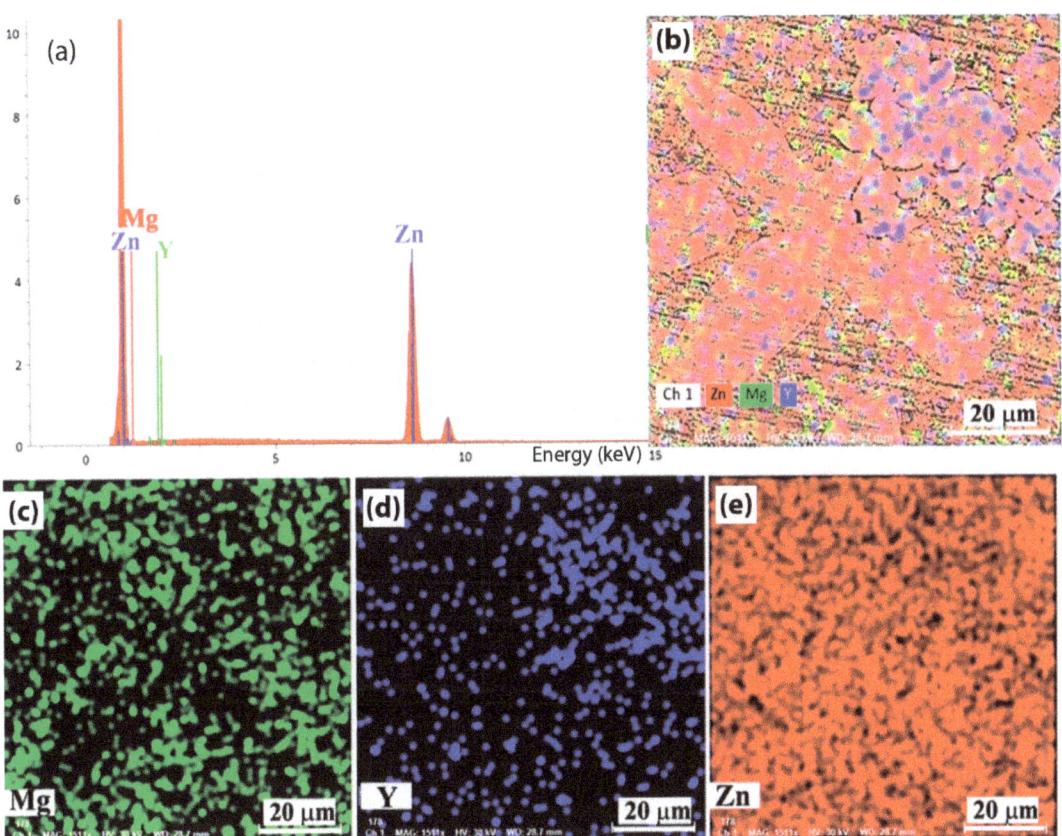

Figure 2. Chemical analysis of experimental alloy (**a**) energy spectrum, (**b**) Zn,MgY elements distribution, (**c**) Mg, (**d**) Y, and (**e**) Zn distributions.

Structurally, five areas were identified as having different shapes denoted x1–x5, the figure from Table 1 and Figure 3c, and their chemical composition was determined and presented in Table 1. The alloy was made of mainly pure Zn and master alloy MgY. For x1 point (a spot with 90 nm diameter), a Zn–Y compound is obtained (Table 1), which was identified also with XRD equipment (Figure 4) that has an 11.5 atomic report between Zn and Y that corresponds to YZn12 phase (chemical formula: Zn24.00Y2.00). Here, the affinity of Y from MgY master alloy was stronger for Zn and the formation of YZn12 compound. Point x2 is situated on a differently morphological compound, a complex one ZnMgY. The third area (point 3) represents the matrix, a solid solution of Zn with more Mg and Y dissolved in. The fourth area analyzed, x4, presents a reduced percentage of Mg and a near proposed percentage of Y in composition with zinc in rest. The last compound, point x5, is a ZnMg without Y element. No MgY compounds (introduced in the metallic bath as master-alloy) were observed on the structure after melting and pouring operations.

Table 1. Chemical composition analysis of structural elements of ZnMgY alloy (Zn3Mg0.6Y).

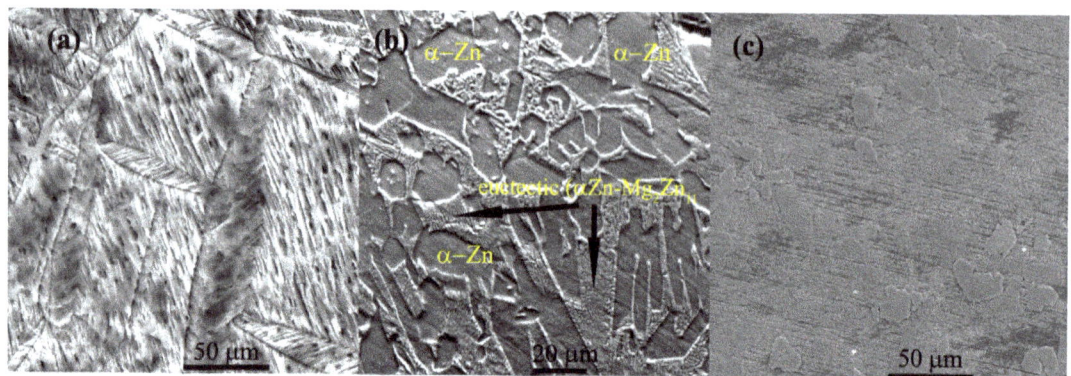

Elements/Analysis Points	Zn		Mg		Y	
	wt%	at%	wt%	at%	wt%	at%
x1	89.11	91.76	–	–	10.88	8.24
x2	99.2	98.4	0.52	1.4	0.28	0.2
x3	96.28	92.16	2.79	7.19	0.93	0.65
x4	99.12	99.19	0.08	0.22	0.8	0.59
x5	99.43	98.48	0.57	1.52	–	–
EDS err.	2.5		0.3		0.5	

Figure 3. Structural aspects of the samples before corrosion tests (a) Zn, (b) ZnMg, and (c) ZnMgY.

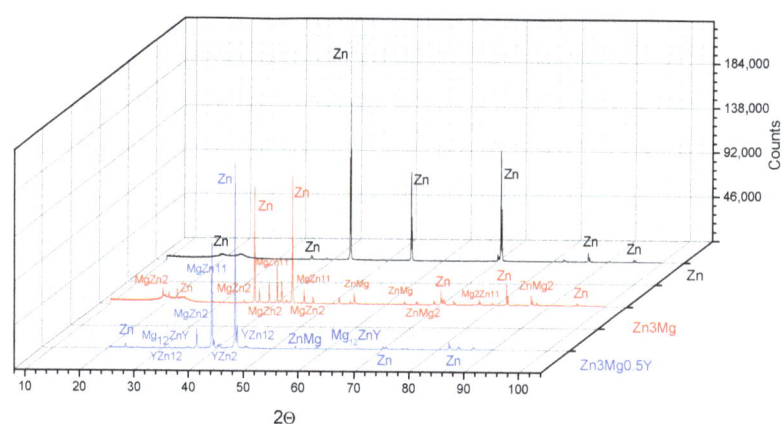

Figure 4. XRD results of Zn, ZnMg, and ZnMgY materials.

After remelting and pouring of the materials in a metallic casting form, experimental ingots were mechanically prepared by eliminating the edges and cutting cylindrical samples, mechanical grinding, and chemical etching with the microstructures presented in Figure 3. In all corrosion resistance tests performed in this article, we used ground and polished samples without chemical etch having a structural state near the ones presented in Figure 3. Chemical composition was determined for each ingot through five determinations and the values presented in Table 2 with standard deviations for each element and EDS detector error.

Table 2. Chemical compositions of the samples (average values from five determinations).

Materials/Elements	Zn		Mg		Y	
	wt%	at%	wt%	at%	wt%	at%
Zn pure	99.99	99.99	–	–	–	–
Zn3Mg	97.02	92.63	2.98	7.37	–	–
Zn3Mg0.4Y	96.8	92.90	2.78	6.74	0.42	0.26
Zn3Mg0.5Y	96.75	92.96	2.75	6.59	0.50	0.35
Zn3Mg0.6Y	96.55	86.93	2.82	5.74	0.63	0.42
EDS error	1.1		0.4		0.1	

StDev: Zn: ±1.5, Mg: ±0.2 and Y: ±0.1.

The samples: $Zn_3Mg_{0.4}Y$, $Zn_3Mg_{0.5}Y$ and $Zn_3Mg_{0.6}Y$ were investigated and compared with pure Zn and Zn_3Mg. In Figure 3, structural aspects of the samples present different orientations of the grains in the case of pure Zn and a dendritic structure of ZnMg alloy. Microstructural investigations of ZnMg alloy in cast state present α-Zn grains (Figure 3b) and a eutectic (consisting of α–Zn and intermetallic Mg_2Zn_{11} phase) located along the grain boundaries [31]. All samples present good structural homogeneity, no high quantities of inclusions on the grain boundaries. Figure 3c presents a good distribution of the ZnY and ZnMg compounds in ZnMgY alloy.

Using a ternary diagram of Mg–Zn–Y predicted from the energy's formation (ternary), eleven stable compounds were determined in this alloy: $Mg_{21}Zn_{25}$, $MgZn_2$, Mg_2Zn_{11}, $Mg_{24}Y_5$, Mg_2Y, MgY, $Zn_{12}Y$, $Zn_{17}Y_2$, Zn_3Y, ZnY, and $MgZnY$. Part of these compounds, based on thermodynamic conditions and chemical composition, formed in alloy and identified by XRD analysis. Besides the usual compounds determined for ZnMg alloy and main peaks of pure Zn, we determined the presence of the YZn_{12} (Figure 4) compound confirming the identification made through EDS detector on Table 1, point x1. For this compound, we identify few 2θ peaks as 40.509 (Int:100, h k l:3 2 1) and 45.497 (Int: 80, h k l:4 1 1). The main characteristics of YZn12 are: a (Å): 8.8750, b (Å): 8.8750, c (Å): 5.1920, Alpha (°): 90.0, Beta (°): 90.0, Gamma (°): 90.0, calculated density (g/cm^3): 7.09, Volume of the cell (10^6 pm^3): 408.95. The spatial group of YZn12 is tetragonal (I4/mmm) with a three-dimensional structure [32,33]. A single Y atom is connected to 20 Zn atoms in a 12-coordinate spatial geometry. The Y–Zn links are positioned at different distances between 3.16–3.41Å. Here, there are three different positions of Zn, first Zn is connected in a denatured geometry to 2 similar Y and 10 Zn atoms. Besides Y-Zn bonds, there are also Zn–Zn links with distances from 2.60–2.88 Å. For the second Zn position, Zn atom is connected to two Y atoms and ten Zn atoms to form a composite of margin, face, and angle corner distribution of Zn in Y2Zn10 (with 2 links of 2.60 Å and 4 of 2.74 Å). For the third position, Zn is linked in a ten-coordinate geometry to one Y and 9 Zn atoms (Zn–Zn link length is 2.59 Å).

A Mg12ZnY intermetallic compound was also identified and exhibits excellent mechanical properties in Mg-based alloys [34]. It is known from the literature that the Mg12ZnY phase has a significant role in enhancing the mechanical properties of alloys [35–37]. This compound presents a hexagonal crystal structure [32] with lattice parameters (nm): a = 0.321 and c = 3.694. These compounds with Y were initially formed pending solidification stage and did not manage to dissolve completely into the α–Zn matrix. To obtain a supra-saturated α-Zn solid-solution, a hardened solution heat treatment is necessary, and to promote this type of precipitation, aging heat treatment at low temperature can be used for obtaining a fine lamellar structure [38]. A higher percentage of solid solution in the experimental alloys will promote a generalized corrosion at the contact with an electrolyte, and a higher number of intermetallic compounds will favor the micro-hardness increase.

3.2. Immersion Tests

Samples were analyzed after the immersion test (37 ± 1 °C, 10xDPBS) by weighing, and the surface by chemical composition acquisitions. Researchers in the field of biodegradable materials agree that the corroding environment has an important influence on the corrosion rate and the behavior of an alloy [39]. Similar to Mg-based alloys big differences were observed in the behavior of Zn-alloys in different electrolyte solutions like simulated body fluids, Hanks's solution, Ringer, Dulbecco, or Dulbecco plus bovine serum [40]. For all samples, an increase in weight is observed after 24, 48, and 72 h (Table 3), except the $Zn_3Mg_{0.5}Y$ alloy at 24 h, which presented a higher added mass than all other materials. The samples show an increase in mass with 1 to 2.6 mg thanks to reactions between materials and DPBS solution. Ultrasound cleaning in alcohol was made to remove unstable compounds from the surface after the immersion test. Only a small part of the compound form on the surface was detached and passed in solution, a fact that shows good stability of the layer formed on top of the alloy during the first three days of immersion. Further research will be conducted to establish the period of layer stability and the degradation starting point by mass loss perspective. When the compound layer formed on the surface with high participation of Zn and Mg will be detached, a further corrosion process will decrease the mass.

Table 3. Weights of samples after the immersion test.

Samples		Initial Weight [g]	Weight After Immersion [g](mg)	Weight After Ultrasound Cleaning [g](mg)
Zn	24H	4.1888	4.1906 (+1.8)	4.1902 (+1.4/−0.4)
	48H	3.9416	3.9431 (+1.5)	3.9428 (+1.2/−0.3)
	72H	4.1369	4.1391 (+2.2)	4.1387 (+1.8/−0.4)
ZnMg	24H	2.6388	2.6410 (+2.2)	2.6401 (+1.3/−0.9)
	48H	3.4378	3.4394 (+1.6)	3.4386 (+0.8/−0.8)
	72H	2.4310	2.4328 (+1.8)	2.4326 (+1.6/−0.2)
$Zn_3Mg_{0.4}Y$	24H	2.2969	2.2981 (+1.2)	2.2979 (+1.0/−0.2)
	48H	2.9427	2.9437 (+1.0)	2.9434 (+0.7/−0.3)
	72H	1.9240	1.9266 (+2.6)	1.9258 (+1.8/−0.8)
$Zn_3Mg_{0.5}Y$	24H	2.4688	2.4755 (+6.7)	2.4753 (+6.5/−0.2)
	48H	2.5858	2.5870 (+1.2)	2.5869 (+1.1/−0.1)
	72H	2.2160	2.2174 (+1.4)	2.2167 (+0.7/−0.7)
$Zn_3Mg_{0.6}Y$	24H	2.2581	2.2605 (+2.4)	2.2604 (+2.3/−0.1)
	48H	2.2529	2.2548 (+1.9)	2.2547 (+1.8/−0.1)
	72H	2.1964	2.1981 (+1.7)	2.1976 (+1.2/−0.5)

It has been proved countless times that electrolyte environment influences the degradation rate and the corrosion compound type. The corrosion rate is relatively lower in 10xDPBS than other mediums like simulated body fluids (SBF) [39]. This behavior is related to the differences in chemical composition, pH values, and different quantities of anorganic salts or aminoacids [41–44].

An important difference that Zn-based materials show, compared to Mg-based alloys, are the normal cathodic reactions that occur [39]:

$$2H_2O + 2e^- \rightarrow H_2 + 2OH^- \tag{1}$$

$$2H_2O + O_2 + 4e^- \rightarrow 4OH^- \tag{2}$$

The release of H$_2$ quantity (reaction (1)) beside O$_2$ reduction (reaction (2)) is much smaller for Zn alloys compared to Mg.

The corrosion Compounds have a partially protective role and decreased the corrosion rate. Mainly, the degradation of Zn is produced by the next anodic reaction (Equation (3)):

$$Zn \rightarrow Zn^{2+} + 2e^- \quad (3)$$

The compounds resulting from these reactions are normally ZnO or/and Zn(OH)$_2$. The growth of Zn(OH)$_2$ is more harmful based on his lower stability compared to ZnO. Moreover, Zn(OH)$_2$ is dissolved in contact with chlorides Equation (4), enhancing the degradation process. In this study, the polarization resistance is increased in 10xDBPS solution with only 8 g/L NaCl.

$$Zn(OH)_2 + 2\,Cl^- \rightarrow Zn^{2+} + 2\,Cl^- + 2\,OH^- \quad (4)$$

In the case of Zn-based alloys, insoluble phosphates Equation (5), or hydrozincite (Zn$_5$(CO$_3$)$_2$(OH)$_6$) and simonkolleite (Zn$_5$(OH)$_8$Cl$_2$·H$_2$O) compounds are more numerous, which affects the general corrosion rate enhancing in the same time the polarization resistance of the material.

$$3\,Zn^{2+} + 2\,HPO_4^{2-} + 2\,OH^- + 2\,H_2O \rightarrow Zn_3(PO_4)_2 + 4\,H_2O \quad (5)$$

Generally, at the first contact, the corrosion rate is slower in 10xDPBS or similar solutions for all biodegradable metals (Mg– or Fe-base) [39]. The corrosion layer formed during immersion is uniform in DPBS and consists of oxides (mainly of Zn) and unsolvable corrosion products made through precipitation, such as metallic oxides/hydroxides, phosphates, and carbonates. The additional mass gained after the immersion tests observed in Table 3 is based on compound accumulation and higher corrosion resistance in the first days of contact with the electrolyte. Comparing the behavior of similar alloys in SBF, the slower corrosion rate can be attributed to the growth of a more stable compound layer, adsorbed amino-acids, and lack of HCl that accelerates corrosion in SBF [39]. The surface of the samples after immersion presents a manly covered area by compounds (Figure 5); most of them passed from the electrolyte solution, interacted with the surface, and produced a resistant corrosion layer. After three days, all samples showed a cover layer on the surface formed from Zn, O, and P (Table 4), small percentages of Mg, Y–oxides and Cl, K as salts from the solution. The morphological structure of the compounds is similar to Zn$_3$(PO$_4$)$_2$ (O$_8$P$_2$Zn$_3$) compound layer for all samples, a layer that covers almost the entire surface [45,46].

Salt compound morphology can be observed as ZnMgY deposits on the surface (Figure 5c, confirmed by Table 4) with Zn$_3$(PO$_4$)$_2$ layer on top. The order of interactions is confirmed by the first formation of the phosphated layer and second by other compounds that passed from the electrolyte to surface. On the surface of Zn$_3$Mg alloy appearance of MgO can be observed on the surface [47]. Uncovered areas were observed on all samples (Figure 5d), and many fine cracks define the deposited layer. The layer formed on the surface has a certain stability after three days of immersion, ultrasound cleaning only removed a part of the compounds (Table 3), but with further contact with the solution, the corrosion will break the phosphated layer, and the material will pass in electrolyte continuing the degradation. The stability of the material in the first three days of immersion can be considered an advantage for biomedical applications.

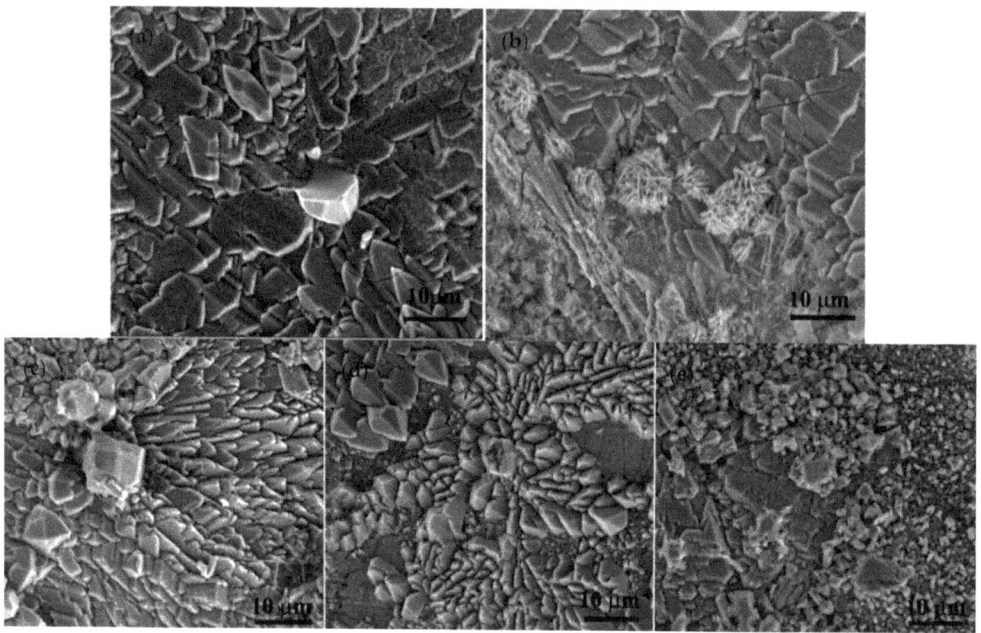

Figure 5. SEM after immersion period of three days and ultrasound cleaning: (**a**) Zn, (**b**) Zn$_3$Mg, (**c**) ZnMg$_{0.4}$Y, (**d**) ZnMg$_{0.5}$Y, and (**e**) ZnMg$_{0.6}$Y.

Table 4. Chemical composition of the surface samples after immersion test (after 3 days and ultrasound cleaning).

Material	Zn		Mg		Y		O		P		Cl		K	
	wt%	at%	wt%	at%	wt%	at%	wt%	at%	wt%	at%	wt%	at%	wt%	at%
Zn pure	55.42	26.97	–	–	–	–	28.76	57.17	12.88	13.23	02.93	02.63	–	–
Zn3Mg	51.60	24.26	1.09	1.38	–	–	29.91	57.45	15.66	15.54	–	–	1.74	1.37
Zn3Mg0.4Y	51.29	23.96	0.67	0.85	0.25	0.09	30.65	58.51	15.70	15.48	–	–	1.44	1.12
Zn3Mg0.5Y	51.54	24.25	0.92	1.16	0.64	0.22	30.18	58.03	15.43	15.32	–	–	1.29	1.01
Zn3Mg0.6Y	53.25	26.80	0.55	0.74	01.36	00.50	25.05	51.50	17.26	18.33	–	–	2.53	2.13
EDS error	1.2		0.2		0.1		0.8		0.75		0.2		0.1	

StDev:Zn: ±1.74, Mg: ±0.41, Y: ±0.57, O: ±2.26, P:1.57, Cl: ±0, K: ±0.91.

All surfaces presented a high percentage of O and P after three days of immersion. Small quantities of Cl and K are also identified. Depending on the stability and thickness of the phosphated layer formed on the surface, the Y element was also identified.

3.3. Potentiodynamic Polarization Curves

The Linear polarization method was used to determine the corrosion rate of Zn, ZnMg with different percentages of Y in 10xDPBS electrolyte. Linear polarization curves are shown in Figure 6a and cyclic polarization curves in Figure 6b. Using electro-chemical polarization methods, we can provide information about the type of corrosion and anodic or cathodic protection. Linear and cyclic potentiometry techniques can be useful for evaluation analysis of the passive region, the passivation possibility of a material in contact with the environment, stability, and the passivation quality (degradation rate).

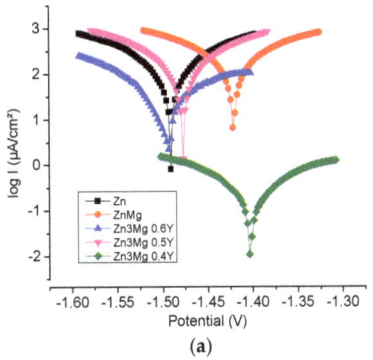 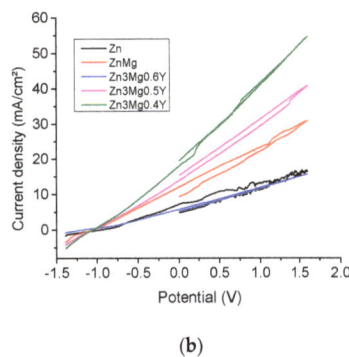

Figure 6. Linear and cyclic potentiometry of the samples. (**a**) Tafel diagram and (**b**) cyclic potentiometry.

The corrosion parameters obtained from the Tafel extrapolation are listed in Table 5. Linear polarization curves show that the corrosion potential values (E_{corr}) were similar with an increase in percentages of Y. This effect can be related to the presence of a passivation process of the surface, as in all metallic materials, and the passivation can last more or less, influencing the degradation rate and the material stability.

Table 5. Linear potentiometry parameters.

Sample	$-E_{cor}$ mV	b_a mV	b_c mV	R_p ohm·cm^2	J_{corr} µA/cm^2	V_{corr} mm/Y
Zn	1491.4	122.2	−141.1	168.97	210.09	3.36
Zn3Mg	1469.8	106.0	−132.0	167.01	253.8	2.85
Zn3Mg0.4Y	1402.3	134.5	−135.6	78.25	575.7	6.46
Zn3Mg0.5Y	1477.9	117.9	−129.1	146.06	311.6	3.49
Zn3Mg0.6Y	1497.0	133.9	−95.8	851.46	66.55	0.74

The tendency of corrosion current density (J_{corr}) values is to decrease along with the increase in percentages of Y from 575.7 µA/cm^2 to 66.55 µA/cm^2. In the case of Zn3Mg$_{0.6}$Y is observed a decrease in the current density (J_{corr}) due to the formation of intermetallic Mg$_2$Zn$_{11}$ and MgZn$_2$ fazes. The corrosion rate of Zn$_3$Mg$_{0.4}$Y is 6.46 mm/Y, much higher than Zn and Zn$_3$Mg 3.36 mm/Y, and 2.85 mm/Y. This is due because the non-homogeneous structure caused by formation of new compounds with Y. A higher concentration of Y homogenizes the composition of the compounds and decreases the corrosion rate to 0.74 mm/Y.

The corrosion potential (E_{corr}) for Zn$_3$Mg$_{0.4}$Y shifted by about 60 mV compared to Zn$_3$Mg, indicating a diminution of the corrosion tendency in thermodynamics. A higher concentration of Y had no further significant impact on E_{corr}.

From cyclic voltammograms presented in Figure 6b, the similar behavior of the alloys and uniform corrosion is observed, reflected by the sharp increase in current density. In the Zn$_3$Mg$_{0.6}$Y case, the cathodic branch of the voltammogram overlaps the anodic branch, indicating advanced corrosion resistance.

3.4. Micro-Hardness Characterization of the Samples

Experiments to establish the microhardness of the samples were realized in laboratory conditions [48,49]. Load-depth variation graphs were obtained in Origin 2021 software (OriginLab Corporation, Northampton, MA, USA). Rockwell microhardness determinations were realized on five areas of the samples. The big difference in the indentation behavior of Zn toward ZnMg and ZnMgY alloys determined us to provide two graphs in Figure 7, with different depth scales.

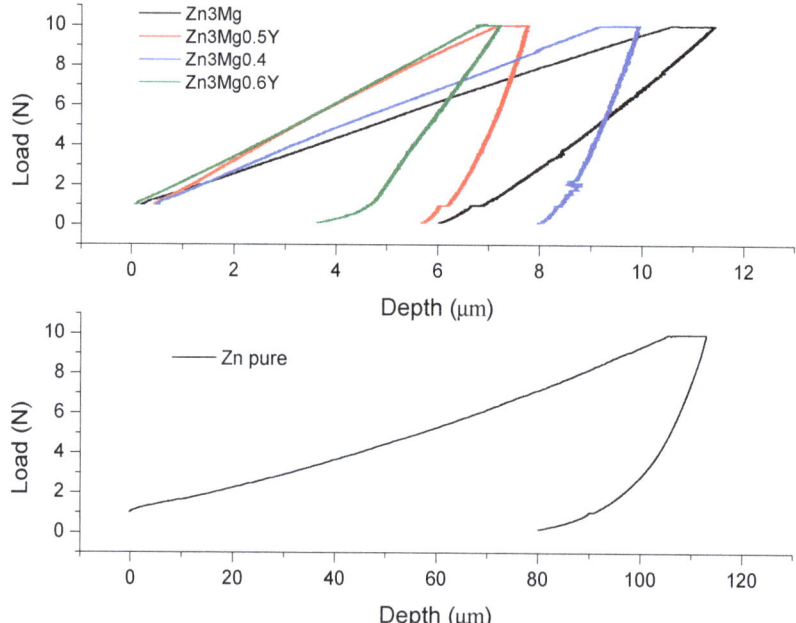

Figure 7. Microhardness experiments pure Zn and Zn-alloy behavior.

The microhardness of a material can give information about the material elongation to fracture, plasticity, and resistance to fracture of materials. The addition of magnesium to zinc clearly improves the microhardness of the material (7.21 times), especially through compounds and phases formed in the microstructure. Furthermore, the addition of the Y element increases even more the microhardness of the alloy till values of 1.25 GPa. The comparative values are presented in Table 6, with a standard deviation of the experimental values. The increase in microhardness with addition of Y element is due to the formation of new compounds YZn_{12} and $Mg_{12}ZnY$. Microhardness seems to increase along with the Y content, but further research in this area is required for confirmation.

Indentation experiments can be useful to analyze the elastic modulus for various depths. From this test, more information about the material can be gathered in the same time based on the load and unload curve. We use average values from separate indentations made into a polycrystalline material and each indent can be situated in different positions regarding the solid solution, intermetallic, or other compounds.

The softer material is pure zinc (higher maximum displacement), and the opposite part is the value for $Zn_3Mg_{0.5}Y$, 6.88 µm. These compounds, based on Mg and Y, have areas under 5000 µm^2 and a homogeneous spread. In almost every indentation case (see Table 6 for contact area values), the indenter tip presses also on a compound at least partially. In case of $Zn_3Mg_{0.5}Y$ alloy, the higher values of indentation Young Modulus and microhardness are based on the fifth measurement point which, probably, was mostly on a harder compound. This affirmation was confirmed by scanning electron microscopy images. Young's modulus values presented an increase with the addition of Mg and a bigger increase with Y, and the $Zn_3Mg_{0.5}Y$ alloy presents the highest value of 81.37 GPa (32.42 times bigger than pure zinc Young modulus). The contact stiffness presents an increase with Y addition based on the increase in material hardness.

Table 6. Microhardness experiment values for samples.

Alloys		Indentation Young Modulus (GPa)	Hardness (GPa)	Contact Stiffness (N/μm)	Maximum Load (N)	Maximum Displacement (μm)	Contact Depth (μm)	Contact Area (μm^2)
Zn pure	Average	2.51	0.14	0.79	8.96	65.48	57.29	7,8807.47
	Std. Dev.	0.51	0.05	0.05	0.02	26.59	26.86	5,3656.28
Zn$_3$Mg	Average	17.10	1.01	1.95	9.01	10.80	7.31	9022.66
	Std. Dev.	1.73	0.06	0.22	0.01	0.46	0.49	599.36
Zn$_3$Mg$_{0.4}$Y	Average	59.64	0.79	7.23	8.79	9.91	8.95	1,1002.22
	Std. Dev.	12.95	0.04	1.47	0.37	0.49	0.43	507.04
Zn$_3$Mg$_{0.5}$Y	Average	81.37	1.22	7.94	9.01	6.88	6.03	7459.77
	Std. Dev.	10.01	0.16	0.69	0.02	0.70	0.69	841.89
Zn$_3$Mg$_{0.6}$Y	Average	39.26	1.25	3.77	9.02	7.51	5.70	7062.93
	Std. Dev.	1.34	0.11	0.36	0.01	0.31	0.35	430.67

4. Conclusions

The effect of Y concentration on the ZnMg alloys properties was investigated. Analyzing the experimental results, we conclude that:

- The Mg and Y addition to pure Zn improves mechanical properties as microhardness, Young's modulus, and contact stiffness.
- The appearance of YZn12 and Mg12ZnY intermetallic phases among ZnMg compounds has the most significant importance concerning the modification of Zn properties.
- The growth of the Zn3PO4 layer on the surface alloys provides protective corrosion resistance effect.
- Y participation in formation of YZn12 intermetallic phase elements gives the new alloy a more noble potential than pure Zinc.
- Formation of YZn12 intermetallic compound has been correlated with a more aggressive galvanic corrosion between Mg2Zn11 and YZn12 with the Zn matrix.

Author Contributions: Conceptualization, R.C. (Ramona Cimpoeșu) and N.C.; data curation, C.P., M.-A.B., and A.-M.R.; formal analysis, R.C. (Ramona Cimpoeșu), N.C., M.-A.B., G.Z., A.-M.R.; R.C. (Romeu Chelariu), and A.S.; funding acquisition, N.C.; investigation, C.P., R.C. (Ramona Cimpoeșu), B.I., N.C., G.Z., A.-M.R., and R.C. (Romeu Chelariu); methodology, C.P., R.C. (Ramona Cimpoeșu), B.I., N.C., A.-M.R., and R.C. (Romeu Chelariu); project administration, R.C. (Ramona Cimpoeșu); resources, C.P., M.-A.B., and A.S.; software, C.P.; supervision, M.-A.B. and R.C. (Romeu Chelariu); validation, G.Z., R.C. (Romeu Chelariu) and A.S.; Writing—original draft, C.P., R.C. (Ramona Cimpoeșu), N.C. and A.-M.R.; writing—review and editing, C.P., R.C. (Ramona Cimpoeșu), N.C., and A.-M.R. All authors have read and agreed to the published version of the manuscript.

Funding: A part of this research was funded by Ministry of Research, Innovation and Digitization, project FAIR_09/24.11.2020, and by the Executive Agency for Higher Education, Re-search, Development and Innovation, UEFISCDI, ROBIM- project number PN-III-P4-ID-PCE2020-0332

Institutional Review Board Statement: Not applicable.

Informed Consent Statement: Not applicable.

Data Availability Statement: Data sharing not applicable.

Conflicts of Interest: The authors declare no conflict of interest.

References

1. Witte, F. The history of biodegradable magnesium implants: A review. *Acta Biomater.* **2010**, *6*, 1680–1692. [CrossRef]
2. Zheng, Y.F.; Gu, X.N.; Witte, F. Biodegradable metals. *Mater. Sci. Eng. R Rep.* **2014**, *77*, 1–34. [CrossRef]
3. Hermawan, H. *Biodegradable Metals from Concept to Applications*; Springer: Berlin/Heidelberg, Germany, 2012.
4. Seitz, J.M.; Durisin, M.; Goldman, J.; Drelich, J.W. Recent advances in biodegradable metals for medical sutures: A critical review. *Adv. Healthc. Mater.* **2015**, *4*, 1915–1936. [CrossRef]
5. Heiden, M.; Walker, E.; Stanciu, L. Magnesium, iron and zinc alloys, the trifecta of bioresorbable orthopaedic and vascular implantation–A review. *Biotechnol. J. Biomater.* **2015**, *5*, 178.
6. Li, H.; Zheng, Y.; Qin, L. Progress of biodegradable metals. *Proc. Nat. Sci. Mater. Int.* **2014**, *24*, 414–422. [CrossRef]
7. Im, S.H.; Jung, Y.; Kim, S.H. Current status and future direction of biodegradable metallic and polymeric vascular scaffolds for next-generation stents. *Acta Biomater.* **2017**, *60*, 3–22. [CrossRef]
8. Hermawan, H. Updates on the research and development of absorbable metals for biomedical applications. *Prog. Biomater.* **2018**, *7*, 93–110. [CrossRef]
9. Bowen, P.K.; Drelich, J.; Goldman, J. Zinc exhibits ideal physiological corrosion behavior for bioabsorbable stents. *Adv. Mater.* **2013**, *25*, 2577–2582. [CrossRef]
10. Vojtech, D.; Kubasek, J.; Serak, J.; Novak, P. Mechanical and corrosion properties of newly developed biodegradable Zn-based alloys for bone fixation. *Acta Biomater.* **2011**, *7*, 3515–3522. [CrossRef]
11. Yang, H.; Wang, C.; Liu, C.; Chen, H.; Wu, Y.; Han, J.; Jia, Z.; Lin, W.; Zhang, D.; Li, W.; et al. Evolution of the degradation mechanism of pure zinc stent in the one-year study of rabbit abdominal aorta model. *Biomaterials* **2017**, *145*, 92–105. [CrossRef]
12. Saghiri, M.A.; Asatourian, A.; Orangi, J.; Sorenson, C.M.; Sheibani, N. Functional role of inorganic trace elements in angiogenesis-Part II: Cr, Si, Zn, Cu, and S. *Crit. Rev. Oncol. Hematol.* **2015**, *96*, 143–155. [CrossRef] [PubMed]
13. Purushothaman, M.; Gudrun, R.; Michal, T.; Bernhard, H. Zinc modulates PPARc signaling and activation of porcine endothelial cells. *J. Nutr.* **2003**, *133*, 3058.
14. Kaji, T.; Fujiwara, Y.; Yamamoto, C.; Sakamoto, M.; Kozuka, H. Stimulation by zinc of cultured vascular endothelial cell proliferation: Possible involvement of endogenous basic fibroblast growth factor. *Life Sci.* **1994**, *55*, 1781. [CrossRef]
15. Little, P.J.; Bhattacharya, R.; Moreyra, A.E.; Korichneva, I.L. Zinc and cardiovascular disease. *Nutrition* **2010**, *26*, 1050–1057. [CrossRef] [PubMed]
16. Li, H.; Yang, H.; Zheng, Y.; Zhou, F.; Qiu, K.; Wang, X. Design and characterizations of novel biodegradable ternary Zn-based alloys with IIA nutrient alloying elements Mg, Ca and Sr. *Mater. Des.* **2015**, *83*, 95–102. [CrossRef]
17. Shi, Z.; Yu, J.; Liu, X.; Zhang, H.; Zhang, D.; Yin, Y.; Wang, L. Effects of Ag, Cu or Ca addition on microstructure and comprehensive properties of biodegradable Zn-0.8Mn alloy. *Mater. Sci. Eng. C Mater. Biol. Appl.* **2019**, *99*, 969–978. [CrossRef] [PubMed]
18. Tong, X.; Zhang, D.; Zhang, X.; Su, Y.; Shi, Z.; Wang, K.; Lin, J.; Li, Y.; Lin, J.; Wen, C. Microstructure, mechanical properties, biocompatibility, and in vitro corrosion and degradation behavior of a new Zn-5Ge alloy for biodegradable implant materials. *Acta Biomater.* **2018**, *82*, 197–204. [CrossRef]
19. Venezuela, J.; Dargusch, M.S. The influence of alloying and fabrication techniques on the mechanical properties, biodegradability and biocompatibility of zinc: A comprehensive review. *Acta Biomater.* **2019**, *87*, 1–40. [CrossRef] [PubMed]
20. Wang, X.; Shao, X.; Dai, T.; Xu, F.; Zhou, J.G.; Qu, G.; Tian, L.; Liu, B.; Liu, Y. In vivo study of the efficacy, biosafety, and degradation of a zinc alloy osteosynthesis system. *Acta Biomater.* **2019**, *92*, 351–361. [CrossRef]
21. Mostaed, E.; Sikora-Jasinska, M.; Mostaed, A.; Loffredo, S.; Demir, A.G.; Previtali, B.; Mantovani, D.; Beanland, R.; Vedani, M. Novel Zn-based alloys for biodegradable stent applications: Design, development and in vitro degradation. *J. Mech. Behav. Biomed. Mater.* **2016**, *60*, 581–602. [CrossRef]
22. Dong, H.; Zhou, J.; Virtanen, S. Fabrication of ZnO nanotube layer on Zn and evaluation of corrosion behavior and bioactivity in view of biodegradable applications. *Appl. Surf. Sci.* **2019**, *494*, 259–265. [CrossRef]
23. Jallot, E.; Nedelec, J.M.; Grimault, A.S.; Chassot, E.; Grandjean-Laquerriere, A.; Laquerriere, P.; Laurent-Maquin, D. STEM and EDXS characterisation of physico-chemical reactions at the periphery of sol–gel derived Zn-substituted hydroxyapatites during interactions with biological fluids. *Colloid Surface B* **2005**, *42*, 205–210. [CrossRef]
24. Su, Y.; Cockerill, I.; Wang, Y.; Qin, Y.; Chang, L.; Zheng, Y. Zinc-based biomaterials for regeneration and therapy. *Trends Biotechnol.* **2018**, *37*, 428–441. [CrossRef] [PubMed]
25. Torne, K.B. Zn–Mg and Zn–Ag degradation mechanism under biologically relevant conditions. *Surf. Innov.* **2017**, *6*, 1–12. [CrossRef]
26. Kambe, T.; Tsuji, T.; Hashimoto, A.; Itsumura, N. The physiological, biochemical, and molecular roles of zinc transporters in zinc homeostasis and metabolism. *Physiol. Rev.* **2019**, *95*, 749–784. [CrossRef] [PubMed]
27. Hunan High Broad New Material, Co.Ltd. Available online: http://www.hbnewmaterial.com/supplier-129192-master-alloy (accessed on 10 June 2019).
28. Baciu, E.R.; Cimpoesu, R.; Vitalariu, A.; Baciu, C.; Cimpoesu, N.; Sodor, A.; Zegan, G.; Murariu, A. Surface analysis of 3D (SLM) Co-Cr-W dental metallic materials. *Appl. Sci.* **2021**, *11*, 255. [CrossRef]
29. Cha, G.H.; Lee, S.Y.; Nash, P.; Pan, Y.Y.; Nash, A. Binary alloy phase diagrams (Ni-Sb; Ni-Sc; Ni-Se; Ni-Si; Ni-Sm; Ni-Sn). In *ASM Handbook. Volume 3. Alloys Phase Diagrams*; ASM International: Russell Township, OH, USA, 1991; pp. 317–318.

30. Ma, S.-Y.; Liu, L.M.; Wang, S.Q. The microstructure, stability, and elastic properties of 14H long-period stacking-ordered phase in Mg–Zn–Y alloys: A first-principles study. *J. Mater. Sci.* **2014**, *49*, 737–748. [CrossRef]
31. Itoi, T.; Suzuki, T.; Kawamura, Y.; Hirohashi, M. Microstructure and mechanical properties of Mg-Zn-Y rolled sheet with a Mg12ZnY phase. *Mater. Trans.* **2010**, *51*, 1536–1542. [CrossRef]
32. Grobner, J.; Kozlov, A.; Fang, X.Y.; Geng, J.; Nie, J.F.; Schmid-Fetzer, R. Phase equilibria and transformations in ternary Mg-rich Mg–Y–Zn alloys. *Acta Mater.* **2012**, *60*, 5948–5962. [CrossRef]
33. Shao, G.; Varsani, V.; Fan, Z. Thermodynamic modelling of the Y–Zn and Mg–Zn–Y systems. *Comput. Coupling Phase Diagr. Thermochem.* **2006**, *30*, 286–295. [CrossRef]
34. Zhu, J.; Chen, X.H.; Wang, L.; Wang, W.Y.; Liu, Z.K.; Liu, J.X.; Hui, X.D. High strength Mg-Zn-Y alloys reinforced synergistically by Mg12ZnY phase and Mg3Zn3Y2 particle. *J. Alloys Compd.* **2017**, *703*, 508–516. [CrossRef]
35. Hagihara, K.; Kinoshita, A.; Sugino, Y.; Yamasaki, M.; Kawamura, Y.; Yasuda, H.Y.; Umakoshi, Y. Plastic deformation behavior of Mg89Zn4Y7 extruded alloy composed of long-period stacking ordered phase. *Intermetallics* **2010**, *18*, 1079–1085. [CrossRef]
36. Hagihara, K.; Yokotani, N.; Umakoshi, Y. Plastic deformation behavior of Mg12YZn with 18R long-period stacking ordered structure. *Intermetallics* **2010**, *18*, 267–276. [CrossRef]
37. Zhang, Z.; Liu, X.; Hu, W.; Li, J.; Le, Q.; Bao, L.; Zhu, Z.; Cui, J. Microstructures, mechanical properties and corrosion behaviors of Mg-Y-Zn-Zr alloys with specific Y./Zn mole ratios. *J. Alloys Compd.* **2015**, *624*, 116–125. [CrossRef]
38. Egami, M.; Ohnuma, I.; Enoki, M.; Ohtani, H.; Abe, E. Thermodynamic origin of solute-enriched stacking-fault in dilute Mg-ZnY alloys. *Mater. Des.* **2020**, *188*, 108452. [CrossRef]
39. Dong, H.; Lin, F.; Boccaccini, A.R.; Virtanen, S. Corrosion behavior of biodegradable metals in two different simulated physiological solutions: Comparison of Mg, Zn and Fe. *Corros. Sci.* **2021**, *182*, 109278. [CrossRef]
40. Liu, X.; Yang, H.; Liu, Y.; Xiong, P.; Guo, H.; Huang, H.H.; Zheng, Y. Comparative studies on degradation behavior of pure zinc in various simulated body fluids. *JOM* **2019**, *71*, 1414–1425. [CrossRef]
41. Wagener, V.; Faltz, A.S.; Killian, M.S.; Schmuki, P.; Virtanen, S. Protein interactions with corroding metal surfaces: Comparison of Mg and Fe. *Faraday Discuss.* **2015**, 347–360. [CrossRef]
42. Gu, X.N.; Zheng, Y.F.; Chen, L.J. Influence of artificial biological fluid composition on the biocorrosion of potential orthopedic Mg-Ca, AZ31, AZ91 alloys. *Biomed. Mater.* **2009**, *4*. [CrossRef]
43. Jamesh, M.I.; Wu, G.; Zhao, Y.; McKenzie, D.R.; Bilek, M.M.M.; Chu, P.K. Electrochemical corrosion behavior of biodegradable Mg-Y-RE and Mg-Zn-Zr alloys in Ringer's solution and simulated body fluid. *Corros. Sci.* **2015**, *91*, 160–184. [CrossRef]
44. Xin, Y.; Hu, T.; Chu, P.K. In vitro studies of biomedical magnesium alloys in a simulated physiological environment: A review. *Acta Biomater.* **2011**, *7*, 1452–1459. [CrossRef]
45. Yua, H.; Xu, J.; Xu, Q.; Cuic, G.; Gu, G. Electrostatic self-assembly of Zn3(PO4)2/GO composite with improved anticorrosive properties of water-borne epoxy coating. *Inorg. Chem. Commun.* **2020**, *119*, 108015. [CrossRef]
46. Gu, W.; Zhai, S.; Liu, Z.; Teng, F. Effect of coordination interaction between water and zinc on photochemistry property of Zn3(PO4)2·2H2O. *Chem. Phys.* **2020**, *536*, 110811. [CrossRef]
47. Kastiukas, G.; Zhou, X.; Asce, M.; Neyazi, B.; Wan, K.T. Sustainable calcination of magnesium hydroxide for magnesium oxychloride cement production. *J. Mater. Civ. Eng.* **2019**, *31*. [CrossRef]
48. Yao, C.; Wang, Z.; Tay, S.L.; Zhu, T.; Gao, W. Effects of Mg on microstructure and corrosion properties of Zn–Mg alloy. *J. Alloys Compd.* **2014**, *602*, 101–107. [CrossRef]
49. Dias, M.; Brito, C.; Bertelli, F.; Rocha, O.L.; Garcia, A. Interconnection of thermal parameters, microstructure, macrosegregation and microhardness of unidirectionally solidified Zn-rich Zn–Ag peritectic alloys. *Mater. Des.* **2014**, *63*, 848–855. [CrossRef]

Article

Immersion Behavior of Carbon Steel, Phosphate Carbon Steel and Phosphate and Painted Carbon Steel in Saltwater

Costica Bejinariu, Diana-Petronela Burduhos-Nergis * and Nicanor Cimpoesu *

Faculty of Materials Science and Engineering, Gheorghe Asachi Technical University, 700050 Iași, Romania; costica.bejinariu@tuiasi.ro
* Correspondence: burduhosndiana@yahoo.com (D.-P.B.-N.); nicanor.cimpoesu@tuiasi.ro (N.C.)

Abstract: The carbon steel is used in many areas due to its good mechanical properties; however, its low corrosion resistance presents a very important problem, for example, when carbon steel carabiners are used in the petroleum industry or navy, the possibility of an accident is higher due to carabiner failure. This phenomenon could occur as a consequence of the corrosion process which negatively affects mechanical properties. This paper study the possibility to improve its corrosion resistance by depositing on its surface a phosphate layer and a paint layer, and also aims to analyze the immersion behavior in saltwater of carbon steel, phosphate carbon steel, and phosphate and painted carbon steel. According to this study, by coating the carbon steel with a phosphate or paint layer, a higher polarization resistance is obtained in saltwater. Moreover, by electrochemical impedance spectroscopy (EIS), it was observed that the corrosion rate decreases with the increase of the immersion time. Meanwhile scanning electron microscopy (SEM) and energy dispersive spectroscopy (EDS) revealed that the main compounds which formed on the sample's surface were iron oxides or hydroxy-oxides, after immersion for a longer period. The overall results show that all types of deposited layers increase the corrosion resistance of C45 steel.

Keywords: seawater; corrosion resistance; paint; immersion behavior; phosphate layer; EIS; carbon steel; EDS; SEM

Citation: Bejinariu, C.; Burduhos-Nergis, D.-P.; Cimpoesu, N. Immersion Behavior of Carbon Steel, Phosphate Carbon Steel and Phosphate and Painted Carbon Steel in Saltwater. *Materials* **2021**, *14*, 188. https://doi.org/10.3390/ma14010188

Received: 9 December 2020
Accepted: 29 December 2020
Published: 2 January 2021

Publisher's Note: MDPI stays neutral with regard to jurisdictional claims in published maps and institutional affiliations.

Copyright: © 2021 by the authors. Licensee MDPI, Basel, Switzerland. This article is an open access article distributed under the terms and conditions of the Creative Commons Attribution (CC BY) license (https://creativecommons.org/licenses/by/4.0/).

1. Introduction

Carbon steels, due to their good mechanical properties [1], as well as their possibility to be processed (welded [2], chipped [3], deformed [4]), are a suitable choice for use in the manufacture of machine parts [5], accessories of fall arrest systems (e.g., carabiners, hooks, and pythons) [6,7], vehicle bodies [8], shipbuilding [9], or in use in buildings, bridges, rails, pipes, [10] etc.

Compared to other metallic or non-metallic materials, carbon steels have a low corrosion resistance [11–13]. According to the literature, various methods have been attempted to improve the corrosion resistance of carbon steel; however, most of them are limited by the high manufacturing costs or complexity of the technologies. Nevertheless, one method which showed promising results was phosphate layer deposition on the surface [14].

According to previous studies [15–18], phosphating is one of the most widely used methods of protection against corrosion in metals [19]. The phosphating process consists of, with the help of a phosphating solution, a layer of zinc, iron, or manganese phosphate (depending on the major constituent of the phosphating solution) on the surface of the metal [20]. This layer is not only deposited on the surface, but also forms bonds with the material, with high adhesion to the substrate [21]. However, the quality and characteristics of the obtained phosphate layer highly depend on multiple parameters, such as the phosphating process parameters (stages, immersion time, and temperature), the characteristics of the substrate surface, the chemical composition of the phosphate solution (solution pH, type, and concentration of the metal ions), etc.

Accordingly, Schmidt D.P. et al. [22] studied the effect of the immersion time on the corrosion resistance of different types of commercial zinc-based coatings. Guenbour A. et al. [23] studied the corrosion resistance properties of steel on the surface, of which a layer of epoxy or chlorinated rubber was deposited, into which zinc phosphate was added. Asadi V. et al. [24] studied the immersion behavior for different periods of coated samples: phosphate with a zinc oxide solution. Moller H. et al. [25] studied the corrosion behavior of carbon steel immersed in natural and synthetic waters, highlighting chemical compounds that appeared on the surface using X-ray diffraction (XRD).

However, to our knowledge, there is no study which approaches the corrosion behavior evaluation of the carbon steel coated with Zn/Fe phosphate layer and phosphate-painted carbon steel immersed in a natural saltwater environment. Based on previous preliminary studies [17,26], the goal of this study is to analyze the influence of immersion time in saltwater of three types of samples: carbon steel, phosphate carbon steel, and phosphate and painted carbon steel, while a new phosphating solution suitable for carbon steel coating was introduced. Moreover, the corrosion mechanism and parameters for different immersion times were analyzed using EIS, SEM, and EDS.

2. Materials and Methods

2.1. Material

The steel used in this study is C45 carbon steel (Anterasteel, Targoviste, Romania), which has the chemical composition according to Table 1.

Table 1. Chemical composition of the experimental low carbon steel.

Element	C	Mn	P	Cu	Cr	Si	Fe
Percent (wt.%)	0.45	0.98	0.02	0.15	0.17	0.22	balance

2.2. Sample Preparation

For the phosphate layer to cover evenly the entire surface of the steel, the phosphating was performed by an immersion comprising three main stages [27]. The first and second stages were performed to prepare the surface for the deposition of the phosphate layer. The surfaces of the samples were prepared by immersing them for 10 min in an alkaline degreasing solution (solution temperature: 85 °C ± 2 °C), after which they were immersed for 20 min in an acid pickling solution (at room temperature). A phosphate solution based on zinc and iron with the composition as shown in Table 2 was used to deposit the phosphate layer. Thus, the samples were immersed in the solution for 30 min at a temperature of 95 °C ± 2 °C. At the end of the phosphating process, the samples were dried for 4 h in an oven at 120 °C [17].

Table 2. Chemical composition of phosphate solution.

Name of the Active Substance	Quantity
Sodium hydroxide (NaOH)	0.75 g
Sodium azotite ($NaNO_2$)	0.45 g
Sodium tripolyphosphate ($Na_5P_3O_{10}$)	0.05 g
Phosphoric acid (H_3PO_4)	10.00 mL
Nitric acid (HNO_3)	4.00 mL
Zinc (Zn)	3.50 g
iron (Fe)	0.038 g

The paint used to cover the phosphate layer is a commercial paint, type KS 1000 (Car System GmbH, Uetersen, Germany). Its deposition was conducted by spraying [28].

In this study, to facilitate expression, the following notations were used:

C45—C45 steel sample;

PS—The C45 steel phosphate in zinc/iron-based solution sample;

PPS—The C45 steel phosphate in zinc/iron-based solution sample and painted;
BSW—Black Sea water, pH = 6.15.

2.3. Methods

The samples, in the form of discs with a diameter of 10 mm and a thickness of 2 mm, were immersed in closed bottles, in which 40 mL of corrosion medium (saltwater) was introduced. Weekly, the solutions were aerated by bubbling air. After 45 or 36 days, and 85 or 75 days, respectively, the samples were removed from the solution and analyzed using the method of electrochemical impedance spectroscopy (EIS) to study the structure of the surface layer and its evolution with the immersion time. The initial studies were performed one hour after immersion in solution (the time required for sample mounting operations in the electrochemical cell and thermostat) [26]. Only samples kept in saltwater for 85 days were used to study the surface morphology. These were removed from the solution, air-dried, and analyzed by scanning electron microscope with the energy of electron gun of 30 kV (SEM—VegaTescan LMH II, SE detector—Scanning Electron Microscope) (VegaTescan, Brno, Czech Republic), evaluated for surface micro-composition by the EDS method (Bruker detector) (Bruker, Berlin, Germany).

The visual analysis of the sample/solution systems indicated that the long contact between the alloy and the corrosion medium produces important changes, both in the surface structure of the samples and in the properties of the corrosion environment.

Some of the corrosion products can be soluble and pass into solution, changing its pH, and other parts, insoluble, remain on the surface or passed into solution if they are not adherent to the metal surface. In the case of the systems studied, the corrosion products were oxy-hydroxides (FeO(OH)) or hydroxides ($Fe(OH)_2$, $Fe(OH)_3$)) of iron. They are insoluble and poorly adherent to the surface, so that in the absence of a reciprocal solution/alloy displacement (liquid jet or alloy displacement through the liquid), part of it passing into the solution and part of it remaining on the surface. The remaining products on the surface are yellow-brown, porous, and slightly adherent to the surface.

The products that passed in the corrosion environment were in the form of reddish-brown flakes, which in time are deposited at the bottom of the vessel, and the soluble products make the color of the seawater turn yellow. To observe, qualitatively, the changes of the corrosion environment, Figure 1 shows the physical aspect (the color) of the corrosion environment, after the samples were kept in solution for a long period. Prior taking the images, the solutions were stirred, in order to disperse the product from the bottom of the vessel.

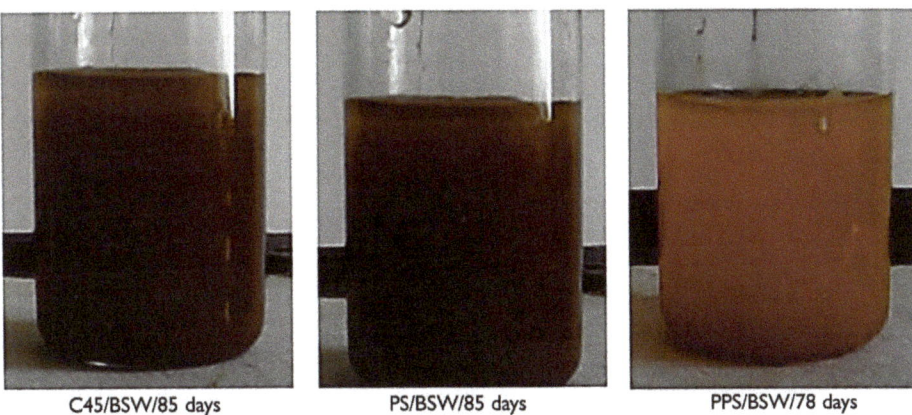

Figure 1. The corrosion environment appearance, after long immersion of alloys.

In the case of seawater, the color intensity is proportional to the amount of insoluble reaction product passed into the solution.

Soluble corrosion products, but also products resulting from the degradation of coatings (phosphate layer, paint), have changed the pH of the corrosion environment, as can be seen from Table 3. The pH measurements were performed with an and pH-METER OP-2641/1 (Radelkis, Budapest, Hungary).

Table 3. Variation of the corrosion environment pH with the immersion time of the samples.

Corrosion Environment	Immersion Time (Days)	C45	PS	PPS
	Initial	6.15	6.15	6.15
pH of BSW	45 or 36	6.25	7.30	7.22
	85 or 78	6.71	7.57	7.76

45 days—C45 and PS; 36 days—PPS; 85 days—C45 and PS; 78 days—PPS.

In all cases, a relatively small solution pH increased with the immersion time of the samples. This phenomenon occurs due to the reduction of H^+ ions, which increases OH^- ion concentration from the solution [29]. Compared to the C45 sample, the PS and PPS samples produce a greater change in pH in seawater, probably because in this corrosion environment, the coating layers (phosphate layer and paint) are damaged.

The structure and composition of the surface layer were analyzed by electrochemical impedance spectroscopy (EIS) and scanning electron microscopy, as well as EDS microanalysis.

The EIS studies were performed with the PGZ 301 potentiometer (Radiometer Analytical SAS, Lyon, France), and the acquisition and processing of experimental data were performed with VoltaMaster 4 software (Version 7, Radiometer Analytical SAS, Lyon, France). The experimental data were processed with the ZSimpWin program (Version 3.5, E-chem Software, Ann Arbor, MI, USA), in which the spectra were interpreted by the fitting process developed by Boukamp, the least-squares method. To process, with this program, the data acquired with the VoltaMaster 4 program, the data were converted using the EIS file converter program. (Version 1.5, Radiometer Analytical S.A, Lyon, France) [30].

In this study, a three-electrode glass cell, C145/170 type (Radiometer Analytical SAS, Lyon, France), with a platinum auxiliary electrode (exposed surface area of 0.8 cm^2), a reference electrode made of saturated calomel, and a 10 mm round sample of 3 mm thickness (exposed surface area of 0.503 cm^2), fixed by Teflon washers on working electrode, was used. Furthermore, all the measurements were performed in the $104\text{--}25 \times 10^{-2}$ Hz frequency range, at a sinusoidal potential amplitude of 10 mV. In the present study, the "ZSimWin" program was used for the analysis of impedance data. The program uses a wide variety of electrical circuits to numerically correlate the measured impedance data. The program can analyze very complicated dispersion data by decomposing the complex response into that of simple subcomponents. This approach, combined with the general procedure of nonlinear least squares correlation, allows the construction of an equivalent circuit whose simulated response has a high degree of correlation with the measured data. Depending on the frequency phase angle, one or more time constants may occur, and these can be used to determine the value of the parameters in the equivalent circuit. The presence of a compact passive film can be indicated from the Bode spectrum if the phase angle is close to 900 over the high-frequency range and if the spectrum has linear portions at intermediate frequencies.

The surface structure, both for freshly ground samples (C45 only) and for electrochemically corroded samples, was studied with a Vega Tescan LMH II electron microscope (30 kV, SE detector, high pressure).

The chemical compositions on various surface areas of the electrochemically untreated alloys and the surface of the corroded samples were evaluated using an EDX QUANTAX QX2 detector manufactured by Bruker/Roentec Co., Berlin, Germany EDS connected to the electron microscope.

To confirm the identification of the main compounds from the corrosion products, the rust crust, formed on the surface of the alloy, the oxides layers from the corroded samples was detached, dried in an oven, and analyzed by spectrophotometry. A BOMEM MB 104 FT-IR spectrophotometer (ABB Training Center GmbH & Co., Berlin, Germany) was used. The dry sample was dispersed in KCl (6% w/w), ground, and compressed into a disk at 10,000 Psi, and then scanned at a resolution of 4 cm^{-1} in the wavelength range 400–4000 cm^{-1}.

3. Results and Discussion

3.1. The Effects of Prolonged Immersion of C45 in Black Seawater

The Nyquist diagram for the C45 sample, after 1 h of immersion in seawater, shows a depressed semicircle in the high-frequency range and an inductive loop in the low-frequency range (Figure 2a). The Nyquist and Bode diagrams for all the samples are presented in Figures A1–A6 from Appendix A. The low-frequency inductive loop is attributed to the adsorption processes of ions or neutral molecules (e.g., inhibitors) or even particles (may also be insoluble corrosion products) in the solution, but may also indicate the existence of intermediates reactions related to the adsorption of aggressive chlorine ions, which produce instability of the electrode surface [31,32].

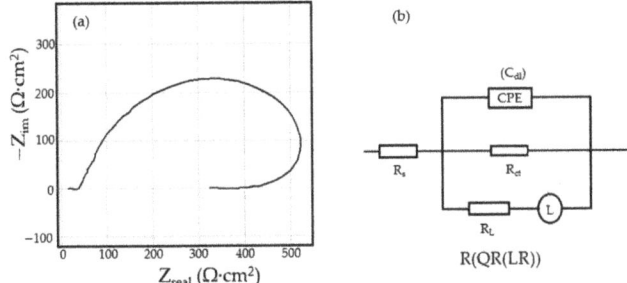

Figure 2. The Nyquist diagram (a) and equivalent circuit (b) for C45 sample, after one hour of immersion in BSW.

The processing of the experimental data for this system was conducted with the equivalent electrical circuit from Figure 2b. Better adjustment (fitting) of the data was obtained by replacing the capacity of the double layer with a constant phase element (CPE), which expresses the ideal behavior of the capacity of the electric double layer (change of capacity with frequency). The equivalent circuit consists of a resistor (R_s) in a series with a parallel combination of a constant phase element (CPE), a resistor (R_{ct}), and a series resistor-inductance combination (R_L, L) [26]. The value of the solution resistance (R_s) is 38.67 $\Omega \cdot cm^2$, the value of the charge transfer resistance (R_{ct}) is 588 $\Omega \cdot cm^2$, and the value of the resistance due to the adsorbed components (R_L) is 1007 $\Omega \cdot cm^2$.

The low value of the R_{ct} indicates a high corrosion rate (compared to phosphate samples, R_{ct} exhibited higher values). For this circuit, the polarization resistance is calculated with the formula:

$$R_p = (R_L \cdot R_{ct})/(R_L + R_{ct}) \quad (1)$$

and the R_p value is 377.23 $\Omega \cdot cm^2$. On the surface of the carbon steel sample, a continuous layer had not formed; instead, this was adsorbed on its surface ions from the saltwater or insoluble corrosion products.

This physical condition is confirmed by the microscopic and compositional analysis performed on a sample maintained for one day in seawater. Figure 3 shows the surface sample micrographs at various degrees of magnification. It was found that the surface of the sample was uniform and small crystalline formations were adsorbed on it, but these were reduced in number.

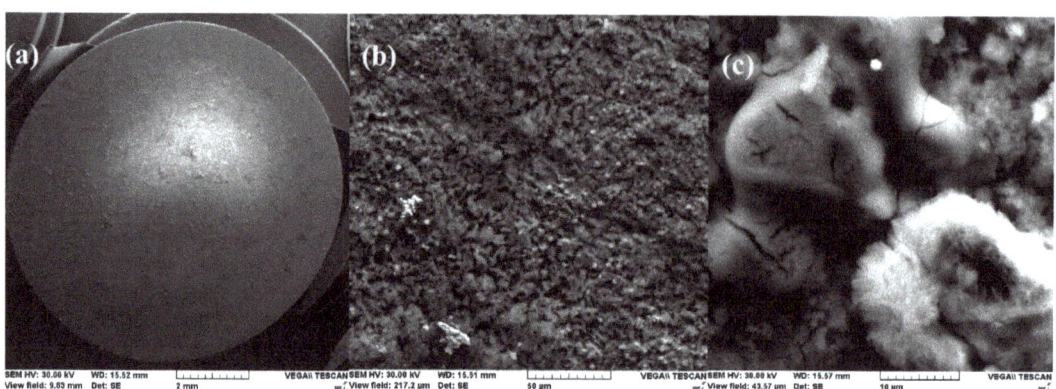

Figure 3. SEM microphotographs for C45 immersed 24 h in seawater (**a**) 100×, (**b**) 1000× and (**c**) 5000×.

A more conclusive figure (Figure 4) was provided by the EDX determinations on a small portion of the surface (850 µm^2), on which the overall surface composition and point compositions on the main table and crystalline micro compounds were measured.

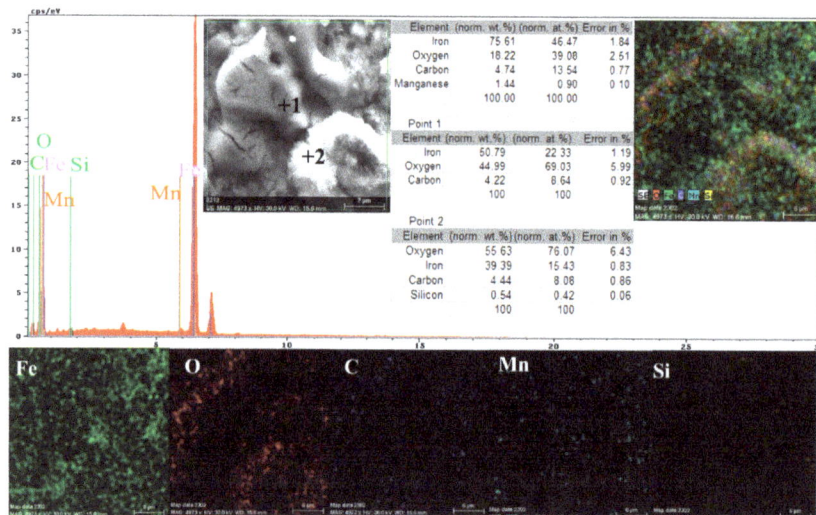

Figure 4. EDS spectrum, compositions and distribution of elements over a portion of the sample surface (Figure 3c).

In the surface composition, oxygen and iron predominate formed a very thin and irregular film on the surface. Due to the very small thickness of this layer, the composition contained very small amounts of manganese and silicon, which were part of the main mass of the alloy. Surprising, however, was that a large amount of carbon (which, in the base alloy, is only 0.45%) and a large amount of oxygen (much more than the corresponding iron oxides) was observed. This anomaly could be explained when it is considered that HO$^-$ ions are involved in the corrosion process (Fe + 2H$_2$O→Fe^{2+} + 2OH$^-$ + H$_2$), in addition to carbon dioxide from the atmosphere. Chen et al. [28] explain the appearance of the inductive loop through a relaxation process, at the level of the surface layer, due to carbon monoxide and some HO⁻ anions in the solution.

The Nyquist diagram, the Bode diagram, and the equivalent circuit used for the C45 sample immersed for 43 days in seawater is shown in Figure 5, as well as for the sample immersed for 85 days in Figure 6. The physical significance suggested by this circuit is the association of a film/electrolyte interface (R_{ext}, C_{ext}) with a kinetic process (R_{ct}, C_{dl}), R_{ct} being the charge transfer resistance through the electric double layer, and C_{dl} being the capacity of this layer. The R_{ext} resistor represents the resistance of the charge transfer through the open pores of the oxide layer, formed on the surface of the alloy, and the constant phase element C_{ext} represents the electrical capacity of this layer.

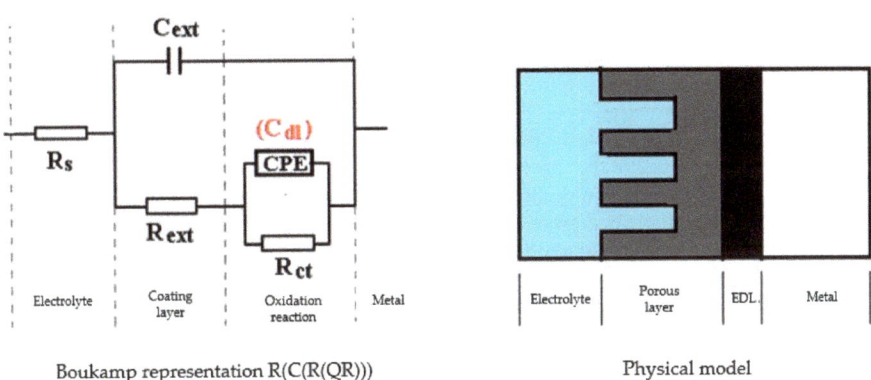

Figure 5. The equivalent circuit for C45 sample, after 43 days of immersion in BSW.

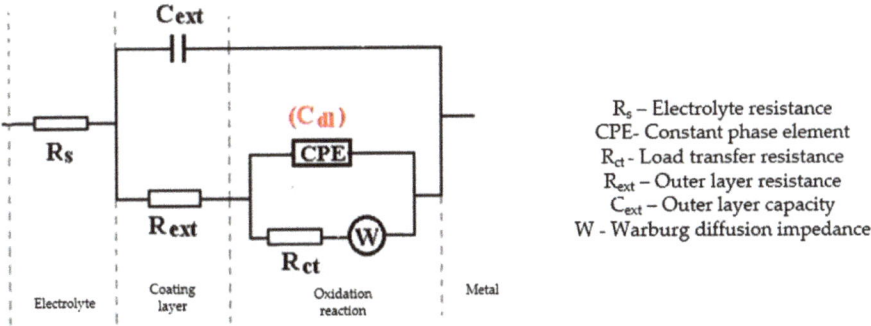

Figure 6. The equivalent circuit for C45 sample, after 85 days of immersion in BSW.

In the case of the sample kept for 85 days in seawater, the bi-layer structure is maintained, as in the case of the sample kept for 43 days in solution; however, for a better adjustment of the experimental data, it was necessary to introduce a Warburg impedance in series with charge transfer resistance. The diffusion impedance can be represented, analogously to the CPE impedance, by the relation [32]:

$$Z_W = \frac{1}{w(j\omega)^{1/2}} \qquad (2)$$

The equation applies to linear diffusion, the unit of measure for W being: $<Q> = \Omega^{-1}\,sn^{\frac{1}{2}}/cm^2 \equiv S\cdot s^{\frac{1}{2}}/cm^2$.

By this, the transfer of charges, to and from the surface of the alloy, is controlled both kinetically (through the electric double layer (controlled by R_{ct} and C_{dl}), and by diffusion through the layer of products deposited on the surface. The values of the circuit elements parameters, for the two time periods considered, are presented in Table 4.

Table 4. Equivalent circuit element values for C45 immersed for various periods in BSW.

Immersion Time	$R_s\,\Omega\cdot cm^2$	CPE		$R_{ct}\,\Omega\cdot cm^2$	$R_{ext}\,\Omega\cdot cm^2$	$W\,Ss^{1/2}/cm^2$	$10^3 \times \chi^2$	ε_z
		$Q\,Ss^n/cm^2$	n					
43 days	39.470	6.92×10^{-3}	0.749	155	5.01	-	2.439	4.94
85 days	43.590	1.25×10^{-2}	0.772	130	145.90	0.065	0.090	0.95

The analysis of the data allows us to advance a series of observations, as follows:

- The resistance of the solution (the electrolyte) increases slightly with the immersion time of the samples in seawater. This is because, in addition to the actual resistance of the liquid, there is also the physical resistance of the layer of corrosion products deposited on the surface. The strength of the product layer is very low because it is porous, non-adherent, and soaked in liquid.
- The resistance of the outer porous layer increases over time, due to the increase of its thickness, and accordingly decreases its electrical capacity.
- The constant Q, from the expression of the constant phase element that replaces the capacity of the double-electric layer, increases greatly with the immersion time ($3.418\cdot \times 10^{-4}$—after 1 h, $6.92\times\cdot 10^{-3}$—after 43 days and $1.25\times\cdot 10^{-2}$—after 85 days). As a result, the impedance of this element increases greatly (according to the calculation relationship of ZCPE (CPE impedance)). The C_{dl} deviation from the idealist (expressed by the value of the exponent n) is large and is due to the increase in roughness due to corrosion.
- The layer of products, deposited on the surface, acts as a screen and causes the charge transfer resistance to decrease over time.
- The constant W, from the expression of the diffusion impedance, has a relatively low value and is manifested only in the case of the sample maintained for 85 days in seawater. As a result, the diffusion impedance is high and opposes the charge transfer along with R_{ct}, thus reducing the corrosion rate over time.

The surface appearance of the sample immersed 85 days in seawater, after removal from solution and drying, is shown in Figure 7 at various magnifying powers.

A thick layer of the reaction product was deposited on the surface of the sample forming a reddish-brown crust, very easily removable from the surface. Increased 1000×, the surface of the sample appears as an agglomeration of small crystalline and non-crystalline formations. The crystalline formations are better visible at a 5000× magnification and appear as an agglomeration of monoclinic crystals, characteristic for iron oxy-hydroxide.

Both the energy spectrum (Figure 8) and the surface composition indicate only the presence of iron and oxygen, along with small amounts of carbon. In this composition, the molar ratio iron/Oxygen is very close to the value of 0.5 that is specific for ferric oxy-hydroxide (FeO(OH)), namely: n(Fe):n(O) = 0.57. Correlated with the type of crystals (monoclinic crystals), it can be considered that the entire surface is covered with a thick layer of FeO(OH). The presence of carbon could be attributed to the carbon dioxide absorbed in the crust. The distribution of the elements in the crust is shown in Figure 9. Areas with thick and large oxides cover parts of the surface (Figure 9), decreasing extensively the substrate iron signal. This layer plays a protective role for a while and after that, based on the continuous process of corrosion, the degradation of the material occurs. The presence of bicarbonate compounds in seawater leads to the formation of a carbon-based compound on the surface, starting from CO_3 reactivity. The carbon presence was observed only on a few common parts of the oxygen distribution (Figure 9), which means that only a part of the surface compounds is complex.

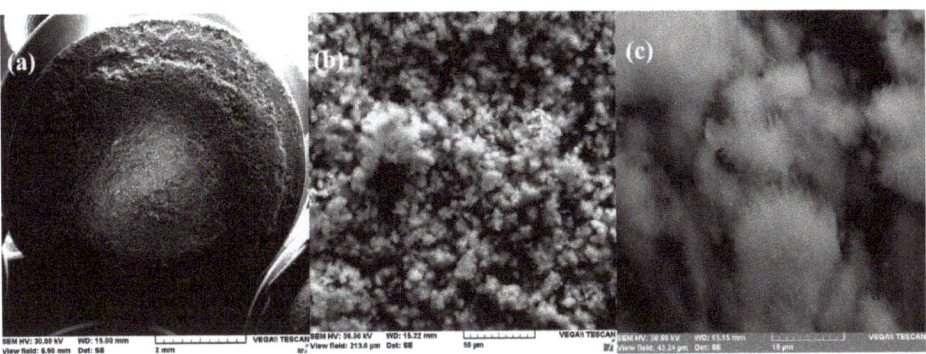

Figure 7. SEM microphotographs for C45 immersed 85 days in seawater (a) 100×, (b) 1000×, and (c) 5000×.

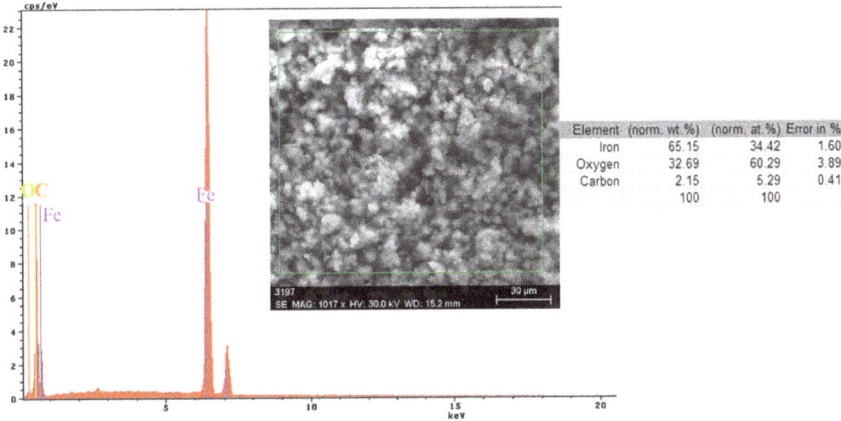

Figure 8. Energy spectrum and crust composition on the C45 surface immersed for 85 days in seawater.

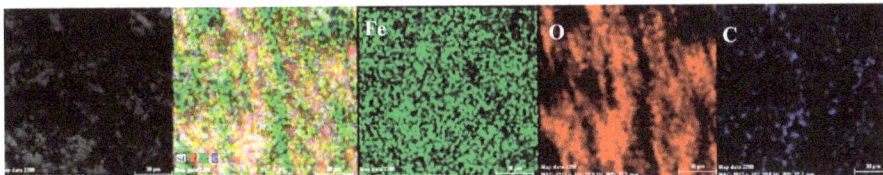

Figure 9. Elemental distribution on the surface of the crust of reaction products in the case of the C45/BSW system, after 85 days of immersion.

An interesting approach is connected to the relationship between the mineral compounds of the corrosion environment (Black Sea water) and corrosion mechanisms of Fe-C alloys elements, with a focus on the effects of chloride and sulfate ions.

In our case, we strictly analyzed the experimental results, but further considerations can be realized. At the same time, we must consider various seawater parameters that affect the corrosion of iron-based elements; other chemical elements were not identified on the surface corroded crust. These parameters include pH value variation, oxygen dissolved in solution, and the number of cations and anions [33]. The influences of all these parameters will be analyzed in further works.

In this FT-IR (Fourier-Transform Infrared Spectroscopy) spectrum (Figure 10) of the corrosion products, the characteristic bands FeOOH [34]: ν (OH) at 3402–3406 cm^{-1}, δ (HOH) at 1633 cm^{-1}, δ (OH) at 1400 cm^{-1}, ν (FeO) at 474.49 cm^{-1} specific to steels were highlighted. The presence of the band 1633.7 cm^{-1} indicates the polymolecular structure of iron oxyhydroxide, formed by the interactions between FeOOH and H$_2$O molecules.

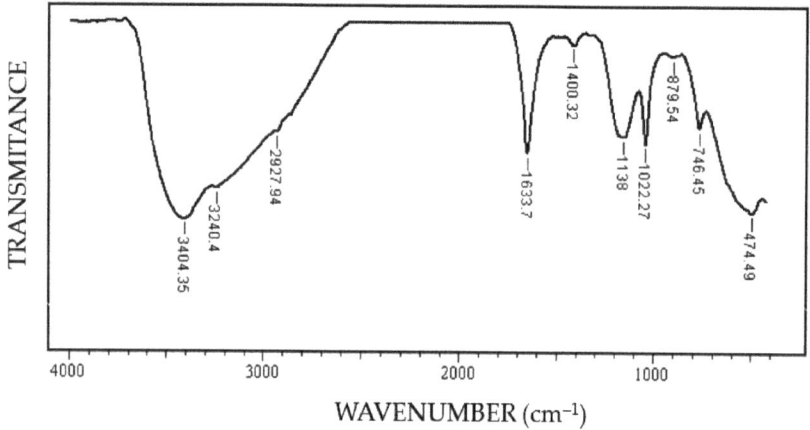

Figure 10. FT-IR spectrum of the corrosion product on the surface of the C45 sample maintained for 85 days in BSW.

3.2. Effects of Immersing C45 Phosphate (PS) in Black Sea Water

The C45 samples were phosphated in a solution based on Zn$_3$(PO$_4$)$_2$ + Fe and were kept in Black Sea water for 43 and 85 days, respectively. After these periods, the samples were removed from the solution and introduced directly into the measuring cell, for the purpose of recording electrochemical impedance spectroscopy data. Another sample, kept in seawater for 85 days, was removed from the solution, dried in air, and analyzed. The measurements were performed after one hour of immersion (after mounting and thermostatic the sample).

The values of the equivalent circuit parameters for the three immersion periods (1 h, 43 days and 85 days) are presented in Table 5.

Table 5. Equivalent circuit element values for PS immersed for various periods in BSW.

Immersion Time	R_s $\Omega\cdot$cm^2	CPE		R_{ct} $\Omega\cdot$cm^2	R_{ext} $\Omega\cdot$cm^2	C_{ext} μF/cm^2	W Ss$^{1/2}$/cm^2	$10^3 \times \chi^2$	ε_z
		Q Ss^n/cm^2	n						
1 h	40.41	2.51 × 10^{-4}	0.613	883.00	11.04	2.870	-	0.694	2.44
43 days	39.47	5.10 × 10^{-3}	0.866	334.20	12.35	2.667	-	0.155	1.24
85 days	41.86	6.63 × 10^{-3}	0.732	280.60	14.55	147	0.042	0.036	0.55

From the data analysis, the following can be highlighted:
- Although a layer (phosphate layer) is already deposited on the surface of the alloy, and after various periods another layer (corrosion products) is deposited on it, only

a circuit with two constants was needed to correlate with the experimental data, considering a bi-layer structure of the surface (a passive layer-electric double layer and an outer layer). The outer layer consists of the phosphate layer and the corrosion products layer, very well intertwined, by occupying the pores of the phosphate layer with corrosion products;

- In this case, the resistance of the electrolyte (Rs) does not change with immersion time, the values obtained oscillating around the average value (40.58 $\Omega \cdot cm^2$), within the limit of experimental errors. This is because the outer layer is very porous (slightly permeable to solution) and has low resistance;
- After the first immersion period (43 days), the strength of the outer layer decreases instead of increasing. This could be due to the fact that the seawater partially damages the phosphate layer, widening its pores, and the corrosion products layer is not yet quite consistent. However, after 85 days of immersion, the resistance of the outer layer appreciably increases, due to the clogging of the pores of the phosphating layer and the thickening of the product layer;
- As in the case of the base alloy (C45) immersed in seawater, the resistance of the double-electric layer (charge transfer resistance) decreases with the immersion time in the corrosion environment, due to the screen effect of the outer layer;
- In the initial moments and in the period of up to 43 days of immersion, the corrosion process is controlled only kinetically, while at long immersion periods, the corrosion is controlled both kinetically and by diffusion;
- The capacity of the double-electric layer (C_{dl}) is replaced here by a constant phase element, which takes into account the variation of the electrical capacity of the double layer with frequency. The values of the frequency exponent fact that in these cases the electric double layer behaves like a non-ideal capacitor.

The appearance of the sample surface and some of its details for various magnifications are illustrated in Figure 11.

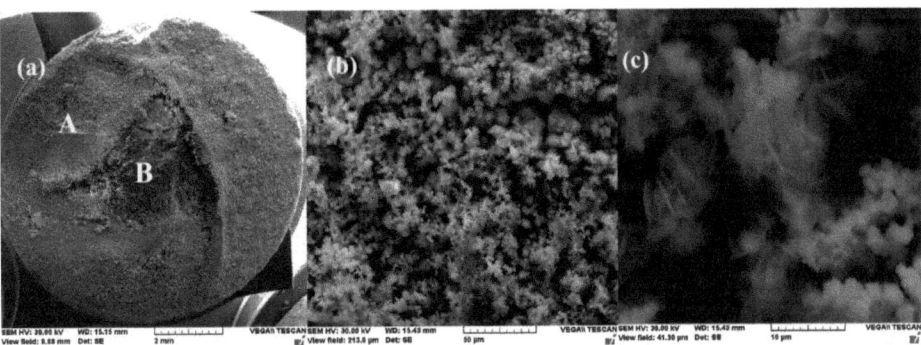

Figure 11. SEM microphotographs of the PS sample maintained for 85 days in seawater (**a**) 100×, (**b**) 1000×, and (**c**) 5000×.

After 85 days of immersion in seawater, a thick layer of corrosion products was deposited on the surface of the sample, which, after drying, formed a fragile crust, easily removable (the A zone). During handling, a portion of the crust was partially detached (the B zone). The microphotographs from the A zone, magnified by 1000× and 5000×, respectively, indicate that at the microscopic level the crust is formed in an agglomeration of crystalline micro formations, a structure that gives the crust a great porosity.

The energy spectrum and surface composition are shown in Figure 12 for the A zone and in Figure 13 for the B zone.

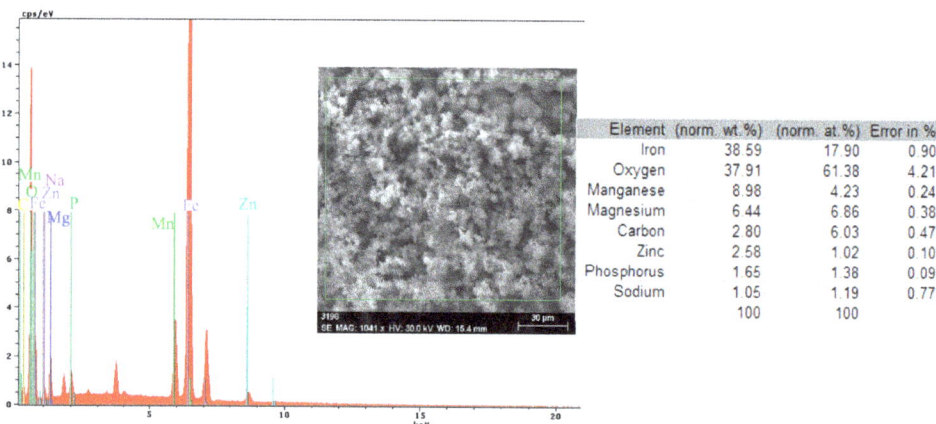

Figure 12. Energy spectrum and surface composition of the crust in the A zone of the PS sample maintained for 85 days in BSW.

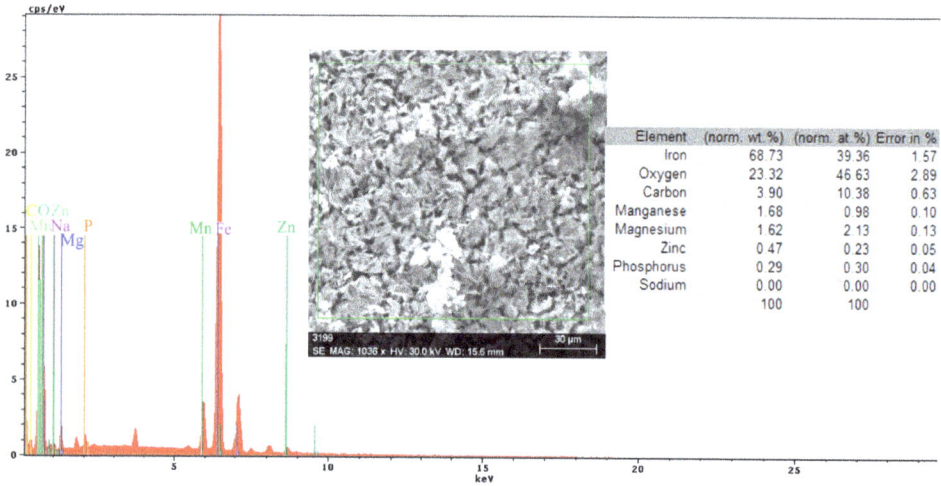

Figure 13. Energy spectrum and surface composition of the crust in the B zone of the PS sample maintained for 85 days in BSW.

In the A zone, the crust had a complex composition, in which the amount of iron was relatively small and the amount of oxygen was much higher than that corresponding to iron oxides or hydroxy-oxides. This means that the crust was not only made up of iron oxides. The composition of the crust contained carbon, magnesium, and sodium, which could come from the seawater left in the surface layer and evaporate on drying. Traces of phosphorus and zinc in the composition of the crust came from the phosphate layer, partially damaged in seawater, especially in the initial periods of immersion. The distribution of the elements in this area is shown in Figure 14.

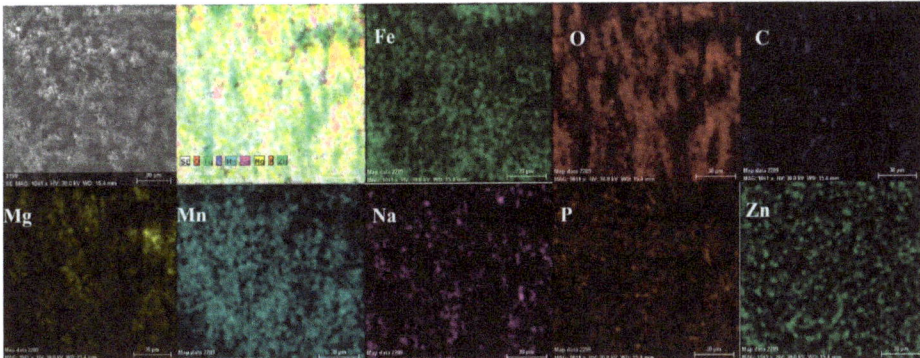

Figure 14. Elemental distribution in the crust on the PS surface, after 85 days of immersion in seawater.

In the B zone, from which a part of the crust came off, the amount of iron is much higher and the molar ratio oxygen: iron is 1:18 and corresponds to ferrous oxide (FeO). The other elements probably have the same origin as those in A zone.

The surface is mainly covered by oxidation products, Figure 14, with areas covered by thicker oxides, Figure 14 (O). Other compounds based on Mg, Mn, P, and Zn are observed on the surface. The oxides distribution is homogeneous on the entire exposed surface. The distribution of C, O, and Na suggests the formation of complex compounds such as carbonate (carbonate CO_3^{2-} with sodium ion to form sodium bicarbonate (NaHCO$_3$) and other bicarbonates than in sodium carbonate (Na$_2$CO$_3$) and other carbonates). The corrosion resistance of the sample depends on the quality of the phosphate layer and the covering percentage (any area of the substrate exposed to direct corrosion environment will facilitate the corrosion).

3.3. Effects of Immersing the Phosphate and Painted C45 (PPS) Sample in Seawater

The C45 phosphate samples were painted and dried prior immersion in Black Sea water for 36 days and 78 days and analyzed by electrochemical impedance spectroscopy and scanning electron microscopy with energy dispersive spectroscopy. The surface microstructure at the initial contact with seawater was evaluated after 1 h of immersion. The data adjustment was made using the equivalent circuit in Figure 15.

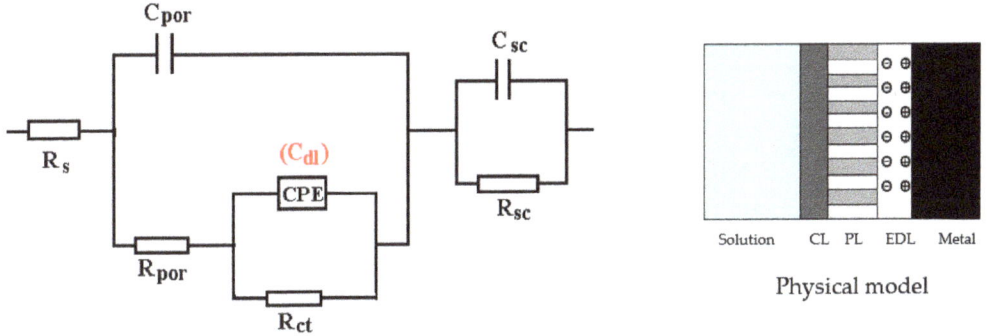

Figure 15. The equivalent circuit for PPS sample, after 1 h of immersion in BSW.

The structure of the surface layer identifies double-electric layer (EDL)—formed by polarizing the metal surface and characterized by R_{ct} and CPE, a porous layer (PL)—representing the porous phosphating layer—characterized by R_{por} and C_{por}, respectively, a compact layer (CL)—representing the paint film—characterized by the R_{SC} and C_{SC} circuit elements. The values of the circuit elements, evaluated with the ZSimp Win software, are presented in Table 6.

Table 6. Equivalent circuit parameters R (C (R (QR))) (CR) for PPS in seawater in the initial moments.

-	R_s $\Omega \cdot cm^2$	C_{por} $\mu F/cm^2$	CPE		R_{ct} $\Omega \cdot cm^2$	R_{SC} $\Omega \cdot cm^2$	C_{SC} $\mu F/cm^2$	$10^3 \times \chi^2$	ε_z
			Q Ss^n/cm^2	n					
-	115	0.307	6.81×10^{-4}	0.68	1121	21.70	1200	0.072	0.846
ε_{EC}	1.36	3.010	0.66	0.38	334.20	4.61	5.15	-	-

The analysis of the results in this table leads to the following observations:

- The chosen equivalent circuit describes very well the experimental data as indicated by the relative error values: $\chi^2 = 7.2 \times 10^{-5}$, impedance measurement error percentage $\varepsilon_z = 0.846\%$ and the relative percentage errors for each circuit element $\varepsilon_{ec} < 5\%$;
- Although the geometry of the measuring cell (distance between electrodes, sample surface) was the same as in the previous measurements, the resistance of the electrolyte (Rs) is approximately three times higher (38.67 for C45 and 0.41 for SP). This is because of the actual resistance of the liquid column in addition to the resistance of the phosphate layer is also added the resistance of the paint film, which is very slightly electrically conductive;
- The resistance of the porous layer (phosphating layer) is low and is practically equal to that found for the PS/BSW system, where only the phosphate layer is present ($R_{ext} \equiv R_{por} = 11.04$ $\Omega \cdot cm^2$);
- Due to the shields produced by the phosphate layer and the paint layer, the charge transfer resistance (R_{ct}) is very high, so the corrosion resistance is good.

The equivalent circuits used for processing the electrochemical impedance spectroscopy data obtained for the painted samples immersed for a long time in seawater are presented in Figures 16 and 17, and the values of the parameters of the circuit elements in Table 7. (Note: the two circuits are equivalent and express the same physical state).

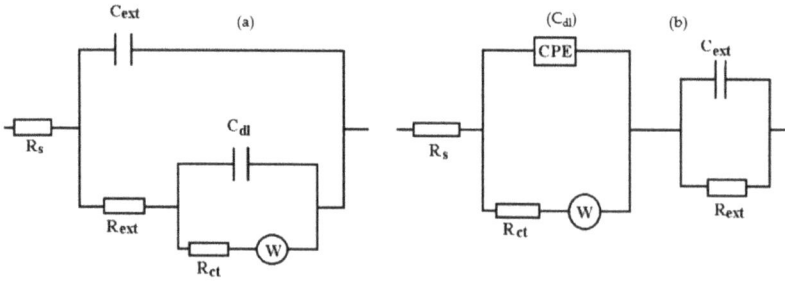

Figure 16. The equivalent circuits used to model EIS data in samples immersed in seawater for (**a**) 36 days or (**b**) 78 days.

Figure 17. SEM microphotographs for the PPS sample maintained for 78 days in seawater (**a**) 100×, (**b**) 1000×, and (**c**) 5000×.

Table 7. Equivalent circuit element values for PPS immersed 36 and 78 days in BSW.

Immersion Time	R_s $\Omega\cdot cm^2$	CPE		R_{ct} $\Omega\cdot cm^2$	R_{ext} $\Omega\cdot cm^2$	C_{ext} μF/cm²	W $Ss^{1/2}/cm^2$	$10^3 \times \chi^2$	ε_z
		Q Ss^n/cm^2	n						
36 days	315.40	$C_{dl} = 1.73$ μF/cm²		176.50	266.90	0.022	5.96 10^{-3}	4.13	6.43
78 days	385.70	6.63×10^{-3}	0.732	145.80	184.60	0.267	6.42 10^{-3}	1.75	2.64

The assumption of the formation of the unit layer seems to be confirmed by the fact that the resistance of this layer is much higher than the sum of the resistances of the phosphate layer and the paint layer. ($R_{por} + R_{SC} = 10.8 + 21.7 = 32.5$).

The resistance of the electrolyte increases to the initial moments and the immersion time in solution is proportional, as a consequence of the clogging of the pores and the deposits of the reaction products.

After the periods of immersion in the solution, it was necessary to introduce the Warburg impedance, which takes into account the diffusion of charges through the product layer. The surface state of the PPS sample maintained for 78 days in seawater is presented in Figure 16, and shown through SEM microphotographs in Figure 17, showing a corroded surface with a thick layer of compounds.

Mainly, the compounds are based on oxides, salts, and carbonates (Figure 18). There are two different corroded areas on the surface, one of them influenced by the gas release under the oxides layer. The main elements identified on the surface (Figure 18) are O, Ca, Fe, C, Na, Mn, Mg, Si, S, and P, and emerged from the substrate and immersion solution. The stability of the material surface and the beginning of the degradation process depend on the complex layer of compounds.

The complexity of the chemical composition result is given by the multiple structures of the surface (metallic substrate, phosphate layer, and paint cover) in contact with the immersion liquid. The presence of oxygen on the surface is highly based on its presence in the paint and from the oxides formed on the surface. The presence of calcium is due to the paint scraps that are mixed with the oxides of the substrate, but also can be explained by the compounds passing from the immersion solution (BSW that contains 146 mg/dm³ of Ca^{2+} component).

Regarding the sulfur element, the paint materials are S free, as identified on the surface based on the compounds formed between the iron-based oxides, as well from the sea solution that has 1305 mg/dm³ of SO_4^{2-}. The magnesium element, as a Mg^{2+} (548 mg/dm³) component, is part of the BSW and passes to the surface as oxide compound or complex compound with zinc (Figure 18).

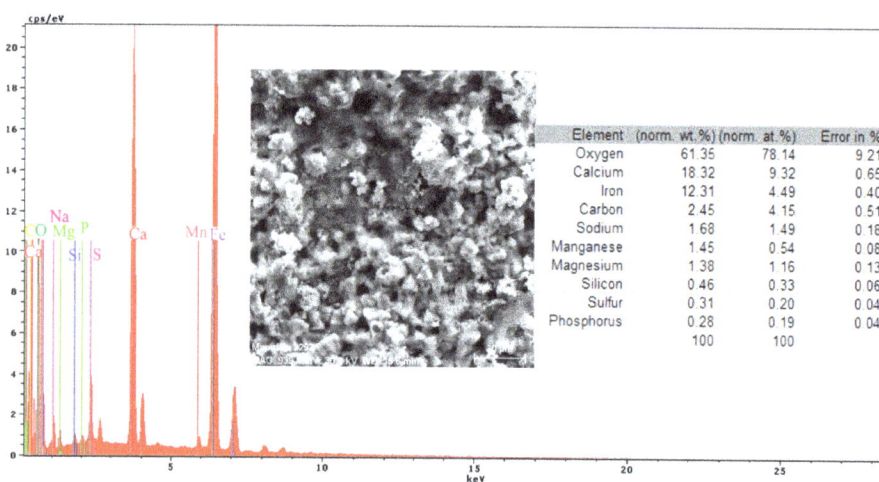

Figure 18. Energy spectrum and surface composition for the crust formed on the PPS surface, after 78 days of immersion in BSW.

A part of the identified chemical elements distribution is presented in Figure 19. The presence of iron is strong in some areas of distribution (Figure 18), which means that both the paint and phosphate layer were penetrated in that part by the corrosion, and the substrate is in straight contact with the liquid environment. The exposed iron areas are near the gas release bubbles.

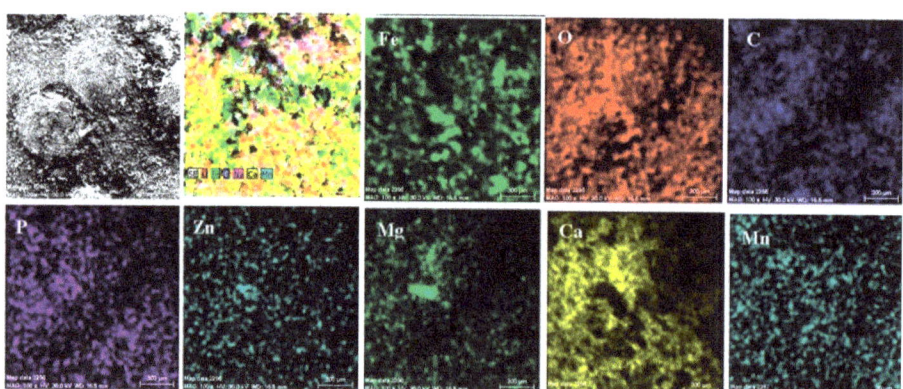

Figure 19. Elemental distribution in the crust on the PPS surface, after 78 days of immersion in seawater.

In the near future, these bubbles will become corrosion pits. During applications of steel metal elements in seawater, two principal arrangements of the corrosion speed-up process appeared: firstly, the old passivation surface layer, represented by partial products, was removed due to the reduction reactions of insoluble ferrous compounds to soluble ferrous ions; secondly, the anodic dissolution rate of steel was promoted because of bio-oxidation of cathodic hydrogen induced by Fe-reducing bacteria existent in seawater.

The calcium ions are specific to the paint layer and the zinc and phosphorus ions are specific to the phosphate layer. As expected, the area from which part of the crust came off is mainly the layer of paint on which a small amount of corrosion product is deposited (Fe(OOH)).

From elemental distribution, we observed an entire surface corroded (oxygen general spread) with areas of the penetrated paint layer. BSW is an extremely corrosive environment and affects most of the surface areas. From the calcium signal, Figure 19, we can observe an unaffected part of the layer where the integrity of the paint is not compromised. The differences of the corrosion resistance are given by the quality of the material's surface and the deposition procedure parameters of the phosphate layer or paint.

Table 8 shows the corrosion rate of the samples calculated according to the method presented in Jafari H. et al., 2011 [35]. Taking into account that the basic material of the samples is carbon steel, the corrosion rate was calculated using its density (7.8 g/cm^3), molar mass (56 g/mol) and valence (2).

Table 8. The corrosion rate of the studied samples.

Corrosion Rate	Immersion Time	C45	PS	PPS
	1 h	0.027	0.011	0.008
CR (mpy)	45 (36) days	0.062	0.029	0.022
	85 (78) days	0.036	0.034	0.030

As can be seen from the table, the trend of the calculated values is in accordance with the results obtained by EIS and with the stated hypotheses regarding the analyzed materials.

4. Conclusions

The immersion behavior, in salt water, of carbon steel, phosphate carbon steel, and phosphate and painted carbon steel was studied. Based on the experimental results, the following conclusions can be highlighted:

C45 sample:

- The immersion time increase results in a significant decrease of the R_{ct} value, from 588 cm^2 to 155 $\Omega \cdot$cm^2, respectively, 130 $\Omega \cdot$cm^2; this aspect indicates a decrease of the corrosion rate with the immersion periods.
- After 1 h of immersion, on the surface of the C45 sample were absorbed ions from saltwater or insoluble corrosion products creating a discontinuous layer. Over time, the thickness of the products layer increase acting as a screen causing the charge transfer resistance decrease.
- The EDS spectra and quantitation data indicate the presence of the iron and oxygen on the C45 surface for all the immersion times. For the C45 sample, immersed for only 1 h, the EDS spectra show small amounts of manganese and silicon, which confirm the hypothesis above.
- For the other sample, the EDS data indicate only the presence of iron and oxygen along with small amounts of carbon. This layer has a protective role for a while and after that, the degradation of the material occurs based on a continuous process of corrosion.
- The FT-IR spectrum confirms the presence of the iron oxyhydroxide on the C45 surface.

Phosphate C45 sample:

- After immersion, the surface of the samples indicated a bi-layer structure (the phosphate layer and the corrosion products layer).
- After 43 days immersion, the phosphate layer is partially damaged, widening its pores, but after 85 days, the resistance of the outer layer increases due to the thickness increase of the product layer.
- The composition of the crust form on the surface sample present not just iron and oxygen, but also phosphorus and zinc from the phosphate layer and carbon, magnesium, and sodium from the seawater composition.

Phosphate and pained C45 sample:

- The EIS data reveal that the phosphate layer resistance is equal compared with the value obtained for the PS sample.
- Even the saltwater damaged the paint layer; this sample has good corrosion resistance.

Author Contributions: Conceptualization, writing and investigation, D.-P.B.-N.; Writing an original draft —reviewing and editing, project administration and scientific supervision, C.B.; Methodology, investigation, data curation and validation, N.C. All authors have read and agreed to the published version of the manuscript.

Funding: This research received no external funding.

Institutional Review Board Statement: Not applicable.

Informed Consent Statement: Not applicable.

Data Availability Statement: The data presented in this study are available on request from the corresponding author.

Conflicts of Interest: The authors declare no conflict of interest.

Appendix A

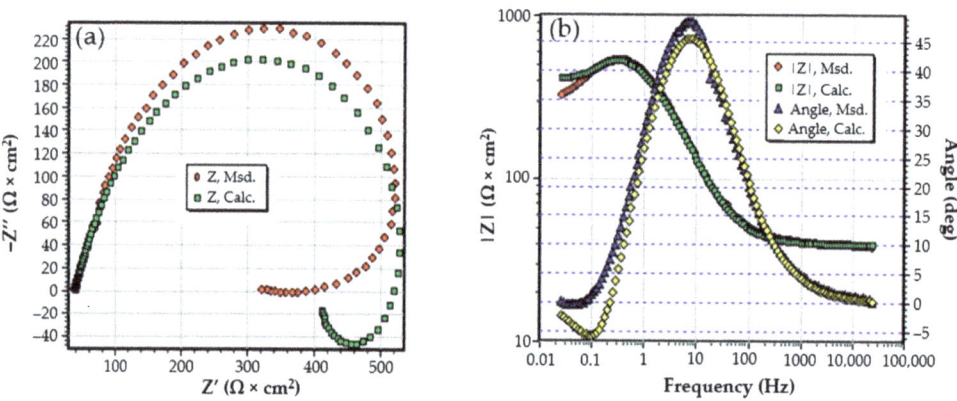

Figure A1. The Nyquist (a) and Bode (b) diagram for C45 sample after 1 h of immersion in BSW.

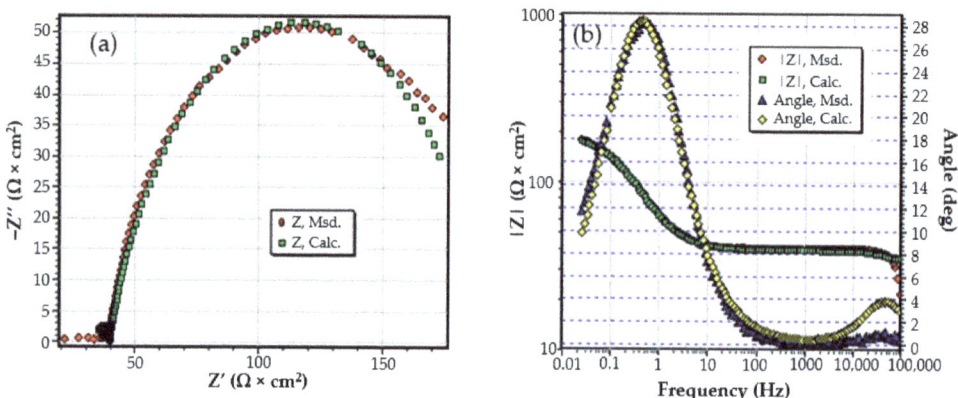

Figure A2. The Nyquist diagram (a) and Bode diagram (b) for C45 sample after 43 days of immersion in BSW.

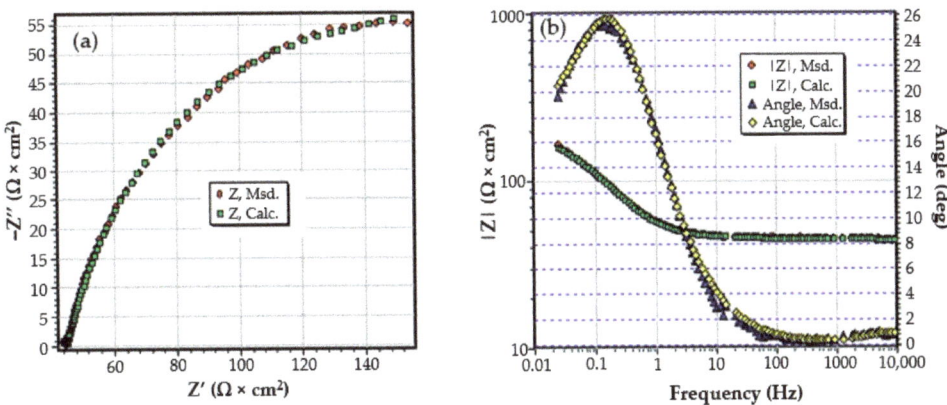

Figure A3. The Nyquist diagram (**a**) and Bode diagram (**b**) for C45 sample after 85 days of immersion in BSW.

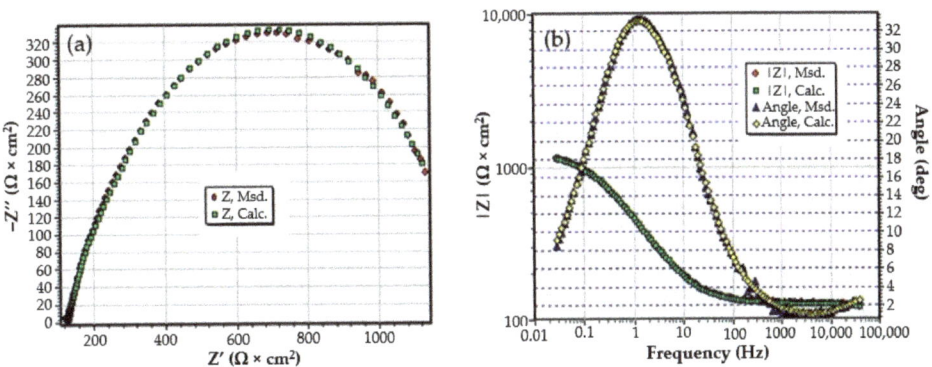

Figure A4. The Nyquist diagram (**a**) and Bode diagram (**b**) for PPS sample after 1 h of immersion in BSW.

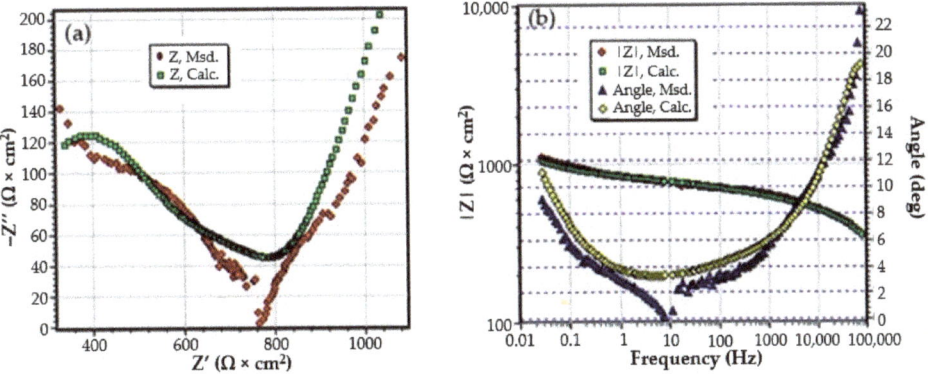

Figure A5. The Nyquist diagram (**a**) and Bode diagram (**b**) for PPS sample after 36 days of immersion in BSW.

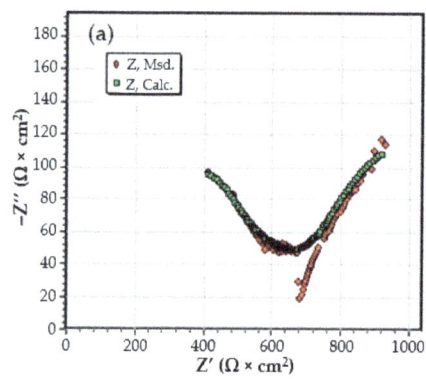 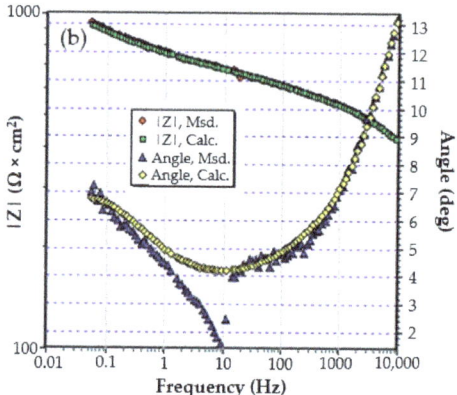

Figure A6. The Nyquist diagram (**a**) and Bode diagram (**b**) for PPS sample after 78 days of immersion in BSW.

References

1. Odusote, J.K.; Ajiboye, T.K.; Rabiu, A.B. Evaluation of Mechanical Properties of Medium Carbon Steel Quenched in Water and Oil. *J. Miner. Mater. Charact. Eng.* **2012**, *11*, 859–862. [CrossRef]
2. Boumerzoug, Z.; Derfouf, C.; Baudin, T. Effect of Welding on Microstructure and Mechanical Properties of an Industrial Low Carbon Steel. *Engineering* **2010**, *2*, 502–506. [CrossRef]
3. Panda, A.; Duplak, J.; Hatala, M.; Krenicky, T.; Vrabel, P. Research on the Durability of Selected Cutting Materials in the Process of Turning Carbon Steel. *MM Sci. J.* **2016**, *2016*, 1086–1089. [CrossRef]
4. Karavaeva, M.V.; Nurieva, S.K.; Zaripov, N.G.; Ganeev, A.V.; Valiev, R.Z. Microstructure and mechanical properties of medium-carbon steel subjected to severe plastic deformation. *Met. Sci. Heat Treat.* **2012**, *54*, 155–159. [CrossRef]
5. Kimapong, K.; Poonayom, P.; Wattanajitsiri, V. Microstructure and wear resistance of hardfacing weld metal on JIS-S50C carbon steel in agricultural machine parts. *Adv. Eng. Forum.* **2016**, *872*, 55–61. [CrossRef]
6. Burduhos-Nergis, D.P.; Baciu, C.; Vizureanu, P.; Lohan, N.M.; Bejinariu, C. Materials types and selection for carabiners manufacturing: A review. In *Proceedings of the IOP Conference Series: Materials Science and Engineering*; Institute of Physics Publishing: Bristol, UK, 2019; Volume 572.
7. Bejinariu, C.; Darabont, D.C.; Baciu, E.R.; Georgescu, I.S.; Bernevig-Sava, M.A.; Baciu, C. Considerations on applying the method for assessing the level of safety at work. *Sustainability* **2017**, *9*, 1263. [CrossRef]
8. Hamidinejad, S.M.; Kolahan, F.; Kokabi, A.H. The modeling and process analysis of resistance spot welding on galvanized steel sheets used in car body manufacturing. *Mater. Des.* **2012**, *34*, 759–767. [CrossRef]
9. Sekban, D.M.; Aktarer, S.M.; Xue, P.; Ma, Z.Y.; Purcek, G. Impact toughness of friction stir processed low carbon steel used in shipbuilding. *Mater. Sci. Eng. A* **2016**, *672*, 40–48. [CrossRef]
10. Usher, K.M.; Kaksonen, A.H.; Cole, I.; Marney, D. Critical review: Microbially influenced corrosion of buried carbon steel pipes. *Int. Biodeterior. Biodegrad.* **2014**, *93*, 84–106. [CrossRef]
11. Nedeff, V.; Bejenariu, C.; Lazar, G.; Agop, M. Generalized lift force for complex fluid. *Powder Technol.* **2013**, *235*, 685–695. [CrossRef]
12. Nica, P.E.; Agop, M.; Gurlui, S.; Bejinariu, C.; Focsa, C. Characterization of aluminum laser produced plasma by target current measurements. *Jpn. J. Appl. Phys.* **2012**, *51*, 106102. [CrossRef]
13. Bacaita, E.S.; Bejinariu, C.; Zoltan, B.; Peptu, C.; Andrei, G.; Popa, M.; Magop, D.; Agop, M. Nonlinearities in drug release process from polymeric microparticles: Long-time-scale behaviour. *J. Appl. Math.* **2012**, *2012*, 1–26. [CrossRef]
14. Dwivedi, D.; Lepková, K.; Becker, T. Carbon steel corrosion: A review of key surface properties and characterization methods. *RSC Adv.* **2017**, *7*, 4580–4610. [CrossRef]
15. Manna, M. Characterisation of phosphate coatings obtained using nitric acid free phosphate solution on three steel substrates: An option to simulate TMT rebars surfaces. *Surf. Coat. Technol.* **2009**, *203*, 1913–1918. [CrossRef]
16. Abdalla, K.; Rahmat, A.; Azizan, A. Influence of activation treatment with nickel acetate on the zinc phosphate coating formation and corrosion resistance. *Mater. Corros.* **2014**, *65*, 977–981. [CrossRef]
17. Burduhos-Nergis, D.-P.; Vizureanu, P.; Sandu, A.V.; Bejinariu, C. Phosphate Surface Treatment for Improving the Corrosion Resistance of the C45 Carbon Steel Used in Carabiners Manufacturing. *Materials* **2020**, *13*, 3410. [CrossRef] [PubMed]
18. Etteyeb, N.; Sanchez, M.; Dhouibi, L.; Alonso, C.; Andrade, C.; Triki, E. Corrosion protection of steel reinforcement by a pretreatment in phosphate solutions: Assessment of passivity by electrochemical techniques. *Corros. Eng. Sci. Technol.* **2006**, *41*, 336–341. [CrossRef]
19. Darband, G.B.; Aliofkhazraei, M. Electrochemical phosphate conversion coatings: A review. *Surf. Rev. Lett.* **2017**, *24*. [CrossRef]

20. Burduhos-Nergis, D.-P.; Bejinariu, C.; Toma, S.-L.; Tugui, A.-C.; Baciu, E.-R. Carbon steel carabiners improvements for use in potentially explosive atmospheres. *MATEC Web Conf.* **2020**, *305*, 00015. [CrossRef]
21. Bejinariu, C.; Burduhos-Nergiș, D.P.; Cimpoeșu, N.; Bernevig-Sava, M.A.; Toma, Ș.L.; Darabont, D.C.; Baciu, C. Study on the anticorrosive phosphated steel carabiners used at personal protective equipment. *Qual. Access Success* **2019**, *20*, 71–76.
22. Schmidt, D.P.; Shaw, B.A.; Sikora, E.; Shaw, W.W. Corrosion protection assessment of barrier properties of several zinc-containing coating systems on steel in artificial seawater. *Corrosion* **2006**, *62*, 323–339. [CrossRef]
23. Guenbour, A.; Benbachir, A.; Kacemi, A. Evaluation of the corrosion performance of zinc-phosphate-painted carbon steel. *Surf. Coat. Technol.* **1999**, *113*, 36–43. [CrossRef]
24. Asadi, V.; Danaee, I.; Eskandari, H. The effect of immersion time and immersion temperature on the corrosion behavior of Zinc phosphate conversion coatings on carbon steel. *Mater. Res.* **2015**, *18*, 706–713. [CrossRef]
25. Möller, H.; Boshoff, E.; Froneman, H. The corrosion behaviour of a low carbon steel in natural and synthetic seawaters. *J. S. Afr. Inst. Min. Metall.* **2006**, *106*, 585–592.
26. Burduhos-Nergis, D.P.; Vizureanu, P.; Sandu, A.V.; Bejinariu, C. Evaluation of the corrosion resistance of phosphate coatings deposited on the surface of the carbon steel used for carabiners manufacturing. *Appl. Sci.* **2020**, *10*, 2753. [CrossRef]
27. Díaz, B.; Freire, L.; Mojío, M.; Nóvoa, X.R. Optimization of conversion coatings based on zinc phosphate on high strength steels, with enhanced barrier properties. *J. Electroanal. Chem.* **2015**, *737*, 174–183. [CrossRef]
28. Burduhos-Nergis, D.-P.; Sandu, A.-V.; Burduhos-Nergis, D.-D.; Darabont, D.-C.; Comaneci, R.-I.; Bejinariu, C. Shock Resistance Improvement of Carbon Steel Carabiners Used at PPE. *MATEC Web Conf.* **2019**, *290*, 12004. [CrossRef]
29. Mobin, M.; Shabnam, H. Corrosion behavior of mild steel and SS 304L In presence of dissolved nickel under aerated and deaerated conditions. *Mater. Res.* **2011**, *14*, 524–531. [CrossRef]
30. Nagiub, A.M. Evaluation of Corrosion Behavior of Copper in Chloride Media Using Electrochemical Impedance Spectroscopy (EIS). *Port. Electrochim. Acta* **2005**, *23*, 301–314. [CrossRef]
31. Chen, M.; Du, C.Y.; Yin, G.P.; Shi, P.F.; Zhao, T.S. Numerical analysis of the electrochemical impedance spectra of the cathode of direct methanol fuel cells. *Int. J. Hydrog. Energy* **2009**, *34*, 1522–1530. [CrossRef]
32. Cao, C.; Zhang, J. *Introduction of Electrochemical Impedance Spectroscopy*; Science Press of China: Beijing, China, 2002; ISBN 7-03-009854-4.
33. Xu, X.; Liu, S.; Smith, K.; Cui, Y.; Wang, Z. An overview on corrosion of iron and steel components in reclaimed water supply systems and the mechanisms involved. *J. Clean. Prod.* **2020**, *276*. [CrossRef]
34. Boily, J.F.; Felmy, A.R. On the protonation of oxo- and hydroxo-groups of the goethite (α-FeOOH) surface: A FTIR spectroscopic investigation of surface O-H stretching vibrations. *Geochim. Cosmochim. Acta* **2008**, *72*, 3338–3357. [CrossRef]
35. Jafari, H.; Idris, M.H.; Ourdjini, A.; Rahimi, H.; Ghobadian, B. EIS study of corrosion behavior of metallic materials in ethanol blended gasoline containing water as a contaminant. *Fuel* **2011**, *90*, 1181–1187. [CrossRef]

MDPI
St. Alban-Anlage 66
4052 Basel
Switzerland
Tel. +41 61 683 77 34
Fax +41 61 302 89 18
www.mdpi.com

Materials Editorial Office
E-mail: materials@mdpi.com
www.mdpi.com/journal/materials

www.ingramcontent.com/pod-product-compliance
Lightning Source LLC
LaVergne TN
LVHW070714100526
838202LV00013B/1090